An Introduction to Biophysics

AN INTRODUCTION TO BIOPHYSICS

C. Sybesma

Laboratorium voor Biofysica
Vrije Universiteit Brussel
Brussels, Belgium

ACADEMIC PRESS New York San Francisco London 1977
A Subsidiary of Harcourt Brace Jovanovich, Publishers

ACADEMIC PRESS, INC.
111 Fifth Avenue, New York, New York 10003

United Kingdom Edition published by
ACADEMIC PRESS, INC. (LONDON) LTD.
24/28 Oval Road, London NW1

Library of Congress Cataloging in Publication Data

Sybesma, C
An introduction to biophysics.

Bibliography: p.
1. Biological physics. I. Title.
QH505.S86 574.1'91 76-13953
ISBN 0-12-679750-1

PRINTED IN THE UNITED STATES OF AMERICA

Contents

Appendix I Elements of Quantum Mechanics and the Electronic Structure of Atoms

Appendix II Elements of Equilibrium Thermodynamics

Preface

For some time there has been a need for a textbook that could be used as a basis for an introductory course in biophysics. This work is an attempt to fill that need. It defines biophysics as an approach to biology from the conceptual viewpoint of the physical scientist. It is not intended to be a textbook on physical chemistry with applications to biology. The emphasis is on fundamental biological problems rather than those related to medical physics and bioengineering.

In Chapters 1 and 2, the relation between the physical sciences and biology is discussed and an approach to biology based on the concepts of the physical sciences is developed. In this context, the fundamental biological unit, the cell and its (macromolecular and membraneous) constituents, is defined. The definition emphasizes the universality of cell structures rather than the diversity of forms in which cells actually occur.

In Chapters 3 and 4, the relation between molecular structure and biological function is described by showing how structures can be determined, how they influence biological function, and, finally, how inter- and intramolecular interactions form the basis of this structure–function relationship.

Chapters 5 and 6 are devoted to the bioenergetics and biophysics of the sensory systems. In these chapters the essential, fundamental processes in living systems are described. Energy conversion is a necessary part of all life processes, and many of these processes are essential for energy conversion. The approach of the physical scientist has been particularly successful in this area. The visual sensory system, so far, has been the only integrated system in which research has progressed substantially. Discussion of other sensory systems has been included, however; it is hoped that this discussion will arouse interest in biophysical research of these systems that are much more difficult to study.

Chapter 7 deals with theoretical biology. This new "subdiscipline" of biology emerged as a result of the physical science approach to biology. Its impact on the growth of biology may be considerable, not only because it aids in our understanding of existing problems, but also because it often leads to and determines the direction of experimental biophysical research. Although it has been possible to discuss only a few facets of theoretical biology, they are presented in such a manner that the reader will appreciate the problems involved and the results that can be expected.

This book evolved from senior and first-year graduate courses in introductory biophysics given for several years at the University of Illinois, Urbana, and the University of Brussels. It has been written with the assumption that the reader is familiar with elementary calculus, quantum mechanics, and thermodynamics. However, to make the book useful for those without such a background, appendixes on quantum mechanics and thermodynamics have been included. Although students of physics and chemistry will probably constitute the major audience for this work, it should provide a satisfactory introduction for others interested in biophysics.

An attempt to name all those who contributed to the realization of this book would undoubtedly result in unforgiveable omissions. I am in great debt to all whose criticism, advice, and devoted assistance made this book a reality. Thanks are also due, of course, to the students and staff of the Biophysical Laboratory in Brussels, who patiently endured the completion of this book. Finally, I thank my family, especially my wife Neri, for their cheerful aid and tolerance without which this work would not have been written.

Biophysics includes everything that is interesting
and excludes everything that is not. K. S. Cole

Chapter

1 Introduction

1.1 What Is Biophysics

What the definition of biophysics should be is a question asked in almost all texts dealing with biophysics or biophysical sciences. Many of the answers given in such texts are as vague as they are negative: biophysics is a discipline without a fixed content; biophysics is not yet an established discipline; its subject matter is not (yet) very well defined; biophysics is more or less what individual biophysicists have made and are making it. Yet many eminent scientists whose work has significantly contributed to our understanding of many biological phenomena have called themselves biophysicists and apparently did so for more or less pragmatic reasons. This indicates that there is an area of scientific activity which, by general consent rather than by definition, is covered by the name biophysics. Therefore, rather than try to define biophysics or emphasize the lack of a definition, it seems to make more sense to identify it by discussing the relation between the physical sciences and biology.

Development of Physics. Physics is an exact science which gradually obtained this characteristic in the days following those of Copernicus and Galileo. Its power is its exactness or, in other words, the fact that it deals with accurately measured quantities which allow, often through abstractions, causal interrelations in terms of a sophisticated conceptuality. Many of man's intellectual endeavors were initially descriptive sciences (limited to a compilation of appearances and events) and later developed into more exact sciences. Chemistry ceased to be a descriptive science when scales entered the laboratory. The introduction of quantitative measurements in chemistry resulted in the important laws of conservation of matter (Lavoisier) and constant and multiple proportions (Boyle), followed by Dalton's atom and Mendeleyev's periodic system of the elements. But, while chemistry used to be a science quite distinct from physics, the more it became an exact science, the closer it moved toward physics. Chemistry and physics now appear to be two aspects of the same thing. A dramatic demonstration of this is the reconciliation of Dalton's chemical atom and the physical atom of Rutherford and Bohr. The stability and valence characteristics of Dalton's atom were inexplicable with the Rutherford–Bohr model. The development of quantum mechanics brought about the connection between the two models and initiated an era of scientific activity in which the fundamentals of physics are not different from the fundamentals of chemistry.

Development of Biology. The development of biology progressed at a slower pace. This science still exists as a collection of numerous subdisciplines. Many of these—such as botany, zoology, entomology, mycology, and ornithology—have long been descriptive sciences and most of them still have that flavor. The development of biology, however, depended on and was connected to that of medicine. Progress in the art of healing is closely linked to the development of disciplines such as anatomy, bacteriology, pharmacology, and physiology. Developments in these areas are responsible for the spectacular advances made in recent times in medicine but also made these areas (especially physiology) more general. A great stimulus for this development was the emergence of biochemistry, after the shattering, by Wöhler's synthesis of urea in the second half of the nineteenth century, of the old notion that organic matter could not arise outside the living organism.

Relation between Physics and Biology. At that time, physiologists also began to incorporate generalized physical concepts such as mechanics, hydrodynamics, optics, electricity, and thermodynamics into their science. Something like a new subdiscipline began to emerge and many people now see this as the beginnings of biophysics. If we try to trace the history of this penetration of the physical sciences into biology we may as well start with Galvani and his work in the eighteenth century on the effects of static electricity on frog muscle. Julius Robert Mayer, in the nineteenth century, was trained in medicine but is well known for his formulation (in 1842) of the law of conservation of energy as a general principle. He was the first to point out that the process of photosynthesis is essentially an energy conversion process. Hermann Ludwig von Helmholtz (also trained as a physician) is probably the most outstanding example of a successful early biophysicist because of his contributions which led to a greater understanding of both biological and physical concepts. He studied muscle contraction, nerve impulse conduction, vision, and hearing, developed instrumentation to analyze the frequencies of speech and music, invented ophthalmometry, and contributed to thermodynamics. A historical example of a trained physicist who contributed to biology is John Tyndall, who studied under Faraday. His contributions to microbiology as an English contemporary of Louis Pasteur are well known.

The incorporation of physical concepts and instrumentation into biology became more firmly established in this century. The development of the technique of structure analysis by X-ray diffraction had a profound influence on biology. Thermodynamics became an essential part of the study of muscle contraction, and the recent development of nonequilibrium thermodynamics has direct implications in biology. Electrophysiology obviously cannot be thought of without the concepts and the technology of electricity and electronics. Spectroscopy is an indispensible tool in many areas of physiology, and quantum mechanics is the basis for the interpretation of structure–function relationships at the subcellular and molecular levels in biological systems. The direct implications for biology of areas such as information theory and cybernetics (emerging from the development of communication technology) are promising and further exploration of their applicability may well lead to new levels of understanding in biology.

This kind of biophysics thus can be considered an offspring of physiology. Its feedback to physiology has helped separate physiology from medicine and make it a science in its own right. Biophysics developed into a major part of biology through incorporation of not only existing physical theories but also the conceptual approaches of the physical scientist. It thus became an interdisciplinary scientific activity which surpasses the original organism-defined departments of biology. One of the most important achievements of biophysics is the recognition that the laws of physics and chemistry *are valid* in living systems; there is no longer any doubt that the fundamental basis of the processes in a living organism is not different from the fundamental basis of the processes of physics and chemistry.

One can, however, also identify another kind of biophysics which did not emerge from (medical) physiology. Its votaries are usually physicists and physical chemists who, in their search for universal principles that will explain the world around us, are impressed by the strange and very special place occupied by living organisms. Thus, this kind of biophysics truly concerns the fundamental principles of biology.

1.2 The Fundamental Principles of Biology

Living versus Nonliving. Almost since man was able to think, he wondered about what makes him and the living world around him distinct from the rest of his environment. Scientists as well as philosophers have developed theories about it, some of them based on religious presuppositions, others devoid of such presuppositions. Until well into the nineteenth century, virtually all these theories were of a vitalistic or animistic nature; in other words, they assumed special, not necessarily physicochemical, "forces" to explain living systems as distinct from the nonliving. Even today such vitalistic or animistic theories, either disguised or not, have strongholds. But in the beginning of the twentieth century a mechanistic view of life began to develop which states that indeed physics and chemistry are all that is necessary to explain the phenomenon of life.

A Biological Complementarity Principle. The more that was known about the molecular processes of living systems, the more this mechanistic view developed. For many eminent scientists, however, it proved difficult to accept a theory based upon known physical principles that would account for the obvious differences between the living and nonliving world. Evolution over a long period of time leading from a relatively simple self-reproducing system to the wide diversity and complexity of the living world of today seems to defy the second law of thermodynamics; also, the faithful replication of species over numerous generations has puzzled many of those who were looking for universal explanations. In 1949 Max Delbrück, a physicist, gave a lecture in which he discussed biology from a physical science point of view. Reflecting on the tremendous enrichment process of evolution, he concluded that the principles on which an organism of today is based must have been determined by a couple of billion years of evolutionary history. He then said ". . . you cannot expect to explain so wise an old bird in a few simple words." Niels Bohr, one of the founders of modern physics, was also impressed by the singularity of living organisms. He saw in biology an uncertainty principle analogous to that of quantum theory, but at a higher level; to find out about life in a living organism one has to make measurements which interfere so strongly with the processes in the organism that life itself is destroyed. In 1933 Bohr proposed a formalization of this dualism in the form of a complementarity principle of biology. Much later still a similar opinion was expressed by

Werner Heisenberg, a founder of quantum mechanics, who stated in 1962 that "It may well be that a description of the living organism that could be called complete from the standpoint of the physicist cannot be given, since it would require experiments that interfere too strongly with the biological functions." These ideas are the basis of a biological theory developed by Walter Elsasser, a theoretical physicist (1958). We shall briefly discuss his theory in Chapter 7.

The Mechanistic View. In the years before the second world war however, physical methods rapidly gained acceptance in biological research and many physical scientists began to show an avid interest in their application. Then in 1943 Erwin Schrödinger, one of the founders of quantum wave theory, gave four lectures in Dublin on the physical aspects of a living cell. Later (1944) these lectures appeared in the book "What Is Life?" These events are seen by many as the beginning of an era in which physical science and physical concepts were being unconditionally accepted in the study of biology. What is remarkable about the book is that slightly more than a decade before the actual discovery of the structure of the gene, the DNA molecule, and the genetic code enclosed in the linear sequence of the nucleotides, these exact ideas were developed. Schrödinger did not start from any fundamental dualism other than that of quantum mechanics itself. He saw the necessity for what he called "aperiodic crystals" as the governing agents of the hereditary "code-script" in the gene. Noticing the smallness of the gene and the tremendous amount of information it must contain to cause the development of a complex living organism, he stated that "A well-ordered association of atoms, endowed with sufficient resistivity to keep its order permanently appears to be the only conceivable material structure, that offers a variety of possible arrangements sufficiently large to embody a complicated system of determinations within a small spatial boundary [p. 65]." Since then the progress of molecular biology, based on pure physical and physicochemical principles, has been quite impressive.

In the following pages of this book we introduce biophysics as an approach to biology that is characterized not only by the description of the physicochemical basis of the processes in living systems, but also by a physical conceptuality of the subject matter. We shall first see why and how a living system can be generalized and then discuss how physical theories and concepts apply to such generalized systems by looking at diverse interactions and processes occurring in these systems, from the molecular level up to more integrated systems. In the final chapter these generalizations will reappear in a discussion of theoretical biology and we shall try to find out what they mean with respect to questions about the fundamental principles of biology.

Chapter

2 The Biological Unit

2.1 The Cell Doctrine

The Cell as the Basic Unit of Life. The study of matter by both physicists and chemists was simplified and unified by the final description, in quantum mechanical terms, of its basic unit the atom. One could ask whether such atomistic units exist in biology. In many biology textbooks one finds the proposition that the *living cell* is such an atomistic unit. The emergence of the cell doctrine after Robert Hooke's discovery of the "little boxes or cells" in cork in 1665, resulted in the general recognition that life has a cellular structure. In spite of the wide variety of size, shape, and function, the cell, as an integral and relatively independent body surrounded by a boundary, is a common feature of all living organisms† and is therefore a good candidate to be such an atomistic unit. One has to be careful, however, not to see this as more than an analogy. The atom and the elementary particles which constitute it are the basic elements of matter as it manifests itself in the nonliving *as well as* living world. We have already argued that there is no fundamental difference between the physics of matter in living and nonliving systems. One could state, therefore, that as much

† A *virus* has no cellular structure. A virus, however, cannot be considered as an independent living organism. It functions only *within* a host cell, which it requires to reproduce itself. Outside a cell a virus behaves like, and can be considered as, lifeless matter.

and as far as the atom is the basic unit of matter it is also the basic unit of life. However, when a cell is broken the fragments, after some time, lose the characteristics they had and the reactions they were able to carry out in the intact cell, although the material components are not changed. In life, therefore, there is something that transcends the basic unit of matter. That "something" could be a not yet well-defined thing called organization. A haphazard heap of all the molecules or even the aggregates which constitute a cell is not living. Only when they are arranged in a very special way and when the interactions and reactions between them are arrayed and controlled do we have a form of matter which is life.

Organization. Organization is found on all levels of life. Life, indeed, consists of a hierarchy of organizational forms, ranging from society via man and multicellular organism to the unicellular organisms, protozoa, algae, and bacteria. If we upset this organization from the top of the hierarchial ladder on down we end up at the cell. Societies can be broken apart, after which new ones are formed (history is full of examples of such events); organs of multicellular organisms can be removed from the organism and kept alive for an indefinite time under the proper conditions. Organs can be dissociated into their component cells and such cells can be cultivated (tissue cultures) and kept functioning for countless generations; even cells dissociated from well-differentiated livers or kidneys from chicken embryos have been shown to develop miniature livers or kidneys. But when a cell is broken apart the components do not develop into a new cell; they can perform specific functions such as fermentation, respiration, primary photosynthetic processes, or even synthesis of the macromolecules which life is making use of, but they die off after some time. The cell, therefore, seems to be the lowest form of *organization* of matter which we can call life.

Cells occur in life in a wide variety, not only of sizes (a nerve cell can be longer than a meter while a pneumococcus is about 0.2 μm in diameter), but of shapes and functions as well. Figure 2.1 shows some shapes of cells from multi- and unicellular organisms. They all have in common, however, a high-precision apparatus for energy-providing, synthesizing, and specific functional reactions, a center for the control of these reactions, and pathways of communication. This machinery is built up along surprisingly uniform lines; the same types of molecules and the same types of structural features are associated with these functions. The all-important synthesis of proteins, for example, takes place according to the same metabolic patterns regardless of the biological origin of the cell. Thus, it seems as if, from a biophysical (biochemical) point of view, we can allow ourselves this generalization in biology and consider the structural and functional features of cells as fundamental for our study of life.

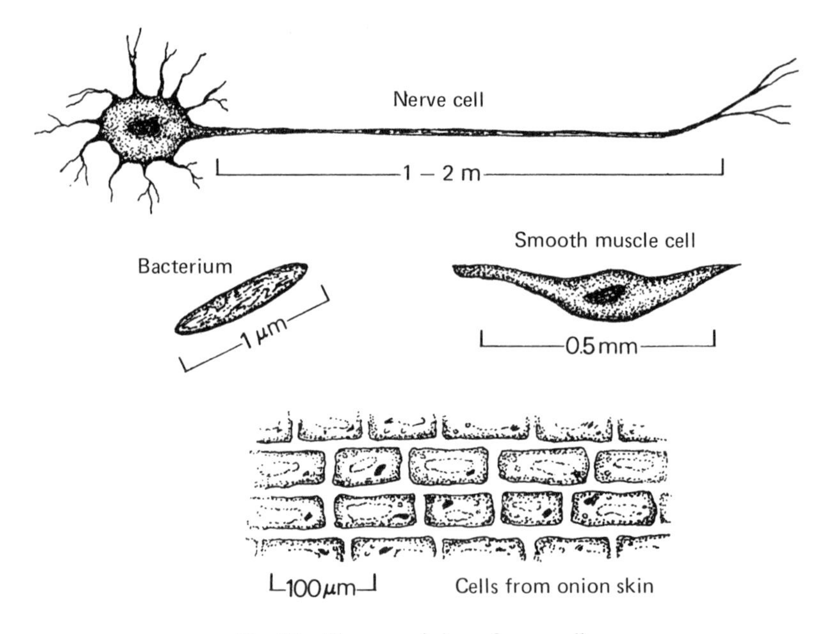

Fig. 2.1 Shapes and sizes of some cells.

Membranes. A common structural feature in all cells is the *membrane*. Membranes are found both as boundaries of the cells as well as more or less elaborate systems within the cell. Many biochemical events take place on or in the membranes. Formerly considered as a static structure with a principal role of simple delineation of the boundaries of cells and subcellular organelles, these membranes are now looked upon as highly dynamic structures essential for many diverse processes. In Section 2.5 we will discuss the most important features of membranes.

The membrane surrounding the cell is called the *cytoplasmic membrane* (or plasmalemma or plasma membrane). It forms a selective barrier which maintains the chemical integrity of the cell. This is done, as will be discussed later, by passive and active transport processes across the membrane which are selective not only in the rate of the movement in or out but also in the kind of molecules that enter or leave the cell. There are other ways for the intake of matter into cells however. Some free cells, an amoeba for example, can take in material by one or both of two processes called *pinocytosis* and *phagocytosis*. Pinocytosis is a process by which the cytoplasmic membrane forms invaginations in the form of narrow channels leading into the cell. Liquids can flow into these channels and the membrane then pinches off pockets that are incorporated into the cytoplasm and digested (Fig. 2.2). Phagocytosis occurs when the cell membrane engulfs a drop

Fig. 2.2 Pinocytosis as occurring at the edge of an amoeba. The cytoplasmic membrane forms channels through which liquid can flow into the cell. In the cell the channels are pinched off as membrane-enclosed droplets; these eventually dissolve in the interior of the cell.

of liquid or a solid particle (a small bacterium for example), and then draws it into the cytoplasm where it can be digested (Fig. 2.3).

The cytoplasmic membrane of a cell can thus be seen as a functional part of the living cell. This view is substantiated by the fact that the internal membrane systems in many cells are shown to be continuations of the cytoplasmic membrane. At the same time, there must be a wide variety in the detailed molecular composition of membranes in order to account for the variations in permeability, functions, and appearances. This variability may have its origin in the great variety of proteins and lipids that could make up the structure of membranes. Recent investigations of many biological phenomena seem to indicate that membranes are a structural feature essential for life processes.

2.2 Prokaryotes and Eukaryotes: Cell Organelles

Advances in the technology of electron microscopy have established the general acceptance of a major morphological distinction between two groups of cells. The majority of cells belong to one group, the so-called *eukaryotes*. Only bacteria and the blue-green algae belong to the other

Fig. 2.3 Phagocytosis as observed in an amoeba. A protrusion of cytoplasm comes in contact with a drop of liquid or a solid particle, surrounds it, and draws it into the cytoplasm where it can be digested.

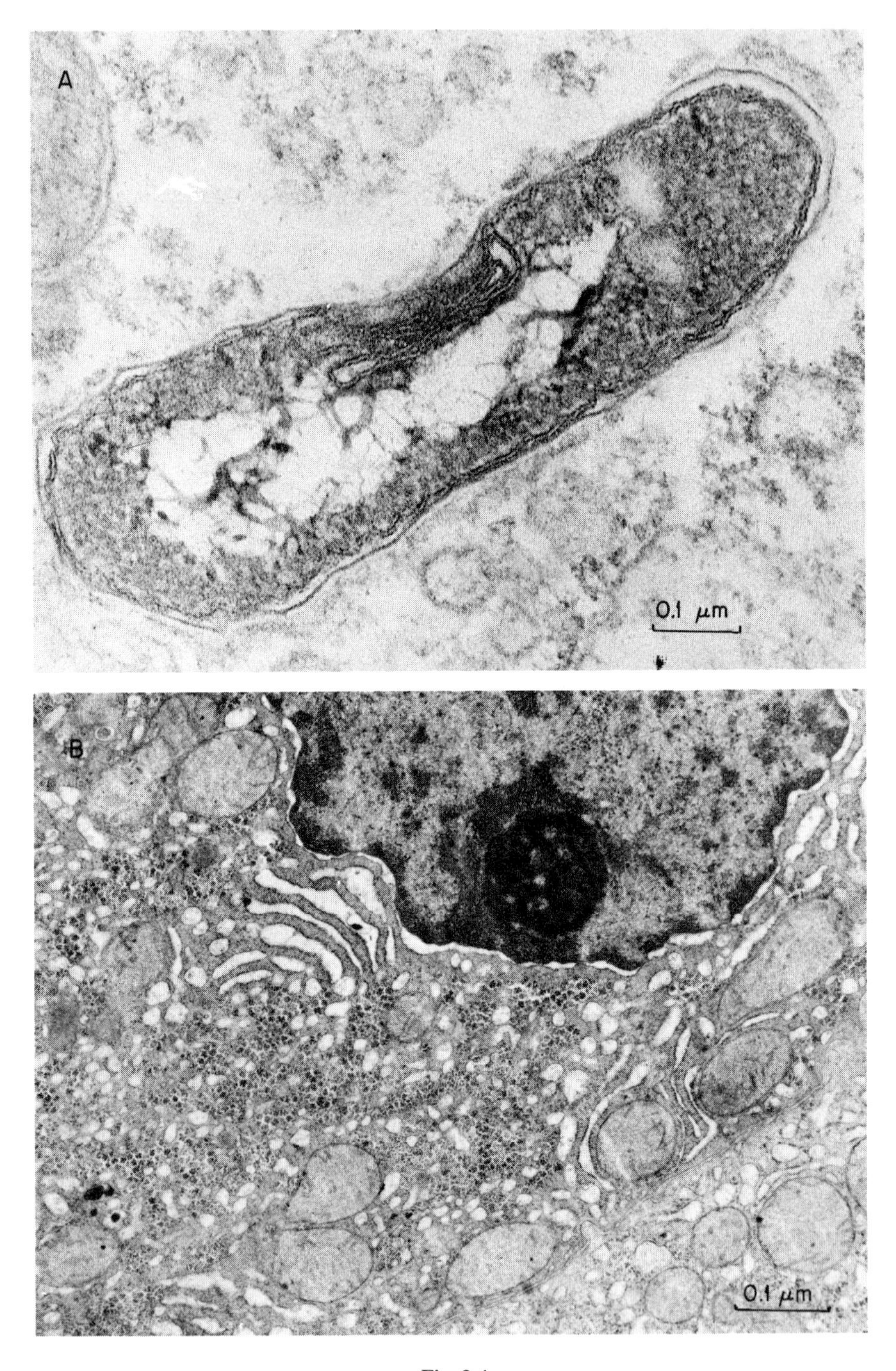

Fig. 2.4

group, the so-called *prokaryotes*. The nomenclature, proposed by E. C. Dougherty in 1957, is derived from the Greek word $\kappa\alpha\rho\upsilon o\nu$ which means *kernel* or *nucleus*. The prokaryon ($\pi\rho o$ = before) is the undeveloped nuclear area in the prokaryotes and the eukaryon ($\varepsilon\upsilon$ = well) is the well-developed nucleus in the eukaryotes. Figure 2.4A shows a cross section of a prokaryotic cell, and Fig. 2.4B shows a cross section of a part of a eukaryotic cell. While in the prokaryotic cell no well-developed internal structure in the form of more or less extensive particulate organelles can be seen, the eukaryotic cell shows a highly differentiated internal structure, with visible particulate organelles. This type of cell has a well-defined nucleus surrounded by nuclear membrane.

In prokaryotic cells a well-defined nucleus cannot be distinguished, although areas with a higher concentration of nuclear material than the rest of the cell can sometimes be recognized. Prokaryotic cells, instead of having systems of particulate organelles, have more or less elaborate membrane systems showing up as either vesiclelike structures or irregular stacks (Fig. 2.4A).

Figure 2.5 shows a schematic representation of two types of *eukaryotic* cells, an "animal" cell (also representative of many eukaryotic protozoa) and a plant cell. Although in view of the diversity discussed above, the designation as "typical cells" should not be taken too literally, the representations in the figure are useful to point out the common features found in real cells.

Eukaryotic Organelles. All eukaryotic cells have a well-defined *nucleus* surrounded by a double-layered membrane. This *nuclear membrane* often shows continuity with membrane systems in the extranuclear part of the cell, called the *cytoplasm*. The nuclear membrane has *pores*, that is small openings at the edges of which the inner and outer membrane are continuous. The material inside the nucleus is a very fine threadlike substance called *chromatin*. The chromatin consists of the nucleic acids [*deoxyribonucleic acid* (DNA) and *ribonucleic acid* (RNA)], a low molecular weight

---

Fig. 2.4 (A) Electron micrograph of a prokaryotic organism *Myobacterium leprae*. The cytoplasmic membrane convolutes sometimes in vesiclelike structures and sometimes into stacks. The nuclear area is not well defined. Courtesy of Dr. W. Jacob and Dr. A. Van Laer, Electron Microscopy Laboratory, Universitaire Instellingen Antwerpen, Antwerp, Belgium. (B) Electron micrograph of part of a human liver cell. The nucleus is a well-defined area bounded by the nuclear membrane. The black round spot in the nucleus is a nucleolus. The circular structures in the cytoplasm are mitochondria. The endoplasmic reticulum is also visible as are many ribosomes. Part of the cytoplasmic membrane is seen in the lower right-hand corner. Courtesy of Dr. W. Jacob and Dr. A. Van Laer, Electron Microscopy Laboratory, Universitaire Instellingen Antwerpen, Antwerp, Belgium.

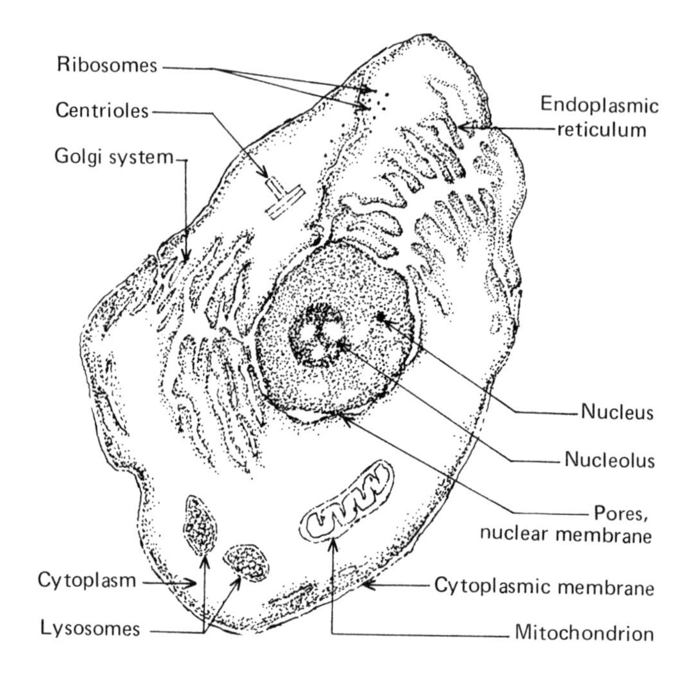

(a)

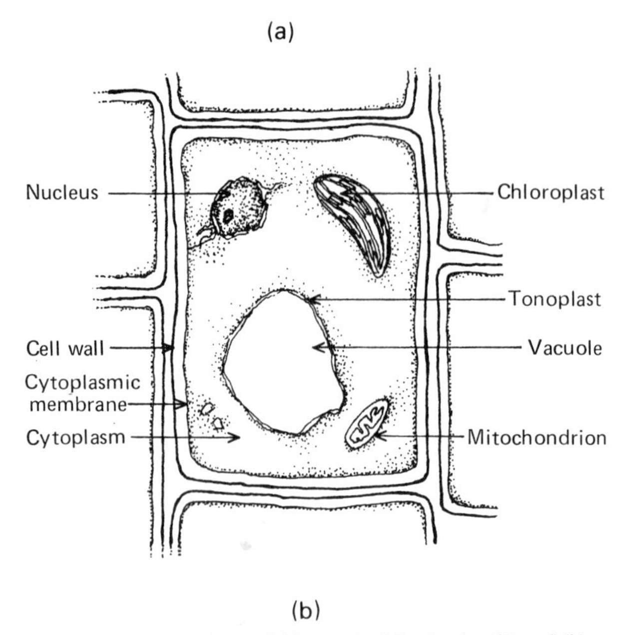

(b)

Fig. 2.5 Schematic representations of (a) a typical "animal cell" and (b) a typical "plant cell." The figures describe the cells and the organelles *in a schematic way*, not as they would appear in reality. Animal cells are bounded by just the cytoplasmic membrane; most plant cells have an additional rigid and sturdy cell wall around them.

protein called *histone*, and some residual protein. A nucleus may contain one or more denser bodies called *nucleoli* (singular: *nucleolus*).

The nucleus is an essential structure of the cells. The DNA contains all the information for the morphology and function of the cell. When the cell divides, the chromatin becomes visible as elongated structures called *chromosomes*. The chromosomes are made up of pairs of so-called *chromatids* which then, in a dramatic and fascinating way,[†] are separated. Each chromatid forms an equal set of genetic material for the two daughter cells. The DNA of the chromatin is exactly duplicated by a process which will be described in Chapter 3.

Plant cells are surrounded by a rigid and sturdy cell wall made of cellulose or celluloselike material (formed by repeating units of glucose). Animal cells do not usually have this feature; their boundary is formed by the *cytoplasmic membrane*. Like the nuclear membrane the cytoplasmic membrane shows continuities with membraneous structures in the cytoplasm. Another structural feature, more characteristic of plant than animal cells, is the large liquid-filled cavity in the middle of the cell called the *vacuole*. The vacuole is surrounded by a unit membrane called the *tonoplast*.

Both types of eukaryotic cells have more or less elaborate membrane structures in their cytoplasm. These channel- or sac-forming (cisternal) structures, known under the name *endoplasmic reticulum*, extend to a variable degree from the nuclear membrane to the cytoplasmic membrane. The type of cell and its state of metabolic activity can be characterized by the appearance of the endoplasmic reticulum. This structure seems to have a variety of functions, an important one being, most probably, that of providing a communication system in the cell; thus, products formed can be segregated and transported either to other parts of the cell or to the outside environment by this system of channels. At least part of the endoplasmic reticulum may also be involved in the synthesis of proteins. Surrounding part of it and sometimes attached to it are numerous little granules called *ribosomes*, giving this part of the endoplasmic reticulum a rough appearance. Ribosomes, as we will see later, are instrumental for the synthesis of proteins.

Another set of membranes seen in both types of cells is the *Golgi apparatus*, sometimes called the *dictyosome*. This peculiar membrane system may act as a packaging and transport system for products and waste by a process which can be seen as a reversed pinocytosis. Specialized cells with a secretory function have a well-developed Golgi system. In such cells products synthesized at or near the endoplasmic reticulum are packaged within membrane-enclosed granules which arise from the Golgi apparatus. These vesicles subsequently migrate toward the cell surface, fuse with the cytoplasmic membrane, and discharge their content into the intercellular

[†] For a detailed description of this process we refer to textbooks in cell physiology.

space by this process of reversed pinocytosis. Such a function may be indicated by, among other things, the appearance of the system; deep inside the cell the vesicles are elongated and flattened and arrayed in a more or less regular parallel stack close to the endoplasmic reticulum. Toward the edge of the cell the vesicles become less flattened and often develop into a large number of irregular but more or less rounded vesicles. The system is not a static one; there seems to be a constant flow of membranes from the inside to the outside of the cell, the replenishment originating from the endoplasmic reticulum.

The above-mentioned *ribosomes*, little granules of about 20–25-nm diameter, are found in every kind of cell. They are intimately connected with the synthesis of proteins, consist of two parts, and contain ribonucleic acid (RNA) and protein. Ribosomes originate in the nucleolus but are assembled in the cytoplasm. Aggregates of them are called *polysomes* and are formed in the process of translation which will be described later.

A cylindrical structure some 300–500 nm long and 150-nm diameter, found more often in animal than in plant cells, is called the *centriole*. When two centrioles are found together, one is generally oriented perpendicular to the other. This organelle is a part of the cell's locomotive system. Centrioles are critical in determining the axes of cell division. The *spindle*, a system of microtubular filaments which appears and attaches itself to the chromosomes when the cell divides, may have its origin in the centrioles. Their chemical composition and exact function, however, remain to be determined.

Lysosomes are vesicle-type organelles with an outer limiting membrane. They contain so-called *lytic* enzymes involved in the breakdown of cellular fragments and large molecules. Lysosomes, therefore, can be considered as disposal units of the cell, removing foreign bodies and cell structures no longer needed. There is strong evidence that the lysosomes originate in the Golgi apparatus. This would be consistent with the functions ascribed to both organelles.

Every kind of eukaryotic cell has at least one, but normally more, *mitochondria* (singular: *mitochondrion*). This extremely important, and probably most studied, organelle is the "power plant" of the cell. Mitochondria range in size from 0.2 to 7.0 μm, vary in shape from spheres to more or less elongated rods, and are surrounded by a smooth membrane wall. On the inside, some 6 nm from the outer membrane, an inner membrane convolutes inward (in the so-called *matrix*) forming thin sheets called *cristae* (see Fig. 2.6). On these cristae are found the enzymes which operate

Fig. 2.6 Electron micrograph of mitochondria from a bat pancreas cell and a three-dimensional drawing of a bisected mitochondrion. The electron micrograph clearly shows the many folds of the inner membrane sticking out to the inside (cristae), thus forming the lamella shown in the drawing. Photograph by K. R. Porter from V. L. Parsegian *et al.*, "Introduction to Natural Science Part 2: The Life Sciences," Academic Press, New York, 1970.

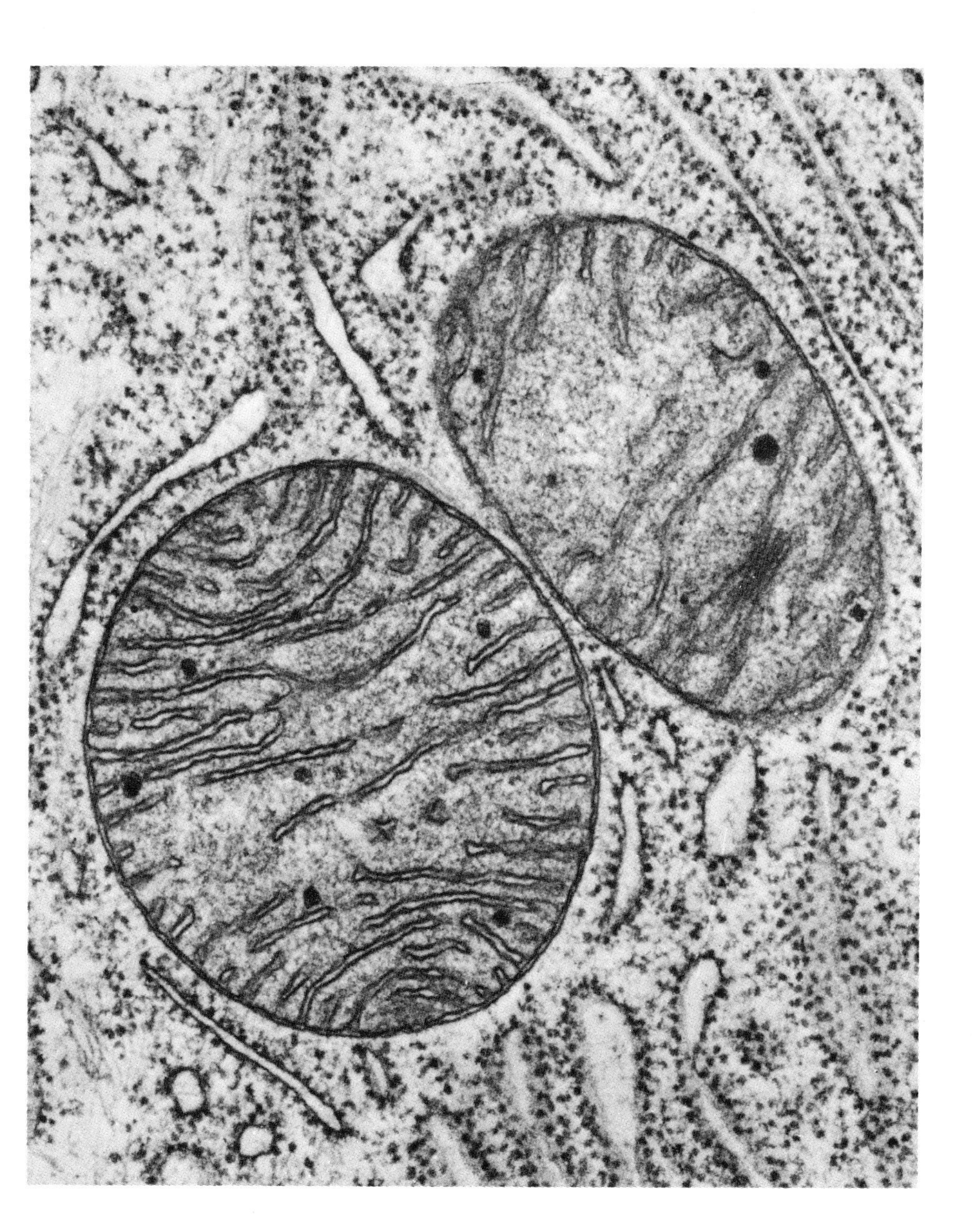

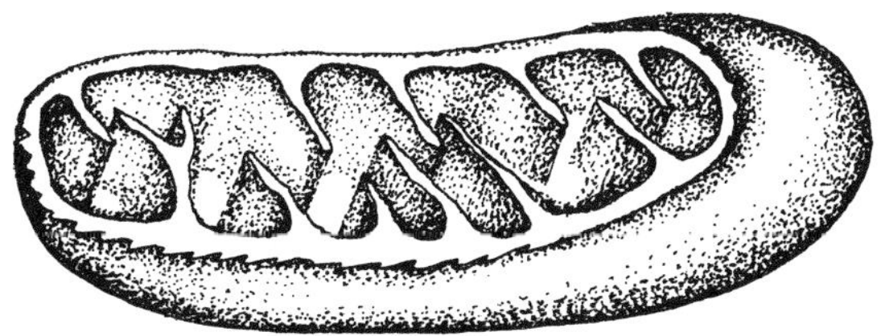

Fig. 2.6

in the conservation of energy obtained in the aerobic part of the breakdown of carbohydrates, fats, and (to a certain extent) proteins, in a process called *respiration*. This process will be described in more detail in Chapter 5.

Cells found in the green parts of plants, especially in the leaves, and in most of the unicellular algae contain *chloroplasts*, organelles which are the site of the *phototrophic* energy conservation. The process in which light energy is absorbed, trapped, and converted into chemical energy is called *photosynthesis* (see Chapter 5). The apparatus for this process is bound to a lamellar system inside the chloroplast, which is again enclosed by a membrane. In many chloroplasts, such as the one shown in Fig. 2.7, the lamellar structure is densely packed at some places (the *grana*) and single-layered at others (the *intergranal lamellae*). The lamellar system is often referred to as the *thylakoid*, the intergranal space as the *stroma*.

2.3 The Relation between Prokaryotes and Eukaryotic Organelles

Prokaryotic and Organellar Similarities. Prokaryotic cells (bacteria and blue-green algae) do not have the specialized organelles just described for the eukaryotic cells. Instead, they have more or less elaborate membrane systems which are linked to a variety of functions. These membrane structures, depending upon the kind of organism and metabolic condition, vary from more or less independent vesiclelike enclosures to stacks of lamellae which seem to be continuations of the cytoplasmic membrane. The common feature of membranous structures in the prokaryotes, however, suggests relationships with the laminar structure inside the energy-conserving mitochondria and chloroplasts. In fact, the inner membrane of mitochondria has many characteristics in common with the bacterial cell membrane, rather than with structures found inside the eukaryotic cell. The discovery of DNA and ribosomes inside mitochondria and chloroplasts, and the fact that mitochondria as well as chloroplasts divide (in some cases concurrent with cell division, in other cases more independently), gave impetus to the speculation that prokaryotic organisms and the energy-conserving eukaryotic cell organelles may have a common origin. Circumstantial evidence tends to support such speculations; the organellar DNA, although incapable of programming for all proteins found in the organelle, suggests an independent existence at least early in evolution. The DNA in the organelle never assumes the form of the chromosomes in the nucleus of the eukaryotic cell; it is much more similar to the bacterial DNA which is present in long closed strands.

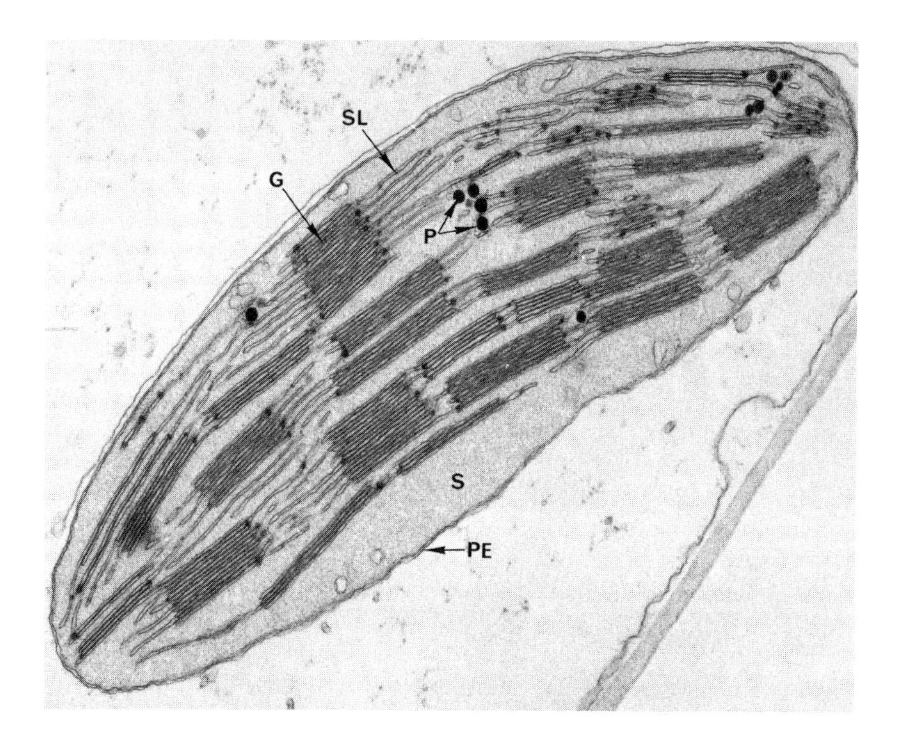

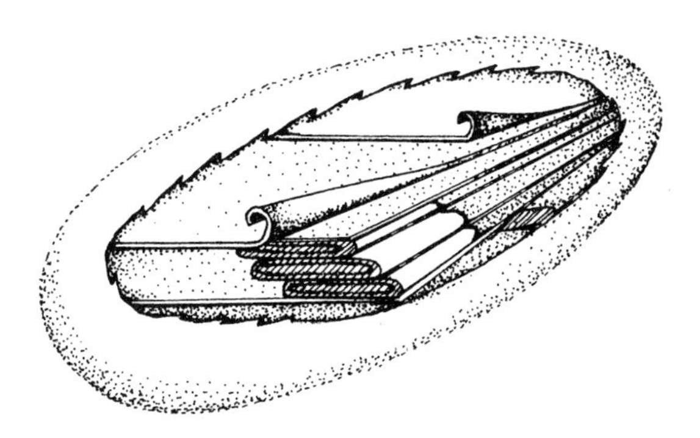

Fig. 2.7 Electron micrograph of a chloroplast from a lettuce leaf cell and a three-dimensional drawing of a cut-open chloroplast. The electron micrograph clearly shows the grana (G) formed by the stacks of flattened vesicles (thylakoids) shown in the drawing. PE is the chloroplast envelope, a double membrane surrounding the chloroplast. SL are the intergranal lamella, S is the stroma, and P are densely stained lipid droplets. From Govindjee, "Bioenergetics of Photosynthesis," Academic Press, New York, 1975, courtesy of C. J. Arntzen.

Endosymbiosis. The demonstration of cellular endosymbiosis (the living of one cell inside the other) may be seen as a supporting indication. Endosymbioses occur with green algae inside certain protozoa; they are easy to detect because the characteristics of a eukaryotic green alga inside a (eukaryotic) host cell cannot be mistaken. Prokaryotic blue-green algae may also become symbionts, although their presence as such inside a host cell is much more difficult to detect. Blue-green inclusions (*cyanelles*) are found in some amoeboid or flagellate eukaryotes but a definite proof that such cyanelles are in fact endosymbiotic blue-green algae can not as yet be given.

Evolution of Cells. Observations such as these suggest that endosymbionts may have been involved in the evolution of the eukaryotic cells. This evolution may have taken place according to one of the diagrams shown in Fig. 2.8. According to Fig. 2.8a the eukaryotic cell may have evolved from a symbiosis of two or more prokaryotic ancestor cells. As shown in Fig. 2.8b, both prokaryotic and eukaryotic cells may have evolved from the same ancestor cell. In the latter case the eukaryotic cells may have come from those cells which developed a cytoplasmic membrane with the ability to take in particulate matter by some kind of pinocytosis. Prokaryotes brought into the cell in this way may have developed into the organelles of the present-day eukaryotes, with the mitochondria and the

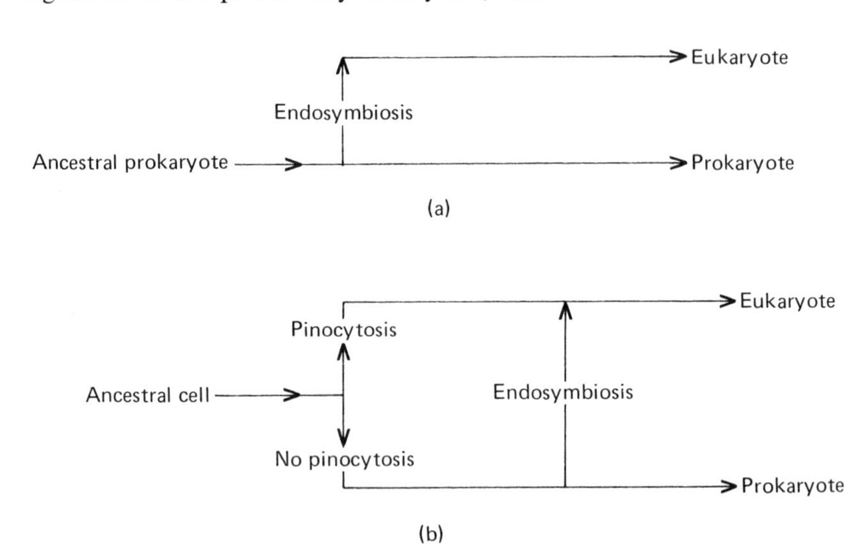

Fig. 2.8 Diagrams showing the evolution of the cell: (a) eukaryotes evolving from an ancestral prokaryote; (b) both eukaryotes and prokaryotes evolving from a common ancestral cell.

chloroplast having retained some of their original independence. The present-day prokaryotes are, according to this suggestion, the progeny of those cells in which the cytoplasmic membrane did not develop this pinocytotic ability. This suggestion was proposed in 1970 by R. Y. Stanier.

There is no evidence to support or reject either one of these courses of events. The fact that in present-day prokaryotic cells the cytoplasmic membrane lacks the plasticity to undergo the complete involutions necessary to take in objects with supramolecular dimensions may support the scheme of Fig. 2.8b; processes like pinocytosis have been observed only in eukaryotic cells.

2.4 The Biological Macromolecules

Many different molecules participate in biological reactions and these reactions, in turn, are of a wide diversity. Moreover, this diversity spreads out over the multitude of different kinds of cells and different stages of development. This diversity, however, is not a random one; on the contrary, it is a manifestation of a highly precise system of regulation and control. The system makes use of features built and organized along surprisingly unified lines. The biological macromolecules are the essential part of these features.

Biopolymers. There are four kinds of biological macromolecules. Three of these are *polymers*, that is, long sequences of relatively small molecules (which are of the same kind but not necessarily identical). These are the *proteins* (polymers of amino acids), the *nucleic acids* (polymers of nucleotides), and the *polysaccharides* (polymers of sugars). The fourth group consists of molecules which are known as *lipids*, loosely defined as that portion of animal or plant tissue which can be extracted with so-called "fat solvents" such as ethanol, ether, chloroform, and benzene. Many of the functions of the macromolecules in living systems are intimately bound to their structure. A discussion of their structure–function relationship is given in Chapter 3. Here, we will look at their composition.

Proteins. A large part of the dry weight of a cell is *protein*. It is a major class of compounds found in all living matter. *Enzymes* are proteins which catalyze and control the rates of many biological reactions; muscular contraction depends on the proteins *myosin* and *actin*; active transport across cell membranes and energy conservation seem to depend on the properties of the protein–lipid complexes in the membranes.

The building blocks of proteins are the *amino acids*. These are relatively small molecules consisting of a carboxyl group (I) and a basic amino group (II) attached to a central carbon atom, the so-called α-carbon. Also attached to the α-carbon is a molecular group R, called the *residue*; the general form of an α-amino acid, is III.

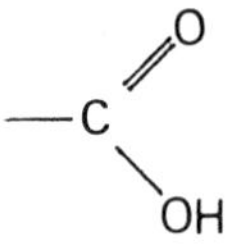

I 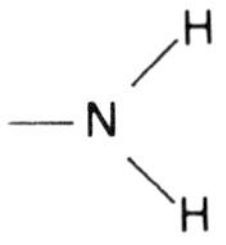 II 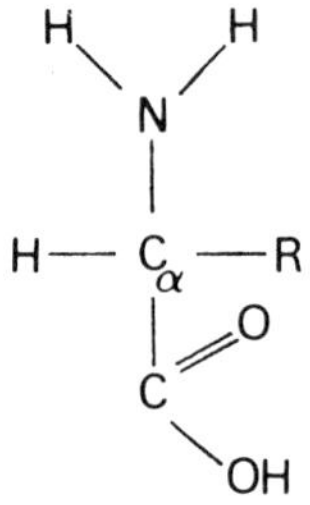 III

Amino acids can react with each other to eliminate, between each two of them, a molecule of water, thus forming a *peptide bond* (2.1).

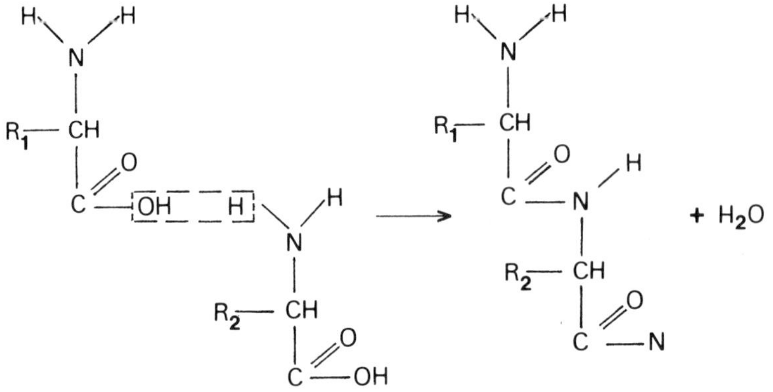

(2.1)

These peptide bonds are covalent bonds and very stable. Long chains of amino acids bound together by peptide bonds are called *polypeptide chains*. When 50 or more amino acids are bound this way, we refer to the chain as a protein.

The specific residues present determine the individual amino acids. Many residues exist and accordingly many α-amino acids can be, and have been, synthesized. In living cells, however, only 20 of them occur. These 20 amino acids together with their names and the abbreviations indicating the residues are given in Table 2.1. They are ordered in three groups according to the reactive properties of the residues. In the first group we have the

TABLE 2.1 Amino Acids

Amino acid	Residue structure	Abbreviation
I. Neutral		
Glycine	—H	gly
Alanine	—CH_3	ala
Valine	—CH(CH_3)$_2$	val
Leucine	—CH_2—CH(CH_3)$_2$	leu
Isoleucine	—CH(CH_2—CH_3)(CH_3)	ile
Serine	—CH_2—OH	ser
Threonine	—CH(OH)(CH_3)	thr
Phenylalanine	—CH_2—C_6H_5 (C, HC=CH, CH, HC—CH ring)	phe
Tryptophan	—CH_2—C=CH—NH— (indole ring: C, C, HC=CH, CH=CH, C, N H)	trp
Methionine	—CH_2—CH_2—S—CH_3	met
Histidine	—CH_2—C=CH (ring: HN, CH, N)	his

TABLE 2.1 *Continued*

Amino acid	Residue structure	Abbreviation
Tyrosine	$-CH_2-C_6H_4-OH$	tyr
Asparagine	$-CH_2-C(=O)-NH_2$	asn
Glutamine	$-CH_2-CH_2-C(=O)-NH_2$	gln
Cysteine	$-CH_2-SH$	cys
Proline	$^-O-C(=O)-C_\alpha$; ring $C_\alpha-H_2C-CH_2-CH_2-NH_2-C_\alpha$	pro
II. Polar, negative		
Aspartic acid	$-CH_2-C(=O)O^-$	asp
Glutamic acid	$-CH_2-CH_2-C(=O)O^-$	glu
III. Polar, positive		
Arginine	$-CH_2-CH_2-CH_2-NH-C(NH_2)=N^+H_3$	arg
Lysine	$-CH_2-CH_2-CH_2-CH_2-N^+H_3$	lys

neutral, or nonpolar residues. These have no net electrical charge and are insoluble in polar solvents such as water. In fact, in their tendency to avoid water the nonpolar chains tend to aggregate with each other. This is particularly true of aliphatic groups which are not as restrained by steric hindrances. The second and third groups contain the predominantly negative and the predominantly positive polar residues, respectively. These groups can interact with other polar groups or molecules, thus forming noncovalent bonds. It is self-evident that these interactions help determine the three-dimensional structure of the protein (see Chapter 3). In an aqueous (water) environment the nonpolar residues, avoiding the water molecules, tend to stay inside the protein while the polar molecules are more on the outside.

One type of residue interaction can lead to other covalent bonds in the protein molecule, for example, the bond between the sulfur atoms in cysteine. The dimer of cysteine, formed by the covalent bond between the sulfur atoms of each after removal of the hydrogen, is called *cystine*. This dimer is responsible for the so-called sulfur bridges which either link two polypeptide chains together (Fig. 2.9a) or make a loop in one polypeptide chain (Fig. 2.9b).

Nucleic Acids. *Nucleic acids* are the biopolymers responsible for the preservation of biological identity. They are bearers and conveyers of all

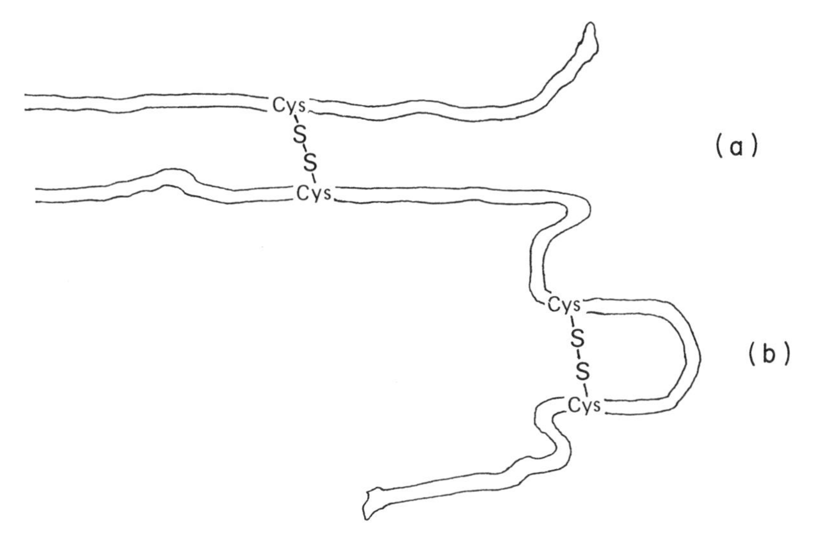

Fig. 2.9 Tertiary structure of protein formed by S bonds.

the information regarding the structure and behavior of the living system. The way in which they accomplish this will be discussed in Chapter 3. Their polymeric character, expressed in sequences of monomeric units in a very stable configuration, is an essential feature for this function. Nucleic acids are polymers of a molecular group called *nucleotides*. A nucleotide consists of a sugar, a base, and a phosphate, and the polymer is formed by covalent diphospho-ester bonds (IV). When the sugar is a five-membered

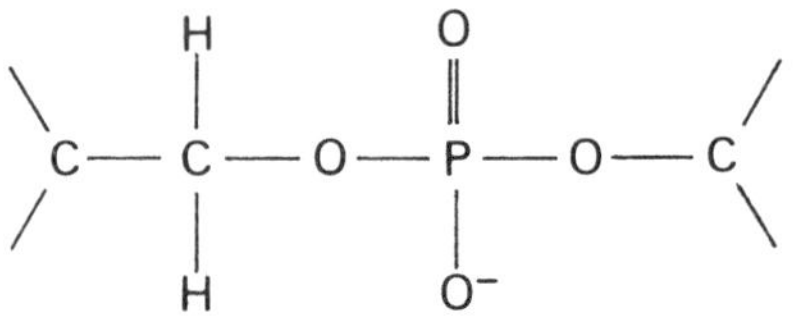

IV

ring (pentose) with the configuration V, it is called D-*ribose* and the nucleic acid is ribonucleic acid (RNA). When the sugar is a five-membered ring (pentose) with the configuration VI, it is D-2-deoxyribose and the nucleic

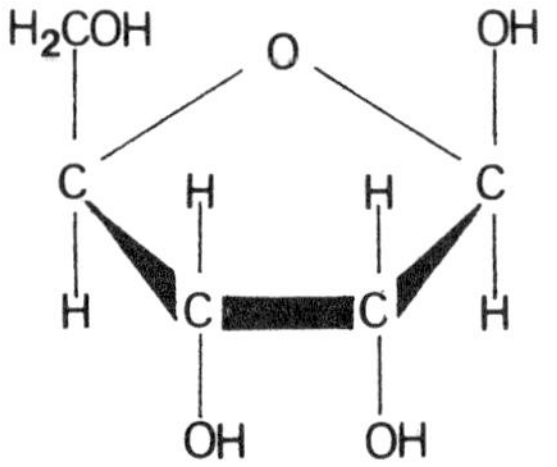

V

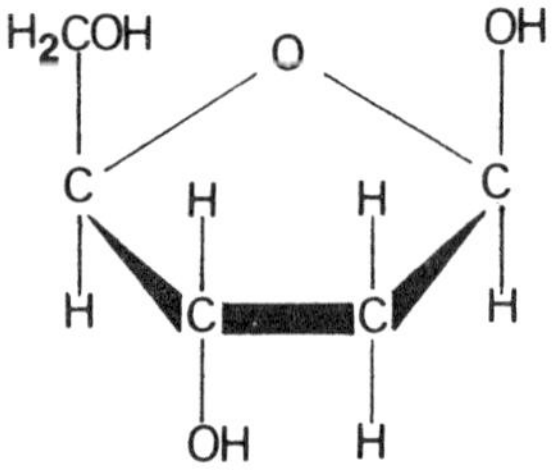

VI

acid is deoxyribonucleic acid (DNA). In both names the D stands for right-turning optical activity.

There are two kinds of bases, those derived from *pyrimidine* (VII), and those derived from purine (VIII). The two pyrimidines found in DNA are *cytosine* (C) (IX), and *thymine* (T) (X). The two pyrimidines found in RNA are cytosine and *uracil* (U) (XI). In both DNA and RNA the purines are *adenine* (A) (XII) and *guanine* (G) (XIII). (In IX–XIII, the carbon atoms of the ring structure are not indicated.) The bases are attached to C_1-carbon of the sugars (see Fig. 2.10). DNA has a double-stranded helical structure with the bases bound to each other by *hydrogen bonds* (see Fig. 2.10), a type of electrostatic interaction which will be discussed in Chapter 4.

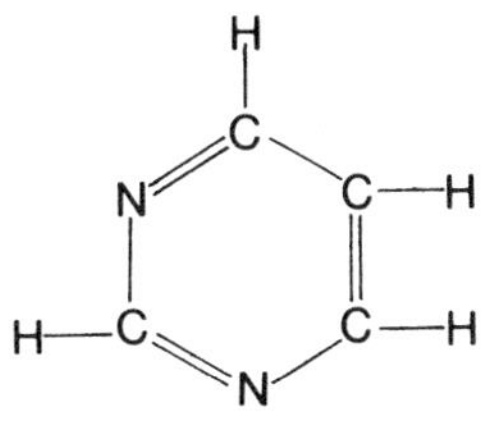

VII

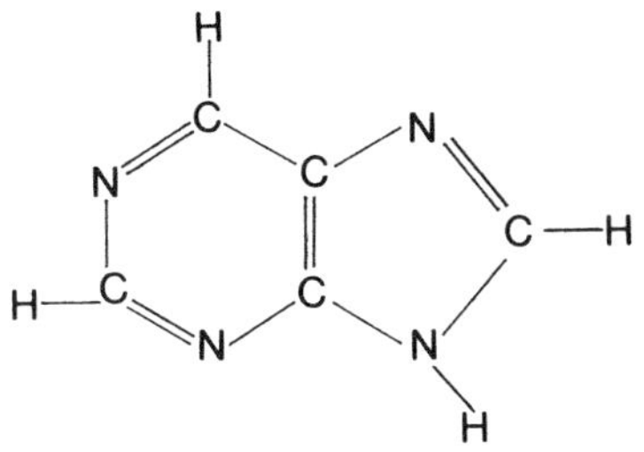

VIII

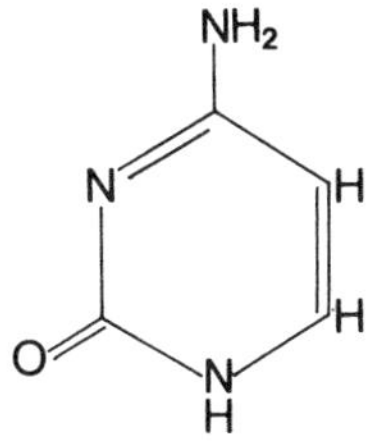

IX

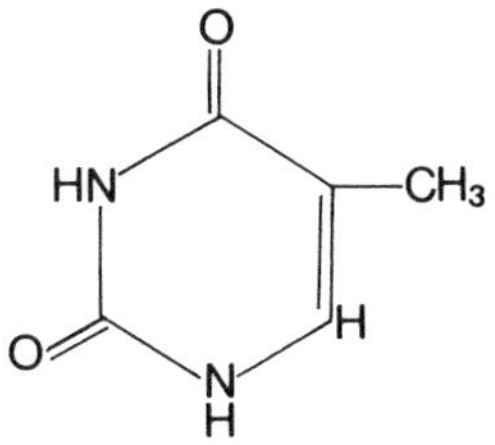

X

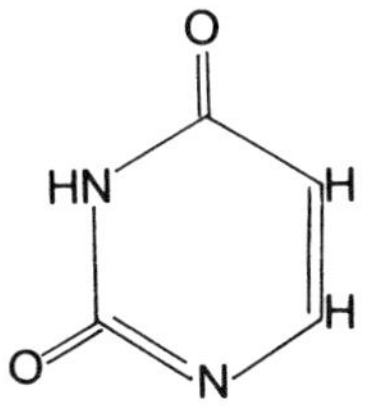

XI

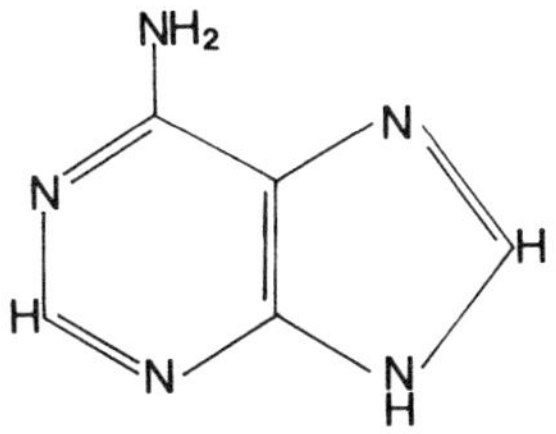

XII

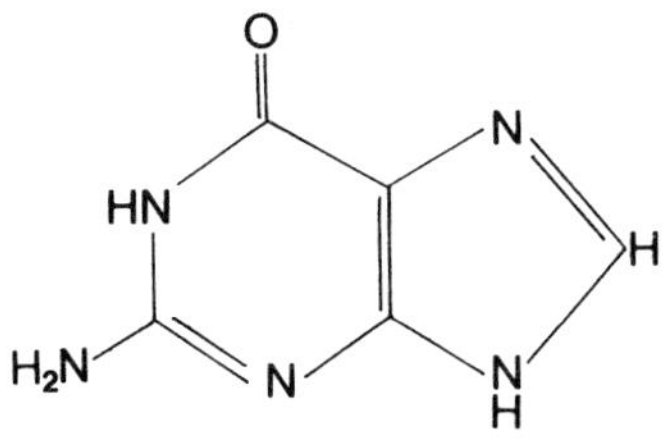

XIII

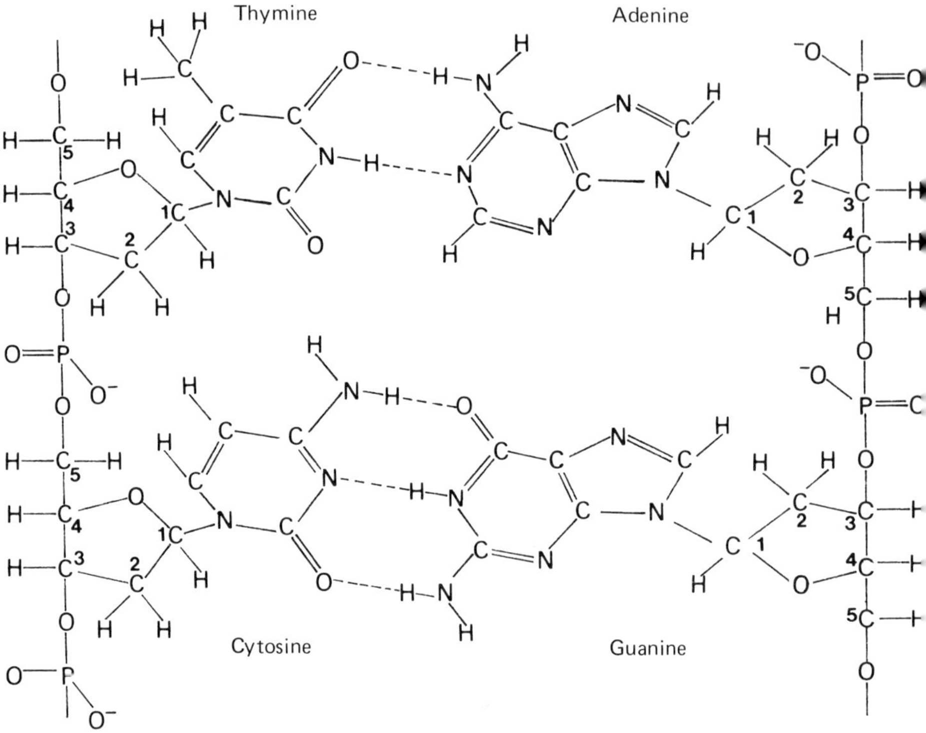

Fig. 2.10 The chemical structure of DNA. The bases thymine, adenine, cytosine, and guanine are attached to deoxiribose–phosphate strands. Two such strands are bound to each other by hydrogen bonds between the bases: two between thymine and adenine, three between cytosine and guanine; thus forming the twisted double-stranded structure. The phosphates in the sugar–phosphate sequence are connected at one side to the 5th carbon atom of the sugar and at the other side to the 3rd carbon of the sugar. In respect to the bases, the sequence runs 5 → 3 in one strand and 3 → 5 in the other strand. The two strands thus are *antiparallel.*

Polysaccharides. Polysaccharides are polymers made up of sugars. The most common sugars making up these long chains are hexoses (six-membered carbon rings). An example is glycogen (XIV), made up of glucose pyranose rings linked together with covalent, so-called glycoside, bonds. Polysaccharides play a role in surface action (cellulose, slime) and in food storage (starch).

Lipids. The *lipids* are a heterogenous group of molecules which can play various roles throughout the cell. As fat (compounds of *glycerol* and *fatty acids*) they can store energy which need not be immediately available. A typical neutral fat is XV. If the third chain is replaced by a phosphoric group, which in turn can be combined with a variety of bases, the compound is a *phospholipid.* An example is *lecithin* (XVI). The molecular group in brackets is a base called *choline.*

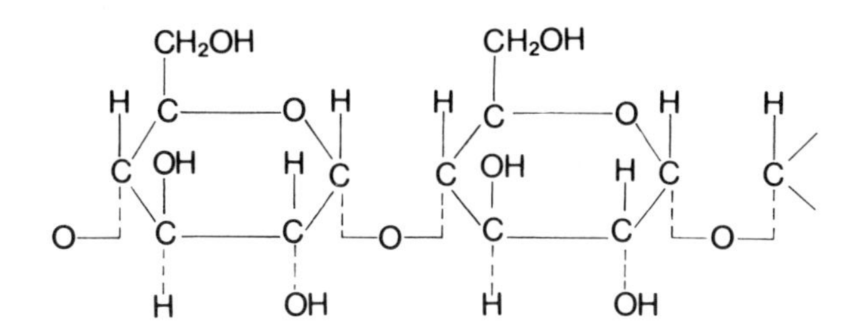

XIV

$$\begin{array}{l} H_2C - O - C(=O) - (CH_2)_{16} - CH_3 \\ | \\ HC - O - C(=O) - (CH_2)_{16} - CH_3 \\ | \\ H_2C - O - C(=O) - (CH_2)_{16} - CH_3 \end{array}$$

XV

$$\begin{array}{l} H_2C - O - C(=O) - (CH_2)_{16} - CH_3 \\ | \\ HC - O - C(=O) - (CH_2)_{14} - CH_3 \\ | \\ H_2C - O - P(=O)(OH) - O - [CH_2 - CH_2 - N^{+}(CH_3)_3] \end{array}$$

XVI

CH_3
C
CH_3
H_2 CH_3
C C C_3H_6 — C — CH_3
H_2 H
C C CH_2 H
H_2 CH_3 H
C C CH_2
H_2C H
C C C
H
C C CH_2
HO H C H
H H_2

XVII

An example of a *steroid lipid* is *cholesterol* (XVII). Cholesterol may have an insulating function in some membrane structures (indicated by the fact that the cholesterol content of the myelin sheath is unusually high). It is most probably a precursor for many if not all steroid hormones.

Lipids, and especially the phospholipids, are important structural factors in biological membranes. Although they cannot form polymers by covalent bonding, they can interact to form sheetlike structures. The long fatty-acid chains form nonpolar tails to the molecules which avoid contact with highly polarized water molecules (they are, therefore, often called *hydrophobic* groups). As a result, these parts tend to attract each other and aggregates can be formed which have these hydrophobic groups sticking to each other on the inside and the polar (or hydrophylic) groups pointing outward. Such structures are called *micelles* (Fig. 2.11a). If the aggregation occurs under well-controlled conditions, structures can be formed in which the hydrophobic tails are more or less parallel to each other on the inside of a prolonged sheet with the hydrophylic heads lining the sheet on the outside. A *lipid bilayer* is just such a structure (Fig. 2.11b). There is evidence that most, if not all, membranes have this structural backbone.

2.5 Membranes

So far there is no generally accepted explanation of biological membrane structure that provides an understanding of how membranes carry out their wide variety of functions. Many membranes (in particular the myelin sheath) show up, in electron micrographs of thin sections of cells, as trilaminar structures consisting of a light (electron transparent) layer 35–40 Å thick, bordered on each side by a dark (electron dense) layer about 20 Å thick. There is little doubt that the electron-transparent part is a *lipid*

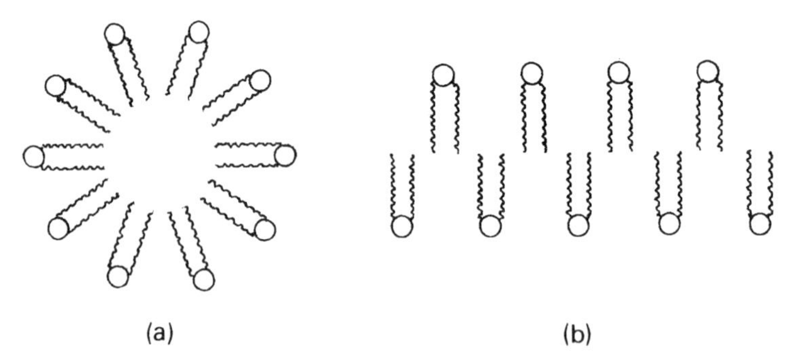

Fig. 2.11 Hydrophobic bonding of phospholipids; (a) in a micelle and (b) in a lipid bilayer.

bilayer while the electron-dense part consists of proteins and/or glycoproteins. The lipoprotein composition seems to be confirmed by many physiological, analytical, physicochemical, and electron-microscopic studies. It is the structural relationship between the lipid and the protein moyeties, much in debate at present times, that must give a clue for the explanation of the wide variety of functions.

Membrane Function. Three broadly defined types of functions seem to be associated with membranes. First, as boundaries of cells and cell organelles, they *create* and *maintain* a definite chemical composition inside which can be quite different from the outside environment. They do this continuously by a combination of selective *passive* diffusion and selective *active* (energy consuming) transport across the membrane. This process will be discussed in more detail in Chapter 5; the mechanisms by which the transport of material across the membranes occurs is not quite known, however. Second, membranes can form a basis on which rapid chemical transformations, requiring an *efficient* supply of reactants and an *efficient* disposal of products and waste, take place. A variety of enzyme systems are associated with, or can be an integral part of membranes. These systems not only govern the transport of ions and/or molecules, but also govern the rate of various biological reactions. A substantial part of the energy-conserving reactions in biological systems requires membranes, thus demonstrating the essential character of membrane systems. Finally, membranes occur as electrical *insulation* material around the fibrous extensions of nerve cells called *axons*. Such an insulation is found in some vertebrate nerve cells and is called the *myelin sheath*. It originates from a satellite cell (called a *Schwann cell*) which winds itself, during the developmental stages, around the axon as indicated in Fig. 2.12. The entire cytoplasmic membrane of the Schwann

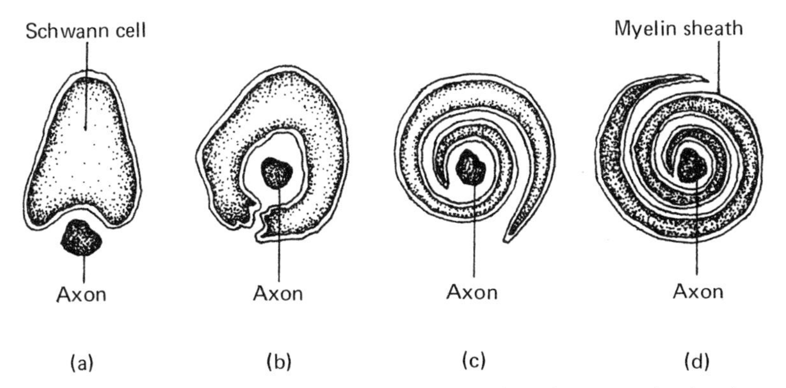

Fig. 2.12 Schematic representation of the myelin sheathing of an axon, showing the progressive envelopment of the axon by the membranes of a Schwann cell.

cell thus forms an elaborate structure serving a specific function in connection with another cell, the neuron.

The myelin sheath is one demonstration of the flexibility of membranes, as their appearance is modified according to a specific function. Another example is the stacking of membrane portions in the form of flat disks as occurs in the rod cells of the retina (Fig. 2.13a) and in the grana of chloroplasts in a leaf cell (Fig. 2.13b). A common feature of these membrane stacks is that they contain light-receptor systems; the retina rod cell is an

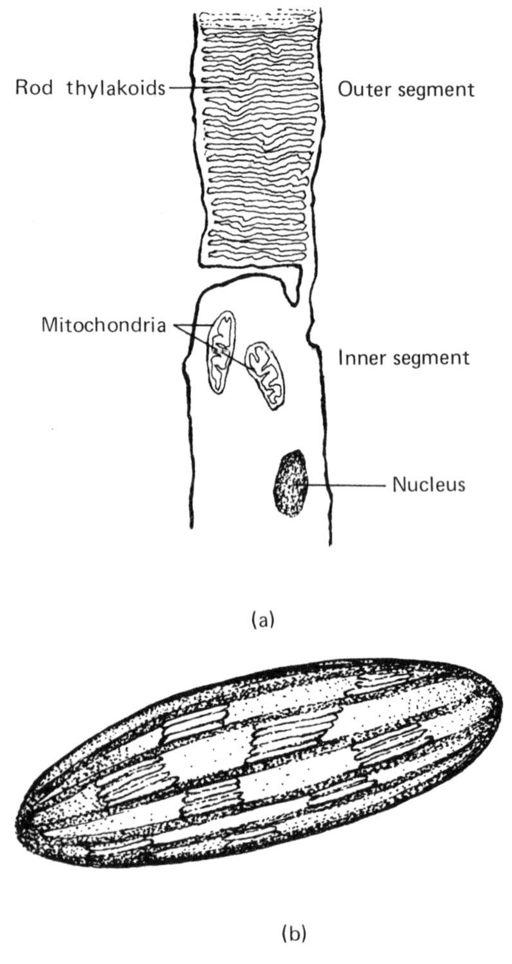

Fig. 2.13 Examples of the stacking of membrane portions in the form of flattened, disklike vesicles called *thylakoids*. (a) The rod receptor of a vertebrate retina; (b) a chloroplast.

essential part of the process of vision and the chloroplast has the necessary apparatus for photosynthesis.

The Unit Membrane. The well-known Danielli–Davson model of membranes provides a central core of lipids in a bilayer with hydrophobic tails opposing each other and hydrophilic heads pointing outward. Globular proteins were postulated to cover the lipids in layers like a sandwich. Later, this model was modified by J. D. Robertson, who proposed the widely accepted *unit membrane* (see Fig. 2.14). In this model the lipid–protein "sandwich" of Danielli was retained but the protein molecules were thought to spread out across the lipid bilayer in uninterrupted sheets with a structure more flattened than the globular configuration of Danielli's model. Electron micrographs of a myelin sheath clearly showed the "railroad track" image consistent with this model, X-ray diffraction patterns also seemed to confirm it. This was held as characteristic, however, for all types of membranes, including those of the intracellular organelles. Only recently has the generality of the unit membrane model again been put in question.

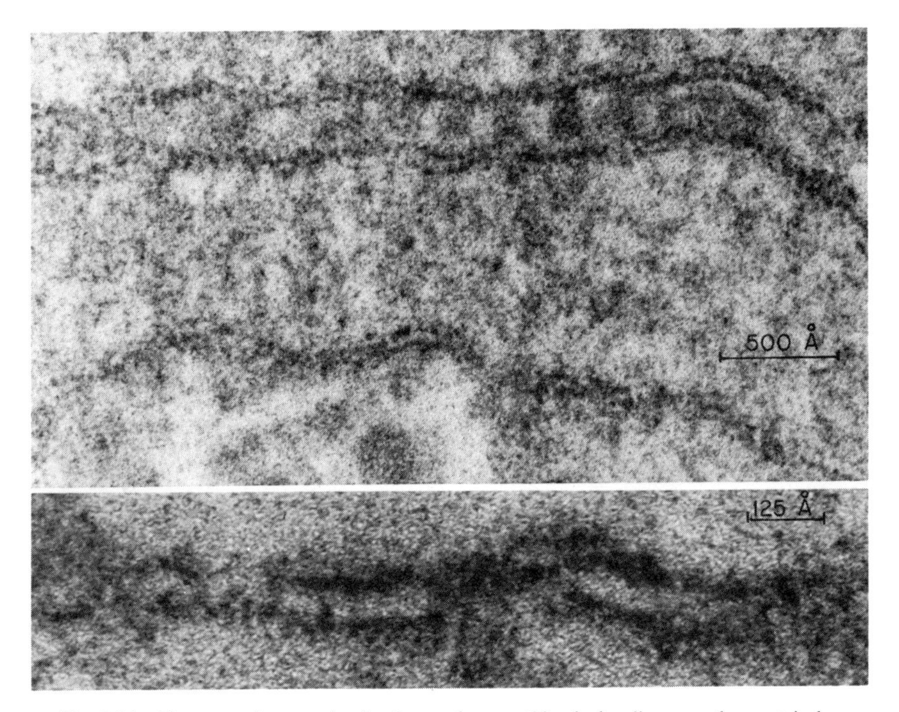

Fig. 2.14 Electron micrograph of unit membranes. The darker lines are the protein layers covering the lighter lipid bilayer. Courtesy of Dr. W. Jacob and Dr. A. Van Laer, Electron Microscopy Laboratory, Universitaire Instellingen Antwerpen, Antwerp, Belgium.

This is mainly due to the availability of more sophisticated optical techniques, such as optical rotary dispersion (ORD) and circular dichroism (CD), for the analysis of membrane proteins. What seems to be firmly established is the presence of a lipid bilayer core. The protein structure, however, is now in debate. The picture which seemed to evolve is a dynamic one in which the amount of "α-helix protein" (see Section 3.5) can be correlated with membrane function. In some cases evidence is found that proteins penetrate deeply into the lipid layer, spanning it rather than covering it in a sandwichlike fashion. The idea that proteins move over and across membranes also seems to have gained ground, especially where transmembranous transport is concerned.

However, more than just the structural aspects of proteins and lipids in membranes are important for the understanding of their functions. In addition to these two major components there are minor ones which may have decisive roles. Carbohydrates (sugars) are found connected to protein, as well as ions and various other small molecules. The function of these small molecules is largely unknown. It seems clear, however, that analyzing the membrane composition is as important as clarification of the structural aspects in order to get a better understanding of membrane functions.

Chapter

3 Biological Structure and Function

3.1 Gross Physical Structure Determination of Biological Macromolecules

Most of the biological macromolecules described in the previous chapter have molecular weights well over 10,000. Since these molecules are in some way or another engaged in the multitude of reactions in a living cell and since the progress and control of these reactions depend on their arrangement and structure, a knowledge of their sizes, shapes, and detailed structure is imperative for understanding their functions and the processes in which they are involved. A number of techniques are used to determine these structures. Many of these techniques involve the observation of deviations due to large size from the idealized behavior of molecules, molecular groups, or particles under certain conditions; osmotic pressure is here given as an example.

Osmotic Pressure. For dilute solutions the osmotic pressure Π obeys the well-known van't Hoff relation

$$\Pi V = nRT \qquad \text{or} \qquad \Pi = CRT \tag{3.1}$$

in which V is the volume, n the number of moles, C the concentration,

R the gas constant, and T the temperature in absolute units. Experience shows that this relation accurately describes situations for which the concentration does not exceed 1%. However, for far more dilute solutions of large molecules (such as rubber in benzene) extensive deviations from the van't Hoff relation are observed.

The questions which then arose were (1) how can the molecular weight at which failure of the van't Hoff law begins be determined and (2) how can the data on osmotic pressure be used to determine the size and shape of large molecules? McMillan and Mayer (1945) showed that the osmotic pressure Π of solutions of nonelectrolytes can be represented by a power series of C, the concentration in grams per cubic centimeter:

$$\Pi = \frac{RT}{M} C + a_2 C^2 + a_3 C^3 + \cdots \tag{3.2}$$

in which M is the molecular weight of the solvent molecules and $a_2, a_3, \ldots$ are coefficients which are independent of C. For fairly dilute solutions the third- and higher-order terms can be neglected so that (3.2) can be rewritten as

$$\frac{\Pi}{C} = \frac{RT}{M} + a_2 C \tag{3.3}$$

Thus, when measurements of the osmotic pressure Π at different values of the concentration C are made, a plot of Π/C as a function of C should yield straight lines at the lower values of C. The molecular weight M can then be determined from the intercept at $C \to 0$.

Light Scattering. Rayleigh worked out a theory of light scattering by isotropic particles whose dimensions were small compared with the wavelength of the incident light. According to this theory the relation between the intensity I of the light scattered at an angle θ, and observed at a distance R_s from the scattering volume, and the intensity I_0 of the incident light is

$$I_\theta R_s^2 / I_0 = R_\theta (1 + \cos^2 \theta) \tag{3.4}$$

in which R_θ, the coefficient of the *Rayleigh ratio*, is given by

$$R_\theta = 8\pi^4 \alpha^2 / \lambda^4 \tag{3.5}$$

in which α is the *polarizability* of the particle (see Chapter 4).

When the dimensions of the particle are no longer small, as compared to the wavelength of the light, the theory becomes more complicated because interference effects of the light scattered at different parts of the particle have to be taken into account. In an extension of the Rayleigh theory,

Debye, in 1947, showed that for dilute solutions of macromolecules the Rayleigh coefficient is given by

$$R_\theta = KCM$$

in which C is the concentration in grams per cubic centimeter; M is the molecular weight; and K is a function of the concentration, the refractive indices of the solvent and the solution, and the wavelength. For more concentrated solutions the relation

$$K\frac{C}{R_\theta} = \frac{1}{M} + \frac{2aC}{M^2}$$

gives a more accurate value for M. The influence of the size of the molecules can be accounted for by a particle scattering factor $P_{(\theta)}$, thus yielding

$$K\frac{C}{R_\theta} = \frac{1}{MP_{(\theta)}} + \frac{2aC}{M^2} \tag{3.6}$$

The scattering data can be obtained from measurements of the scattered light at different values of the scattering angle θ and the mass concentration C, extrapolated to zero angle and zero concentration. Thus, from Eqs. (3.4) and (3.6) values for M can be obtained. Measurements of light scattering are also used to determine the diffusion constant (see Section 3.2).

Other Methods. There are many other methods for the determination of gross physical appearance of the macromolecules (molecular weight, classification of shapes, etc.); a summary is given in Table 3.1. In the following we will discuss, in somewhat more detail, methods based on the motion of molecules due to their thermal energy and to outside forces such as gravity and centrifugal forces.

3.2 Diffusion

Friction and Viscosity. When large molecules move through a liquid with a velocity v they undergo a frictional force

$$F = -fv \tag{3.7}$$

The frictional coefficient f is related to the molecular size and shape, and to the viscosity η of the liquid. If we can determine f, and the constituents and concentration of the molecules are known, we can calculate the molecular weight, provided that the gram-molecular volume can be determined

TABLE 3.1 Physical Techniques Used for Studying Large Molecules[a]

Technique	Information obtained	Purity needed
Osmotic pressure	Molecular weight	Very high
Diffusion	Molecular volume if shape and bound water are known	High or low
Sedimentation velocity	Molecular volume if shape and bound water are known	Intermediate
Sedimentation equilibrium	Molecular weight	High
Viscosity	Shape if bound water is known	High
Flow birefringence	Shape if bound water is known	High
Electric birefringence	Shape if bound water is known	High
Dielectric dispersion	Shape if bound water is known	High
Rotational dispersion	Helical fraction of molecule	High
Light scattering	Molecular weight, anisotropy	Very high
X-ray scattering	Size and shape	Very high
Electrophoresis	Relates charge to size and shape	Intermediate
Electron microscopy	Size and shape	High
Electron scattering	Intermolecular spacings	High
X-ray crystal analysis	Atomic positions	Very high
Electromagnetic absorption	Intramolecular forces and orientation	High
Charged particle irradiation	"Sensitive volume" and its location	Low
Photon irradiation	Structural coherence	Low

[a] From R. B. Setlow and E. C. Pollard, "Molecular Biophysics," Addison-Wesley, Reading, Massachusetts, 1962.

from the shape parameters. The latter is rigorously possible only for spherical molecules for which Stokes's law

$$f = 6\pi\eta a \tag{3.8}$$

a being the radius of the molecule, is valid, and for ellipsoid molecules for which a similar relation can be derived. For more complicated shapes, approximations have to be applied in order to arrive at an estimate of the gram-molecular volume.

Diffusion. In order to determine f we must have the molecules move through a medium. This can be accomplished by making use of the process of *diffusion* in which the driving force for the movement is provided by a concentration gradient, or by *sedimentation* in which the driving force is gravity or centrifugal force.

Diffusion is a direct result of the random motion of the molecules. If there is a region of high concentration in a vessel, the net random movement will be out of that region and into regions of lower concentration. Fick's law states that if dn/dt is the number of molecules that in one dimension pass through a cross section A per unit time

$$dn/dt = -DA\, \partial c/\partial x \tag{3.9}$$

in which $\partial c/\partial x$ is the (one-dimensional) concentration gradient and D is the *diffusion constant*. This diffusion constant is related to the friction coefficient

$$D = kT/f \tag{3.10}$$

in which k is the Boltzmann constant and T is the absolute temperature (see also Section 5.5). Table 3.2 lists values of D for some protein molecules.

The diffusion constant can be measured by making use of the setup illustrated in Fig. 3.1. Two containers, one filled with a solution at the initial concentration $c = c_0$ and the other with the solvent ($c = 0$), can be moved so as to bring the solutions in contact with each other at time $t = 0$. Since the number of diffusing molecules in such an experiment is conserved, the

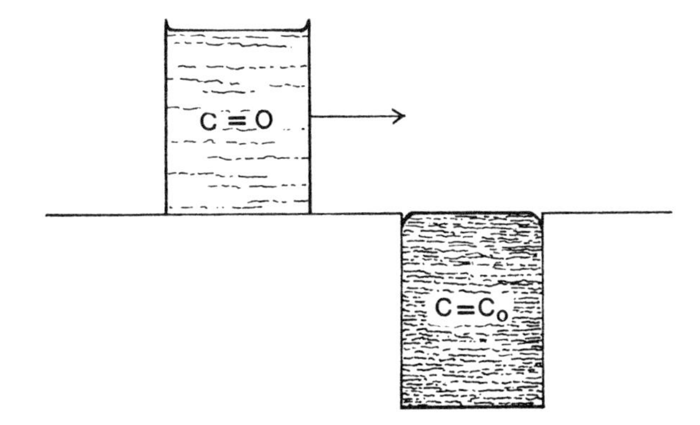

Fig. 3.1 Schematic representation of a diffusion experiment. Two containers, one containing a solvent and the other containing a solution with concentration c_0, are moved so as to bring the two solutions into contact at $t = 0$, thus forming a sharp initial boundary.

increase in concentration within a volume of cross section A and length dx is the excess of material diffusing into that volume over that diffusing out. This leads to the diffusion equation

$$\partial c/\partial t = D\, \partial^2 c/\partial x^2 \tag{3.11}$$

The boundary conditions (for $t = 0$), determined by this experiment, are $c = 0$ for $x < 0$ and $c = c_0$ for $x > 0$. A solution is the probability integral

$$c = \tfrac{1}{2}c_0\left[1 + \frac{2}{\sqrt{\pi}} \int_0^{x/(4Dt)^{1/2}} e^{-\xi^2}\, d\xi\right] \tag{3.12}$$

the values of which are widely tabulated. From (3.12) it follows, for the concentration gradient, that

$$\frac{\partial c}{\partial x} = \frac{c_0}{(4\pi Dt)^{1/2}} \exp(-x^2/4Dt) \tag{3.13}$$

Figure 3.2 shows the concentration (a) and the concentration gradient (b) as functions of x at different values of t. The diffusion could be followed in the experimental setup illustrated in Fig. 3.1 by optical methods, and the diffusion constant D could be calculated from the tables of the probability integral. However, recent developments in light scattering spectroscopy (photon correlation spectroscopy; see, for instance Chen and Yip, 1974) which came about with the development of laser technology have made the determination of the diffusion constant much more easy. The linewidth Γ of scattered (originally coherent) laser light is equal to

$$\Gamma = 2D|\mathbf{k}|^2$$

in which $\mathbf{k}$ is the scattering vector. This vector is given by

$$\mathbf{k} = 2\mathbf{k}_0 \sin \tfrac{1}{2}\theta$$

in which $\mathbf{k}_0$ is the wave vector of the incident laser light and θ is the scattering angle. Thus, by measuring the linewidth Γ of the scattered laser light at different scattering angles and plotting Γ against $|\mathbf{k}|^2$, a direct and unique determination of D is possible. There is no longer a need to establish a concentration gradient in order to initiate diffusion or to avoid complications from mechanical disturbances, temperature gradients, and electrical charges on the molecules or particles.

3.3 Sedimentation

Gravitation. The masses of particles or large molecules are usually determined by observing their motion under the influence of outside forces

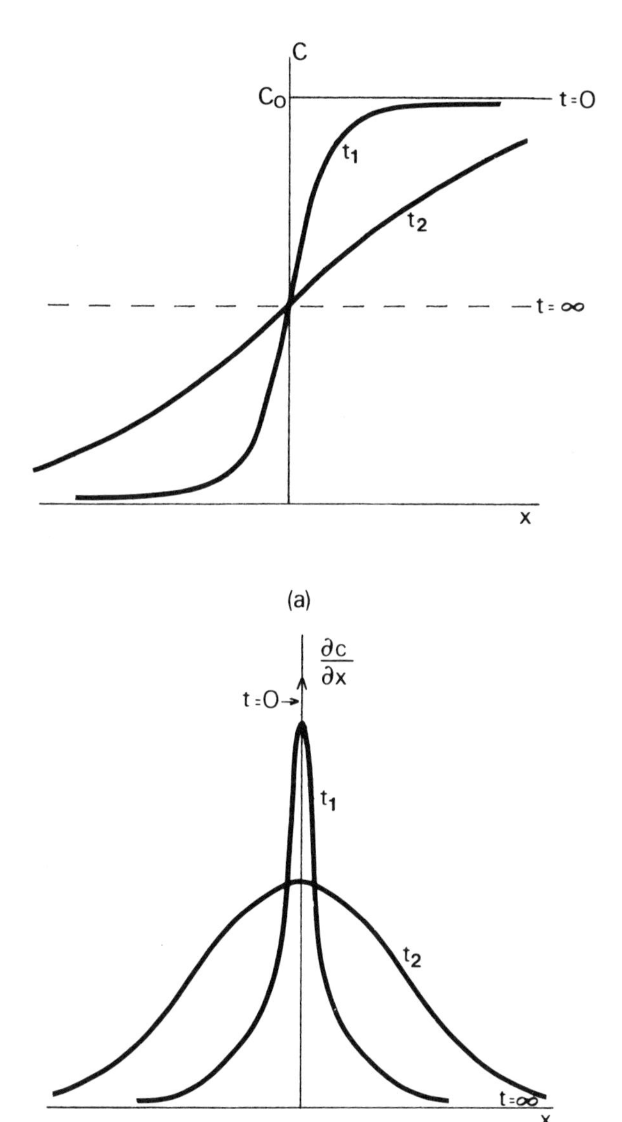

Fig. 3.2 (a) The concentration and (b) the concentration gradient of a diffusing solution as a function of distance at different times after the time t_0 at which the solution came into contact with solvent.

such as gravity or centrifugal force. In a gravitational field the terminal velocity of a particle is determined by the balance between the force of gravitation and the frictional resistance of the fluid medium to the motion of the particle. The force of gravitation, which is the difference between the weight of the particle and the buoyant force due to the displacement of the fluid medium, is $mg - V\rho_0 g$, where m is the mass of the particle, V its volume, and ρ_0 the density of the fluid medium. Since $V = m/\rho$, where ρ is the density of the particle, the force of gravitation can be written as

$$mg\left(1 - \frac{\rho_0}{\rho}\right)$$

When this force balances with the frictional force fv, the uniform velocity v at which the particle sediments will be

$$v = \frac{mg}{f}\left(1 - \frac{\rho_0}{\rho}\right) \tag{3.14}$$

In general, the weakness of the gravitational field restricts the observation of the sedimentation of particles in such a field to quite massive particles.

Centrifugation. The development of centrifugal techniques that can generate "gravitational fields" up to $300{,}000g$ makes this method suitable for large molecules such as proteins, as well as smaller molecules. In this case g in Eq. (3.14) has to be replaced by $\omega^2 r$, in which ω is the angular velocity of the centrifuge and r is the distance between the particle and the center of rotation. The *sedimentation constant*

$$s = v/\omega^2 r \tag{3.15}$$

is a characteristic constant for a given molecular species in a given solvent. It is expressed in *Svedberg units* after The Svedberg, who was instrumental in the development of the technique. One Svedberg unit is equal to 10^{-13} sec. Combining (3.14) and (3.15) then yields

$$s = \frac{m}{f}\left(1 - \frac{\rho_0}{\rho}\right) \tag{3.16}$$

Since f is determined only for spherical and ellipsoid particles we would like to eliminate it, for instance by an additional measurement. When the diffusion constant is known, we can make use of relation (3.10) giving, for the mass m,

$$m = \frac{skT}{D[1 - (\rho_0/\rho)]} \tag{3.17a}$$

and for the molecular weight,

$$M = \frac{sRT}{D[1 - (\rho_0/\rho)]} \tag{3.17b}$$

TABLE 3.2 Characteristic Constants of Some Protein Molecules at 20°C[a]

Proteins	$1/\rho$ ($cm^3\ g^{-1}$)	s (10^{-13} sec)	D ($cm^2\ sec^{-1}$) $\times 10^7$	$M \times 10^{-3}$
Myoglobin (beef heart)	0.741	2.04	11.3	16.9
Hemoglobin (horse)	0.749	4.41	6.3	68
Hemoglobin (man)	0.749	4.48	6.9	63
Hemocyanin (octupus)	0.740	49.3	1.65	2800
Serum albumin (horse)	0.748	4.46	6.1	70
Serum albumin (man)	0.736	4.67	5.9	72
Serum globulin (man)	0.718	7.12	4.0	153
Lysozyme (egg yolk)	(0.75)	1.9	11.2	16.4
Edestin	0.744	12.8	3.18	381
Urease (jack bean)	0.73	18.6	3.46	480
Pepsin (pig)	(0.750)	3.3	9.0	35.5
Insulin (beef)	(0.749)	3.58	7.53	46
Botulinus toxin A	0.755	17.3	2.10	810
Tobacco mosaic virus	0.73	185	0.53	31400

[a] From W. Moore (1972), "Physical Chemistry," 4th ed., p. 939. Reprinted by permission of Prentice-Hall, Inc., Englewood Cliffs, New Jersey, U.S.A.

in which R is the gas constant. Table 3.2 lists the sedimentation constant, diffusion constant, molecular weight, and the inverse of the density (or specific volume) of some molecules at 20°C.

Sedimentation Rate and Equilibrium. The sedimentation rate of a molecule can be determined by rotating a suspension (solution) of the molecule in an analytical centrifuge (Fig. 3.3) and observing the moving boundary. By making use of a special optical setup known as *Schlieren optics* one can record the boundary in terms of a concentration gradient. One can thus obtain photographs of typical Schlieren patterns such as those shown in Fig. 3.4 for a preparation of *E. coli* ribosomes.

At equilibrium the rate at which the particles (molecules) move from the rotation axis just equals the rate at which they move in the opposite direction as a result of the concentration gradient. The rate at which the particles sediment (per unit area) is equal to the velocity of sedimentation times the concentration:

$$\frac{dn}{dt} = vc = c\frac{M\omega^2 r}{RT}D\left(1 - \frac{\rho_0}{\rho}\right)$$

The rate at which the particles diffuse back (per unit area) is given by Fick's law:

$$\frac{dn}{dt} = -D\frac{\partial c}{\partial r}$$

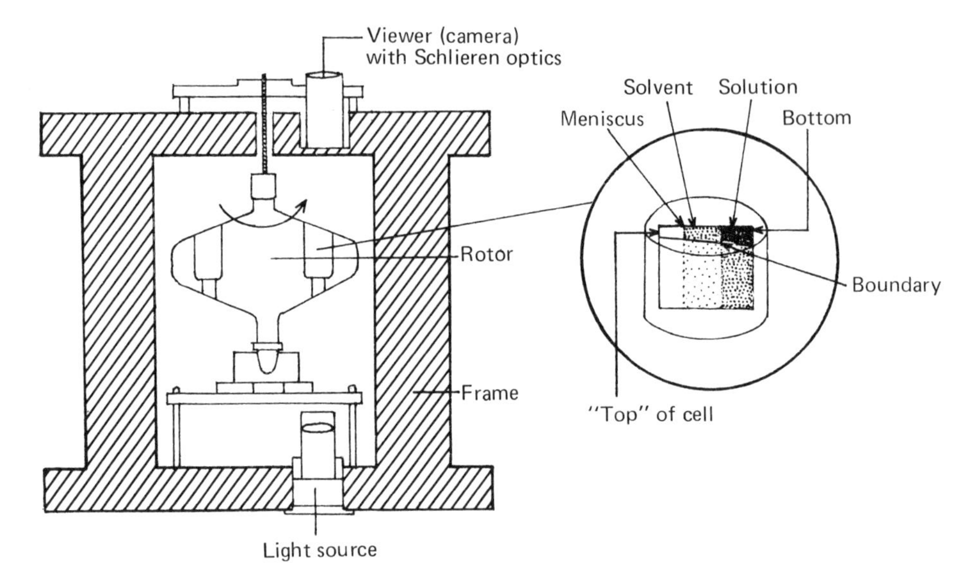

Fig. 3.3 Schematic representation of an ultracentrifuge and the centrifuge cell.

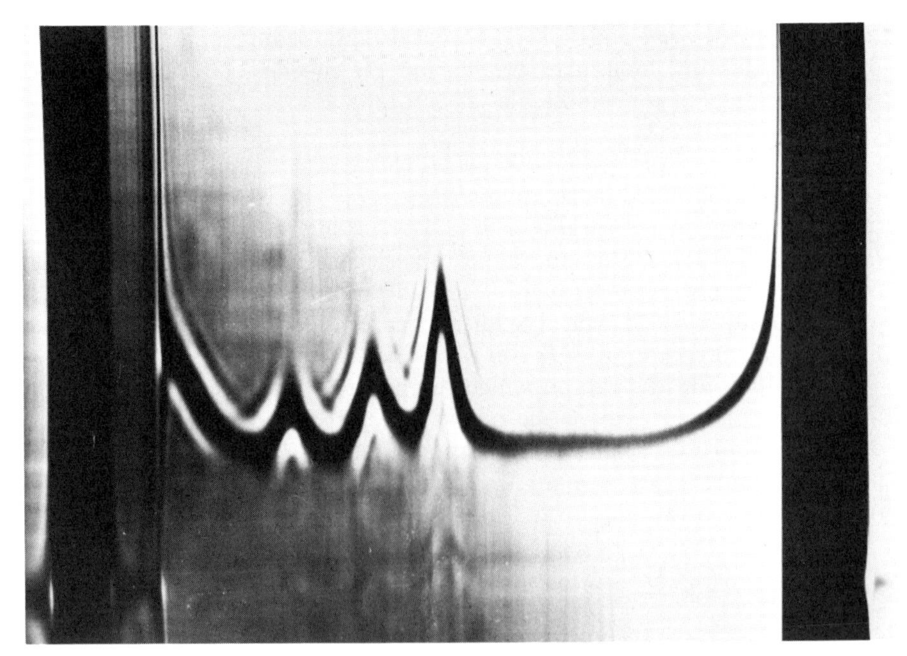

Fig. 3.4 A photograph of a Schlieren pattern of a suspension of ribosomes from the colon bacterium *E. coli*. The three peaks correspond to (from left to right) the sedimentation rates of the two subunits, 30 S and 50 S, and that of the whole ribosome, 70 S.

At equilibrium the two rates are equal and we obtain

$$\frac{dc}{c} = \frac{M\omega^2}{RT}\left(1 - \frac{\rho_0}{\rho}\right) r\, dr \tag{3.18}$$

Integration of the differential equation (3.18) between r_1 and r_2 yields

$$M = \frac{2RT \ln(c_2/c_1)}{[1 - (\rho_0/\rho)]\omega^2({r_2}^2 - {r_1}^2)} \tag{3.19}$$

This expression (3.19) for the molecular weight is independent of the size and shape of the particles (molecules). The sedimentation equilibrium method, therefore, does not need an independent measurement of D in order to fix the molecular weight. The time required to reach the equilibrium condition, however, is so long that the method is not practical for substances having a molecular weight greater than 5000. A modification of the method makes use of the fact that at the meniscus and at the bottom of the cell there cannot be any net flux. Since the equilibrium condition states that there is no net flux at all times across any plane in the solution, the measurement of the concentrations at the plane of the meniscus and at the plane of the bottom of the cell shortly after the centrifuge is brought to speed could be used to give the equilibrium values.

Density Gradient. If a solution of a substance of low molecular weight (e.g., sucrose) is centrifuged, there will be a density gradient across the cell at equilibrium. If we add a substance of high molecular weight to the cell it should float in this solution of varying density at the particular position at which its buoyant density equals the density of the solution. If the substance is made up of various fractions of different molecular weight, each fraction should separate out in a band at a particular plane of the solution. This method of separation is called the *density-gradient method.* It is widely used in biochemical and biophysical research and has, for instance, been very successful in the separation of macromolecular components of the two photochemical systems in higher-plant photosynthesis (see Chapter 5).

3.4 X-Ray Crystal Structure Analysis

X-Ray Diffraction. Diffraction of X rays can occur when the spacing of the scattering units is small enough. Since a crystal is a regular array of a unit structure (the unit cell), diffraction of X rays by crystals occurs when the wavelength of the X rays is of the same order of magnitude as the distances between the atoms in the crystal. In 1912 the group of Max von Laue in Germany showed that this indeed was the case. Thus, if, as in Fig. 3.5, an X-ray beam is incident on a column of scattering units (atoms for example) in a crystal with a repeat distance a, a diffraction pattern results with maxima in the direction ϕ given by

$$a \sin \phi = k\lambda \tag{3.20}$$

The Bragg Relation. In three-dimensional arrays of scattering units the resulting pattern of maxima is too complicated to be analyzed in terms of a three-dimensional extension of the theory described. W. H. Bragg and his son W. L. Bragg developed a method which, essentially, reduces the three-dimensional problem to a two-dimensional one. They showed that the scattered X rays could be seen as being *reflected* from planes in the crystal made up of the individual scattering units.

In Fig. 3.6 such planes are indicated by lines connecting rows of scattering units. Incoming X rays, making an angle θ with these planes, are reflected at an equal angle θ. If the distance between two adjacent planes is d, constructive interference occurs when the difference in optical pathlength is again an integer multiple of the wavelength. Hence, from Fig. 3.6, it follows that maxima are observed in the direction θ given by

$$2d \sin \theta = n\lambda \tag{3.21}$$

the so-called *Bragg relation*. The essential difference between Eqs. (3.20)

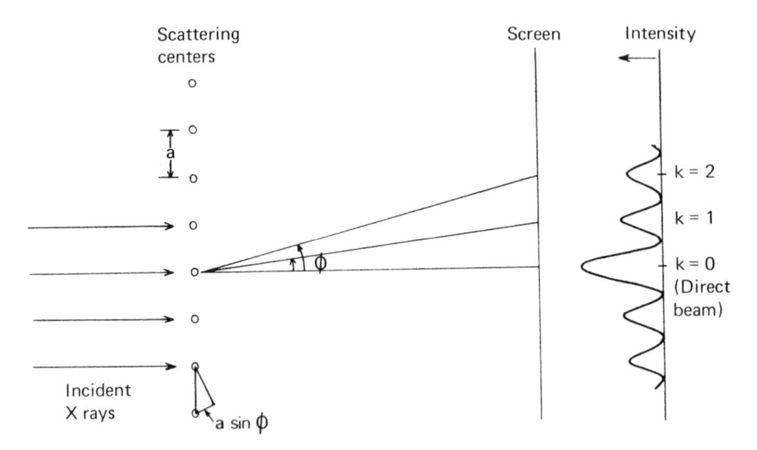

Fig. 3.5 Scattering of X rays from a single row of atoms. The intensity maxima occur as a result of constructive interference in the direction given by $a \sin \phi = k\lambda$, where a is the repeating distance between the scattering atoms.

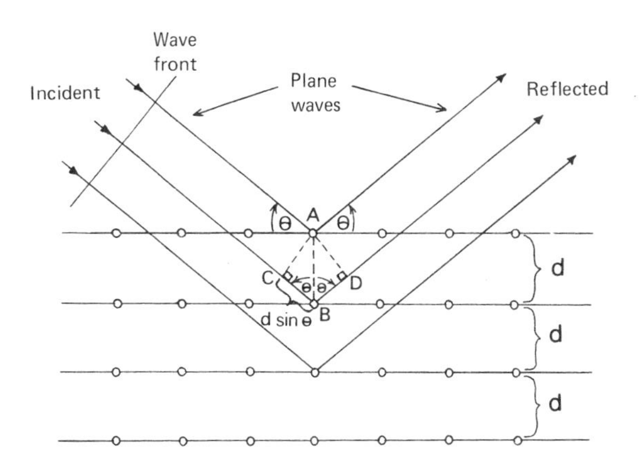

Fig. 3.6 Reflection of X rays from planes formed by scattering atoms. Constructive interference occurs in the direction given by an incident angle (which equals the reflection angle), satisfying the Bragg relation $n\lambda = 2d \sin \theta$, in which d is the distance between the reflecting planes.

and (3.21) is that in (3.21) the quantities d, θ, and λ are fixed in a resting system with monochromatic X rays, whereas, according to (3.20) a maximum will occur at a given scattering angle ϕ at any incident angle and at any value of a and λ, provided that the latter two values are compatible. According to (3.21) however, a maximum occurs only when the incoming X rays make an angle θ with a set of planes, satisfying the Bragg relation. Thus, although planes can be constructed in the crystal in a variety of directions, at a fixed direction of incidence no maxima will, in general, occur.

X-ray diffraction pattern from crystals can be obtained in one of the following ways.

(a) By using nonmonochromatic X rays the value of λ in Eq. (3.2) is made variable. The resulting diffraction pattern is known as a Laue pattern. Since the values of λ are not defined, the information obtainable from a Laue pattern is limited.

(b) One may use monochromatic X rays (obtained by selecting a line of the characteristic radiation, such as the K_α-line of Cu, with a suitable absorbing filter, such as Ni) on a mass of finely divided crystals with random orientations (a powder). In such a random-orientation arrangement some of the tiny crystals have orientations satisfying the Bragg relation (3.21). Since the reflected beam then makes an angle 2θ with the incident beam but the angles 2θ are oriented randomly around the incident beam, the maxima for each set of reflecting planes outline a cone. A photographic plate perpendicular with the direction of the incident beam will, therefore, show the maxima as concentric circles. These so-called Debye–Scherrer rings give direct information about the reflecting plane spacings.

(c) By rotating a single crystal in a fixed monochromatic X-ray beam the angle θ can be made variable. θ is measurable from the distance between the maxima showing on a cylindrical photographic film around the crystal. If the film is moved back and forth with a period synchronized with the rotation of the crystal, the positions of the spots on the film immediately indicate the orientations of the crystal. This technique is widely used for precise structure determinations. Since the value of λ is fixed, the plane distances d of the different sets of reflecting planes can be determined and thus, information can be obtained about the crystal structure. Also, the intensity of the spots, as will be seen later, contains important information about the fine structure of the crystal.

Miller Indices. In order to define the different sets of reflecting planes in the crystal, a set of indices has been developed. These indices, known as Miller indices, are defined *as the reciprocals of the fractional intercepts* of the planes with the a, b, and c axis of the unit cell. In Fig. 3.7 some planes in a cubic crystal are shown, illustrating the Miller indices. Planes which are parallel to one or two of the axes intercept at infinity. The Miller index then is 0 (Figs. 3.7b and 3.7c).

Diffraction Patterns. When a crystal is rotated in a beam of monochromatic X rays about an axis parallel to one of its unit cell sides, the

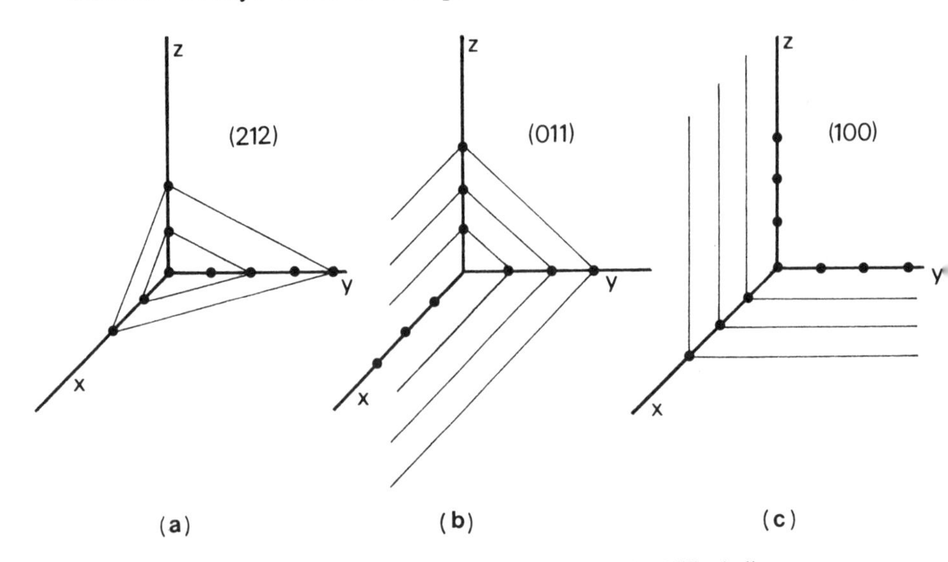

Fig. 3.7 Three examples of sets of planes given by Miller indices.

reflected radiation, according to Eq. (3.20), must outline a cone $\sin \phi =$ constant. These cones cut lines on a cylindrical film around the crystal. The maxima thus produce spots on the film which lie on these so-called *layer lines* and the distance between them is given by

$$\sin \phi_k = k\lambda/x$$

in which x is length of a unit cell axis (see Fig. 3.8). Thus, measurements of the layer line distance after rotation of the crystal about an axis parallel to the unit cell axes (a, b, or c) will yield the dimensions of these unit cell axes. The spacings between the spots on the film are determined, of course, by the Bragg relation (3.21) and indicate the reflecting plane distances d.

Intensity of Diffraction Maxima. As an example of the way in which the maxima can be described, consider a structure with scattering elements in all points of the simple cubic lattice illustrated in Fig. 3.9a. Monochromatic X rays scattered from one (100) plane will be exactly in phase with those from successive (100) planes when the angle of incidence satisfies the

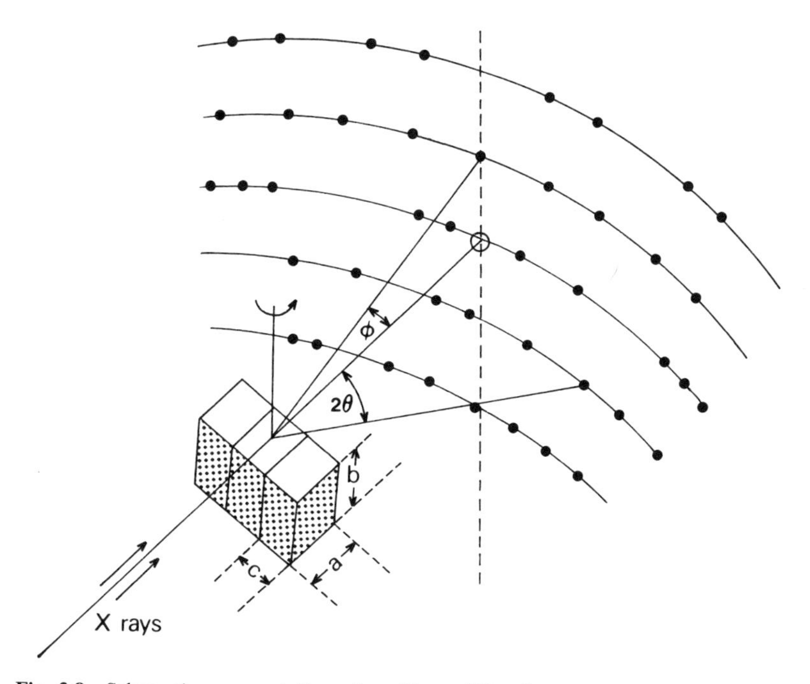

Fig. 3.8 Schematic representation of an X-ray diffraction experiment. The maxima are situated on layer lines. The vertical deflection of the layer lines is given by ϕ and the various diffraction maxima have deflections 2θ.

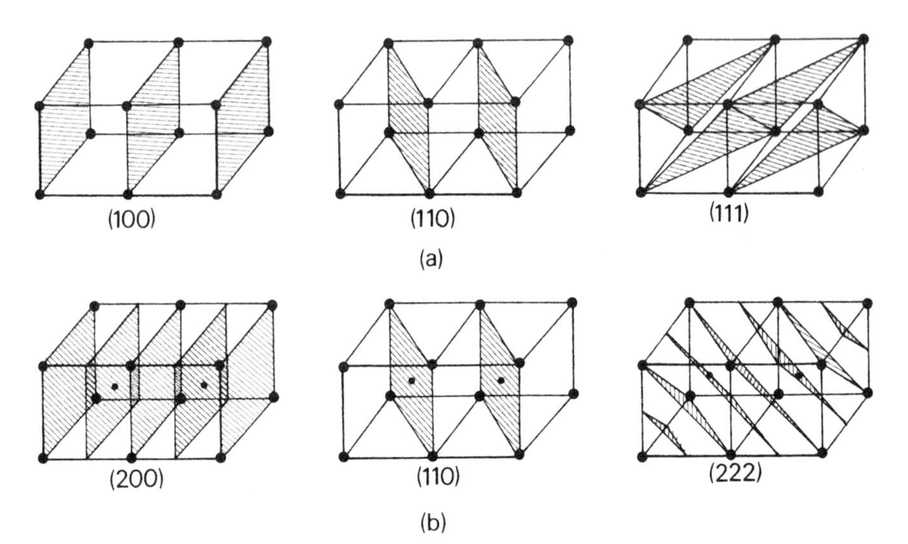

Fig. 3.9 Reflecting planes in (a) a simple cubic lattice and (b) a body-centered cubic lattice.

Bragg relation. This is observed because all (100) planes contain all the scattering elements of the structure. The same is true for the (110) and the (111) planes. The lattice illustrated in Fig. 3.9b is one of a body-centered structure. In this case the (100) planes do not contain all the scattering elements of the structure. Hence, while X rays coming in at the Bragg angle and reflected at the (100) planes reinforce each other, X rays reflected from the interleaved planes (which are exactly out of phase with the others) will reduce the intensity. If the scattering elements all have the same scattering power the resultant intensity will be reduced to zero and no first-order (100) maximum will appear. If, however, the scattering elements are different, the first-order (100) maximum will still appear but will be reduced in intensity. The (110) planes of the body-centered structure do contain all the scattering elements and, therefore, a strong first-order (110) maximum will show up; also, the second-order reflection at the (100) planes will show a strong maximum. This is the exact equivalent of the first-order reflection at the (200) planes which again contain all scattering elements of the structure.

Diffraction by Helical Structures. The application of diffraction techniques for a helical structure is shown in Fig. 3.10. In the figure the helix has a pitch of a and an integral number (six) of scattering elements per pitch. The first-order (100) reflection will be zero (when each of the scattering elements has equal scattering power) since each of the interleaving planes

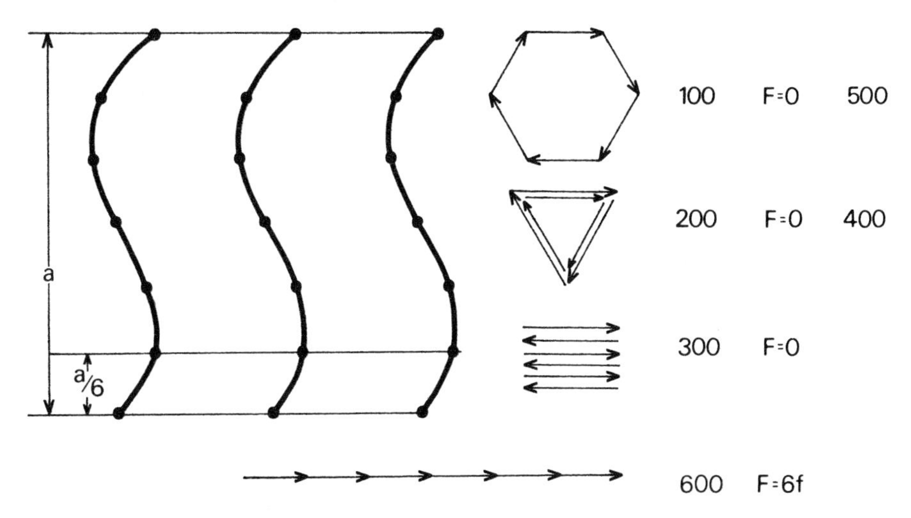

Fig. 3.10 A helical structure with pitch a and six scattering elements per pitch. The vector diagrams show how the maxima of a diffraction pattern are formed.

containing scattering elements will cause a reflection which is $\frac{1}{6}$ out of phase. The same is true for the second-, third-, fourth-, and fifth-order (100) reflections [of course equivalent to the first-order (200), (300), (400), and (500) reflections, respectively]. The first-order (600) reflections will then show a strong maximum. This is illustrated quite clearly when the vector description is used; the scattered radiation has amplitude and phase and, therefore, can be represented by vectors. In the case of the helix of Fig. 3.10 the vector from each scattering element has a $2\pi/6$ phase difference with the preceding one. The resulting diffraction pattern will be that shown in Fig. 3.11.

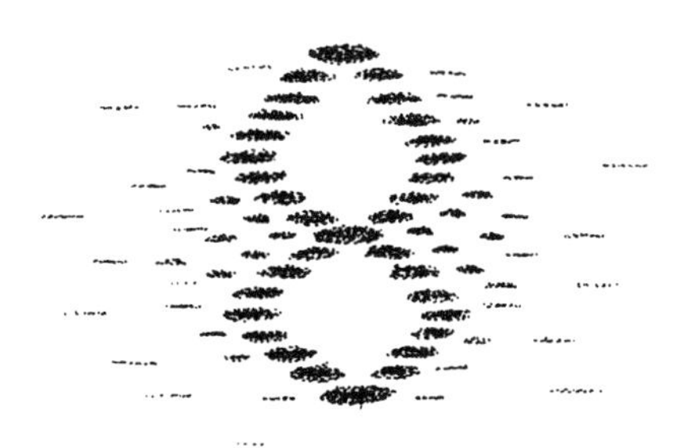

Fig. 3.11 A sketch of an X-ray diffraction pattern from a helical structure as the one showed in Fig. 3.10.

Phase Difference. As can be seen from the preceding example, the intensity of a diffraction maximum is determined not only by the amplitude of the scattered X rays but also by their phase differences. In an actual crystal the lattice points are usually occupied not by a single scattering element but by a group (actually, an electron-density distribution). As an example consider (see Fig. 3.12a) a structure formed by replacing each point in a lattice by two atoms, a black one and a white one. If a set of reflecting planes is drawn through the black atoms another parallel, but slightly displaced, set can be drawn through the white atoms. When monochromatic X rays are incident on the set of planes at the Bragg angle the reflections from all the black atoms will be in phase and the reflections of all the white atoms will be in phase. The radiation from the black atoms, however, is slightly out of phase with the radiation from the white ones, so that the resulting amplitude will be diminished by interference (Fig. 3.12b).

To find a general expression for the phase difference, consider the two-dimensional cross section illustrated in Fig. 3.12c. The black atoms are placed in the corners of a unit cell with sides a and b, and the white atoms have displaced positions. Take the coordinates of the black atoms as $(0, 0)$ and of the white atoms as (x, y). A set of planes hk is shown for which the Bragg condition is fulfilled. The spacings a/h along a, and b/k along b

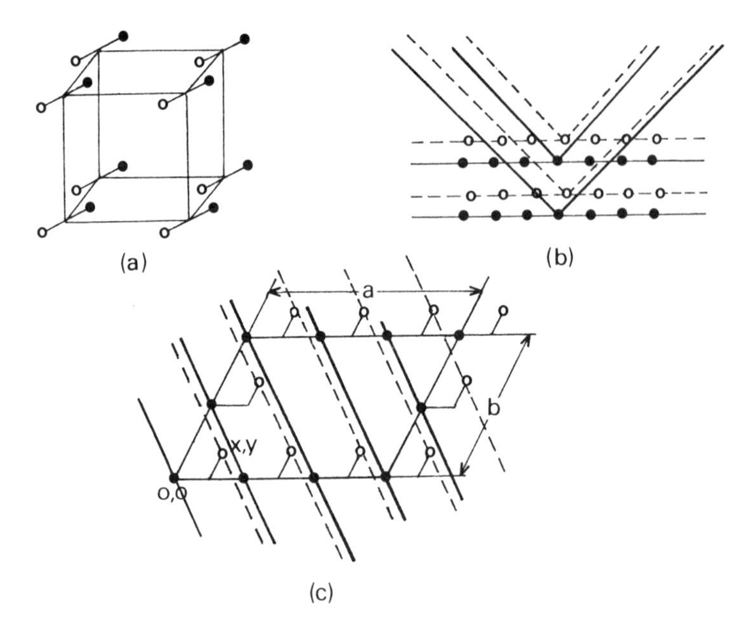

Fig. 3.12 (a) Simple cubic lattice with two scattering elements at each point; (b) Bragg reflection at a set of planes from the cubic lattice of (a); (c) two-dimensional cross section of a simple lattice.

correspond to positions for which the phase difference is 2π rad, i.e., the scattering from these positions is exactly in phase. The phase difference between these planes and those going through the white atoms is proportional to the displacement of the white atoms. The phase difference ϕ_x for the displacement x in the a direction is given by $x/(a/h) = \phi_x/2\pi$ or $\phi_x = 2\pi h/(x/a)$. The total phase difference for the displacement in both the a and the b directions becomes

$$\phi_x + \phi_y = 2\pi[h(x/a) + k(y/b)]$$

and by extension to the three-dimensional case,

$$\phi = 2\pi[h(x/a) + k(y/b) + l(z/c)] \tag{3.22}$$

Superposition of waves with different amplitudes and different phases is best accomplished by vector addition. If f_1 and f_2 are the amplitudes of the waves scattered by atoms 1 and 2, respectively, and ϕ_1 and ϕ_2 are their phases, the resulting amplitude will be $F = f_1 e^{i\phi_1} + f_2 e^{i\phi_2}$. For all the atoms in the unit cell this becomes

$$F = \sum_j f_j e^{i\phi_j} \tag{3.23}$$

When (3.23) is combined with (3.22) we obtain the resulting amplitude of the waves scattered from the (hkl) planes by all the atoms in the unit cell

$$F(hkl) = \sum_j f_j \exp 2\pi i\left(\frac{hx}{a} + \frac{ky}{b} + \frac{lz}{c}\right) \tag{3.24}$$

This quantity is called the *structure factor* of the crystal. It is dependent on the positions of the atoms and the *atomic scattering factors* f_j. The values of the atomic scattering factors are given by the scattering angle and the number and distribution of the electrons in the atom; it is the electrons in a crystal structure that scatter the X rays.

Fourier Analysis. A crystal examined by X rays may be regarded as a periodic three-dimensional distribution of electron density, $\rho(xyz)$; such a density distribution can be expressed as a Fourier series. It can be shown that the Fourier coefficients are then the structure factors divided by the volume V of the unit cell. Thus,

$$\rho(xyz) = \frac{1}{V}\sum F(hkl) \exp -2\pi i(hx/a + ky/b + lz/c) \tag{3.25}$$

The summation is carried out over all values of h, k, and l, so that there is one term for each set of planes (hkl) and, hence, for each spot on the X-ray diffraction pattern.

The Phase Problem. Equation (3.25) summarizes the whole problem of structure determination. Since the crystal structure is simply the electron-density distribution $\rho(xyz)$, positions of individual atoms are peaks in ρ with heights proportional to the atomic numbers (numbers of electrons). If $F(hkl)$ is known, the structure could be immediately plotted. The problem is, however, that we only know the intensities of the spots, which are only proportional to $|F(hkl)|^2$; we know the amplitudes but we have lost the phase differences in taking the X-ray scattering pattern. The only way to obtain the structure data is to assume a trial structure and calculate the intensities. If the assumed structure is approximately correct, the most intense observed reflections should correspond to large calculated intensities. We then compute the Fourier series using the *observed* values of F and the *calculated* (from the assumed structure) phases. The Fourier summation should then give new positions for the atoms from which new values of F can be calculated, allowing more of the phases to be correctly determined. The procedure can sometimes be simplified by the introduction of a heavy atom into the crystal structure at a known position. The large contribution of the heavy atom makes it possible to determine the phases of many of the structure factors. This technique is called *isomorphic displacement.*

In spite of the long and tedious work involved, the elucidation of the structures of a number of proteins has been successful. Notable successes have been the determination of the structure of myoglobin (Kendrew), hemoglobin (Perutz), and vitamin B_{12} (Hodgkin). The method has been employed in a dramatically successful way by Crick and Watson and by Wilkins in the discovery of the double helix structure of DNA, as a result of which the molecular mechanism of heredity could be explained (see Section 3.6).

3.5 Structure and Function of Proteins

Astbury Structures. Shortly after the discovery of X-ray diffraction by crystals the technique was used to investigate biological macromolecules, in particular proteins which could be crystallized and nucleic acids. Around 1930 Astbury tried to interpret the diffraction pattern obtained from crystallized protein and nucleic acid fibers. He showed that protein fibers give rise to two types of patterns which he called α and β. He recognized that the α pattern was due to a more folded and dense structure (Fig. 3.13a) while the β pattern was from a more stretched structure (Fig. 3.13b). The structures in his models are held together by electrostatic interaction between the more negative double-bonded oxygen and the more positive

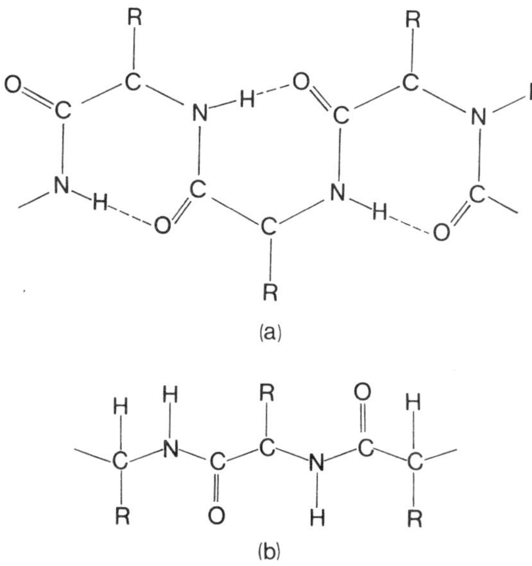

(a)

(b)

Fig. 3.13 Protein structures according to Astbury; (a) α-structure and (b) β-structure.

hydrogen on the nitrogen. This type of electrostatic interaction is called *hydrogen bonding* (see Chapter 4) and is indicated in Fig. 3.13 by dotted lines.

The α Helix. Although the importance of Astbury's models for the understanding of protein structure cannot be overemphasized, they did not give a completely satisfactory explanation of the diffraction patterns. The helical structure proposed by Pauling and Corey in 1951 did fit the diffraction patterns of many fibrous proteins much better. Astbury and others have also tried helical models but they never thought of fitting a noninteger number of amino acid residues into one turn. Taking into account the known bond distances and bond angles, Pauling and Corey showed that a helical structure with 3.6 amino acid residues per turn, a diameter of about 6.8 Å, and a distance between turns of about 5.4 Å (Fig. 3.14a) would be a definite possibility. This α-helical structure fits the diffraction patterns observed for synthetic polypeptides very well and there are also many aspects pointing to the α helix in the diffraction patterns of many naturally occurring fibers of the α-type. Pauling and Corey also made pleated sheet models similar to Astbury's β structure but they did not restrict the peptide bonds to one plane (Fig. 3.14b). Both structures, the α helix and the pleated sheets, retain the idea of an α structure being folded and a β structure being stretched out.

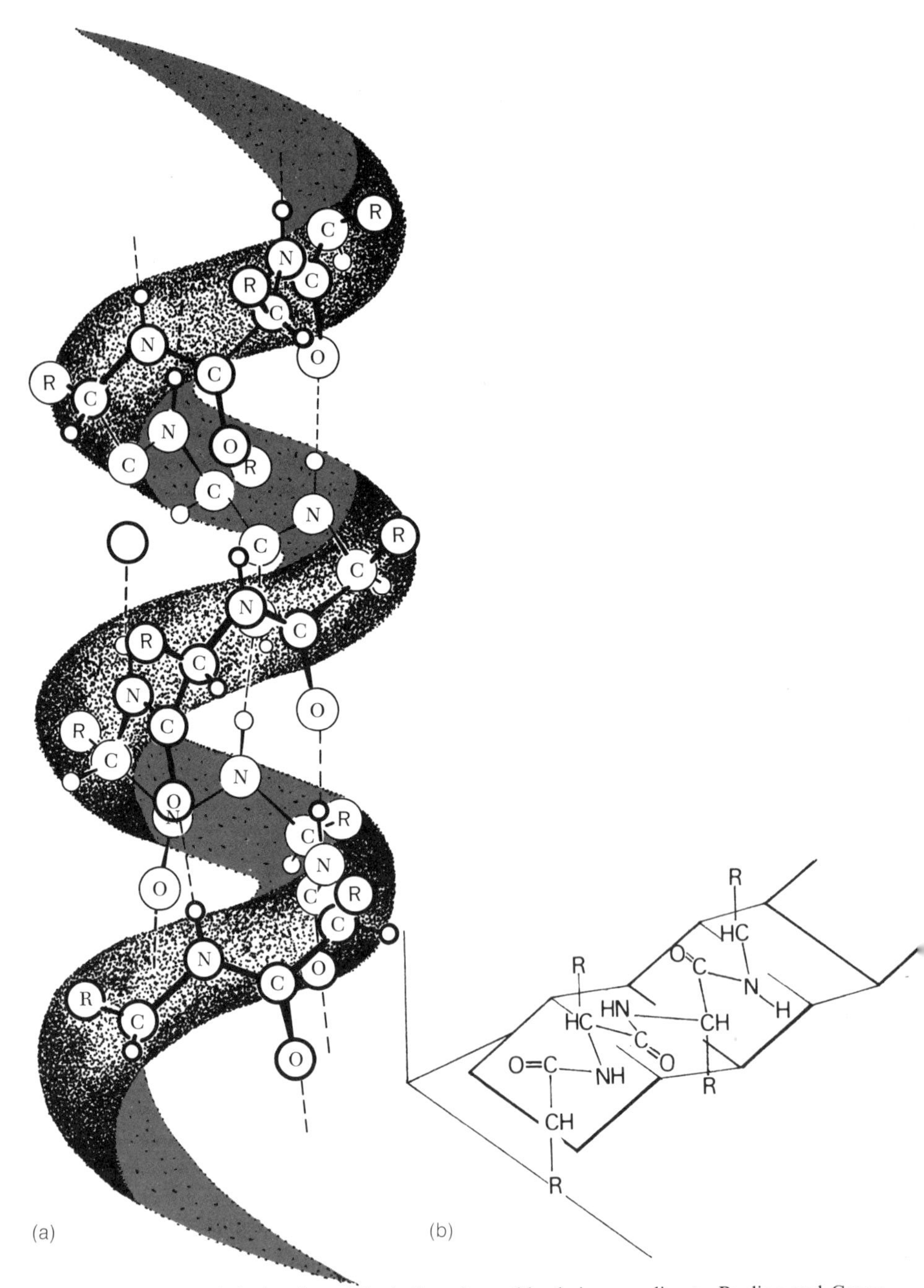

Fig. 3.14 (a) A drawing of the helix polypeptide chain according to Pauling and Corey. The small circles without a letter represent hydrogen atoms; the dotted lines designate hydrogen bonds. R are the amino acid residues. The helix has a radius of 3.4 Å and an interturn distance of 5.4 Å. There are 3.7 residues per turn. (Parsegian *et al.*, 1970). (b) A pleated sheet configuration of a polypeptide chain, such as that used as a basis for a model for keratin.

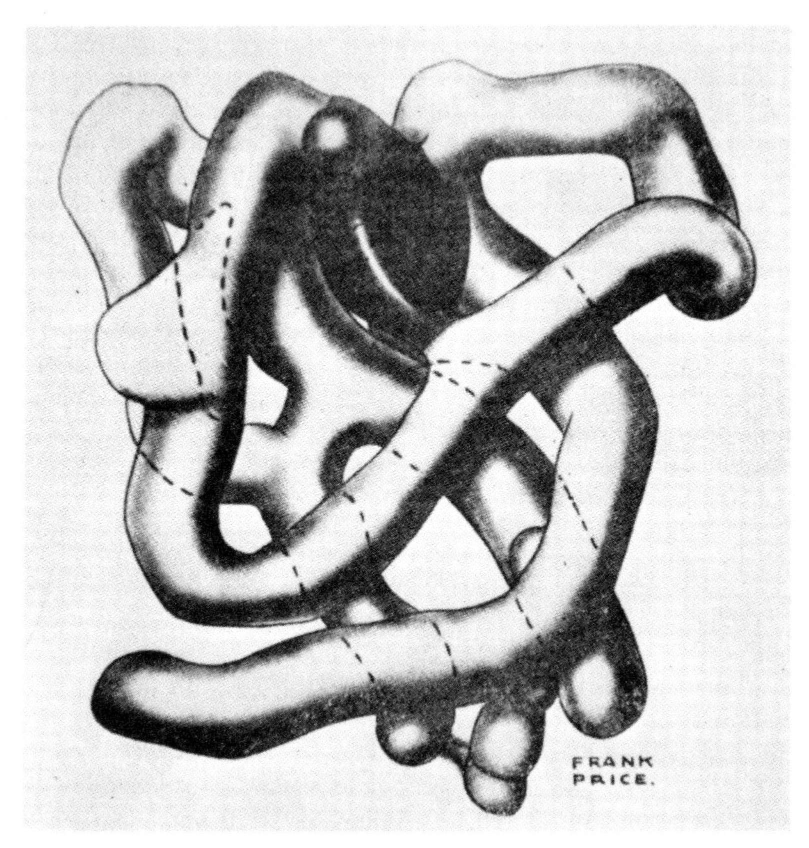

Fig. 3.15 Kendrew's model of sperm whale myoglobin. The black area is the heme group. [From J. C. Kendrew *et al.* (1958), *Nature* **181,** 662.]

Myoglobin and Hemoglobin. Although it is not at all certain that the α helix is a *general* structural feature of proteins, it has been proven convincingly, by X-ray diffraction, that it is a structural factor in the globular proteins *myoglobin* (Kendrew) and *hemoglobin* (Perutz). Myoglobin is a red pigment found in muscle; it probably functions by buffering the oxygen concentration in muscle. It has a molecular weight of 18,000 and contains 153 amino acid residues. Thus, there are some 1200 atoms other than hydrogen in the molecule and one would need to measure the intensity and initially guess the phases of some 20,000 diffraction spots to locate all these atoms. The analysis of about 400 spots showed that there are two molecules in a unit cell. The polypeptide chain and the iron-containing *heme group* (see Chapter 4) could also be located. The further analysis of about 10,000 spots has shown the electron density of the molecule with a

Fig. 3.16 Perutz's model of horse hemoglobin. The white units are an identical pair as are the black units. The disks are the heme groups, one in each unit (two are hidden at the other side of the molecule). The resolution was 5.5 Å. [From M. F. Perutz (1960), *Brookhaven Symp. Biol.* **13,** 165.]

resolution of about 2 Å. Although this is not quite sufficient to indicate the separate atoms, it is sufficient to confirm that a major portion of the molecule consists of α helices. In order to fit the single polypeptide chain into a globular protein the chain must be bent and twisted, and where this occurs the α-helical form is lost for some 2 or 3 residues. There is also a group of about 13–18 residues which is not in the α-helical form. Figure 3.15 shows the form of the polypeptide chain as revealed at this resolution. The straight portions of the chain are right-handed α helices and the molecule has one continuous polypeptide chain, confirming the chemical analysis.

Similar studies were carried out for hemoglobin, the protein which transports oxygen through the blood. This protein has a molecular weight of 65,000 and it was shown that it consists of four subunits which form two identical pairs. Each of the subunits contains a heme group. The model,

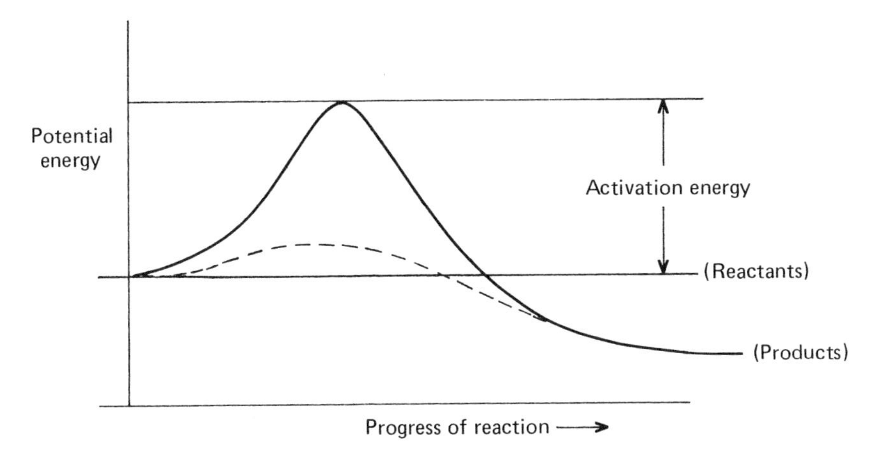

Fig. 3.17 The activation energy barrier. Although the potential energy of the products of a reaction is lower than that of the reactants, the reaction cannot proceed at measurable speed because it has to go through a state of high potential energy. When the reaction is catalyzed (by an enzyme, for example) the potential energy barrier is substantially lowered (dashed line).

based on 5.5-Å resolution studies, is shown in Fig. 3.16. Each subunit is again a continuous polypeptide chain folded around itself in a form very similar to that of the myoglobin molecule.

Today, the X-ray diffraction patterns of more proteins have been studied. The importance of these structure studies is that they show the close relationship between structure and function. Most known proteins are *enzymes*, biological catalysts with a sometimes very pronounced specificity. Enzymes make so-called *substrates* react. Their specificity is based on a recognition process between the enzyme and the substrate. The structural features play an important role in this process.

Enzymes. Enzymes speed up the rate of attaining equilibrium but do not change the equilibrium itself. They enter the reaction but are in the same state after the reaction as they were before. The oxidation of glucose, for example, can be represented by the equilibrium equation

$$C_6H_{12}O_6 + 6O_2 \rightleftharpoons 6CO_2 + 6H_2O$$

The equilibrium strongly favors the CO_2 and H_2O, because the potential energy of the carbon dioxide and the water together is much lower than that of the sugar and the oxygen together. However, when glucose and oxygen are brought together, for instance in a solution, no reaction will take place because the reaction rate is too slow. This is the case since going from the left-hand side of the equation to the right-hand side the system has to go through an intermediate state with a potential energy much higher than even the initial state (as illustrated in Fig. 3.17). The difference between the initial potential energy and that of the intermediate state is called the *activation energy*. At room temperature or at slightly elevated temperatures, such as that inside a human body, only very few molecules have enough energy to overcome this barrier. Heating often results in a speeding up of the reaction by increasing the initial energy of the molecules (one can "burn" sugar). Enzymes have the effect of lowering the activation energy. They do this by forming noncovalent complexes with the substrate.

There are many kinds of enzymes which are grouped according to the kind of reaction they catalyze. The *hydrolases* catalyze hydrolysis reactions (the splitting of a molecule, adding H to one part and OH to the other), the *transferases* catalyze the transfer of a piece of one molecule to another, the *isomerases* accelerate isomerizations, the *carboxylases* remove CO_2, and the *respiratory and photosynthetic enzymes* catalyze the oxidation–reduction reactions of respiratory and photosynthetic processes.

Many enzymes consist of a protein molecule as well as a smaller molecule. In this case the large protein part is called the *apoprotein* (or *apoenzyme*). The smaller part, dependent upon its nature, is called a *coenzyme* (a large organic molecular group which separates easily from the enzyme), a *cofactor* (a loosely bound small inorganic ion), or a *prosthetic group* (small nonprotein parts of enzymes which are attached so firmly that they cannot be removed easily without irreversibly changing the enzyme). An example of a prosthetic group is the heme group, containing iron, in the oxidative enzymes as hemoglobin, myoglobin, and the cytochromes.

Enzyme Kinetics. The fact that the action of enzymes is based on the formation of a substrate–enzyme complex was already established by Michaelis and Menten, who concluded this from the reaction kinetics of enzymatic reactions. If the reaction rate of an enzymatic reaction is plotted against the substrate concentration, one obtains curves such as those illustrated in Fig. 3.18. The rate starts to increase linearly but levels off to a plateau. The height of the plateau depends on the concentration of the enzyme. Thus

$$\mathrm{E} + \mathrm{S} \underset{k_2}{\overset{k_1}{\rightleftharpoons}} \mathrm{E}\cdot\mathrm{S} \xrightarrow{k_0} \mathrm{P} + \mathrm{E}$$

in which E is the enzyme, S the substrate, and P the reaction products. After a while, if [S] > [E] (which is usually the case), the reaction rate

$$d\mathrm{P}/dt = k_0[\mathrm{E}\cdot\mathrm{S}] \tag{3.26}$$

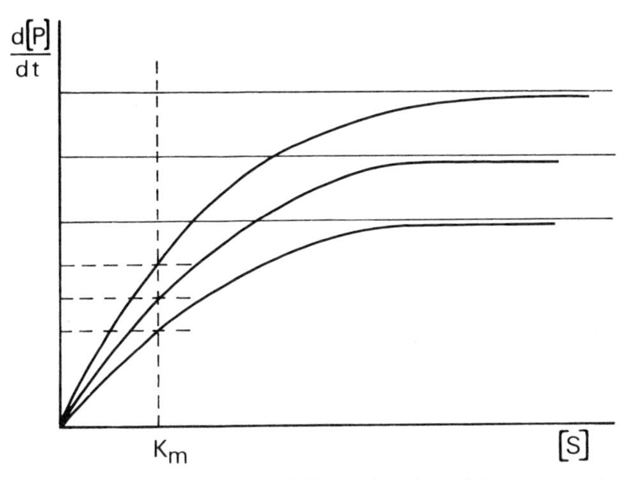

Fig. 3.18 A plot of the reaction rate $d[P]/dt$, as a function of the concentration of substrate [S] at different enzyme concentrations. At half of the saturation rate the substrate concentration is equal to the Michaelis constant K_m.

becomes constant, so

$$d[E \cdot S]/dt = 0 = k_1[E][S] - k_2[E \cdot S] - k_0[E \cdot S] \tag{3.27}$$

Since the concentration of free enzyme [E] is equal to the maximum concentration $[E]_0$ minus the concentration of the complex,

$$[E] = [E]_0 - [E \cdot S]$$

Eq. (3.27) becomes

$$k_1\{[E]_0 - [E \cdot S]\}[S] = (k_2 + k_0)[E \cdot S] \tag{3.28}$$

We now define

$$K_m = \frac{k_2 + k_0}{k_1} \tag{3.29}$$

From (3.28) and (3.29)

$$K_m = \frac{[E]_0 - [E \cdot S]}{[E \cdot S]}[S]$$

and

$$[E \cdot S] = \frac{[E]_0[S]}{K_m + [S]} \tag{3.30}$$

The constant K_m is called the *Michaelis constant* and describes, in a relatively simple way, the enzyme reaction kinetics. If

$$[S] \gg K_m$$

Eq. (3.30) becomes

$$[E \cdot S] = [E]_0 \qquad \text{and} \qquad dP/dt = k_0[E]_0$$

Thus at high substrate concentration the rate is limited by the maximum amount of enzyme. All of the enzyme is bound in the complex and there is no free enzyme left. The rate is independent of the substrate concentration. If, however,

$$[E] \ll K_m$$

Eq. (3.30) becomes

$$[E \cdot S] = \frac{[E]_0[S]}{K_m}$$

and the rate is

$$\frac{dP}{dt} = \frac{k_0[E]_0[S]}{K_m}$$

The reaction is limited by the rate at which the enzyme and the substrate can form the complex

At high substrate concentration the reaction is of the zeroth order (rate is constant). This is the plateau in the curves of Fig. 3.18. Since

$$\frac{dP}{dt} = k_0 \frac{[E]_0[S]}{K_m + [S]}$$

the reaction rate is at half maximum when $[S] = K_m$.

Enzyme Specificity. The specificity of the enzymatic reactions is contingent upon the complex formation between enzyme and substrate. Since the complex is formed by noncovalent interactions, there must be a relatively large number of points at which this interaction can take place in order to ensure a certain amount of stability for the complex. Moreover, the noncovalent interaction must be at close range, in other words there must be many places at which atoms of the substrate can come very close to atoms of the enzyme. Thus, a structural correspondence must exist between the substrate and the enzyme. It is this very fact that lies at the basis of the recognition process (this is illustrated in Fig. 3.19). Although

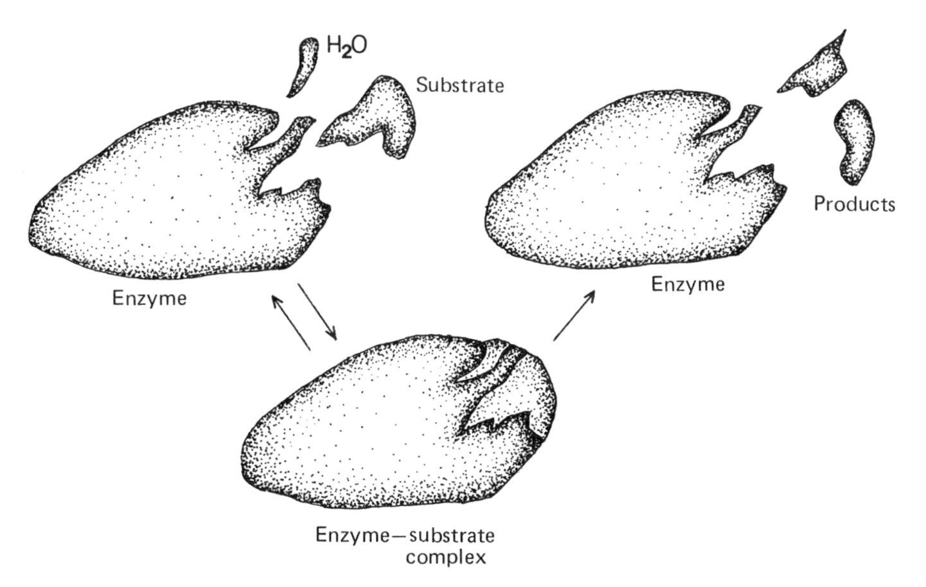

Fig. 3.19 A representation of enzyme action. The substrate can form a complex with the enzyme because its configuration "fits" at the active site of the enzyme. The substrate can then react (for instance by hydrolysis), thus forming the products and free enzyme. The drawn shapes have no relation with actual configurations.

the drawn shapes should not be taken literally, the key–keyhole concept explains the enzyme reaction kinetics excellently. According to this concept the action of competitive inhibitors (Fig. 3.20) can also be analyzed.

Allosteric Effect. Another structure-related functional effect is the allosteric effect. If, for instance, the rate of the binding of oxygen by hemoglobin is plotted as a function of the partial pressure of the oxygen, a sig-

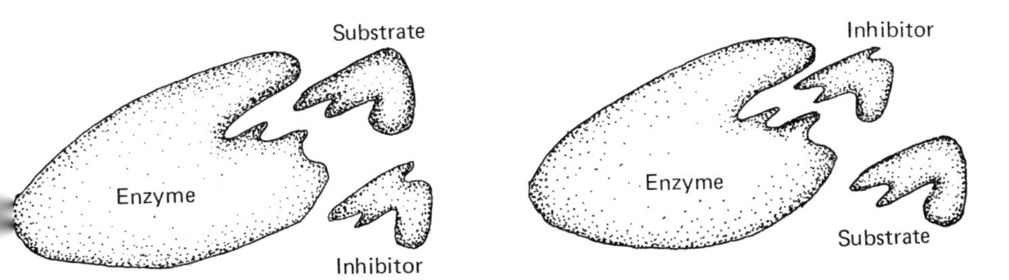

Fig. 3.20 Competitive enzyme reaction inhibition. The inhibitor has a configuration which is similar to that of the substrate. The inhibitor can bind to the enzyme, thereby blocking the formation of the enzyme–substrate complex.

moidal (S-shaped) curve is obtained (Fig. 3.21). This means that the affinity of hemoglobin for oxygen is increased as more oxygen is bound by the molecule. Heme–heme interaction cannot explain this effect because of the relatively large distance between the heme groups in the subunits. The interaction must be indirect, that is through the binding sites of the subunits. Conformational changes in the protein structure which alter the affinity for specific substrates result from this allosteric effect.

Allosteric effects now have been found in many enzymes. The effect is very important for the *regulation* of enzymatic processes; sometimes the effect is stimulating (as with the case of hemoglobin), sometimes the effect is inhibitory; sometimes the products or their degradation products exert the allosteric effect, sometimes the precursors exert the effect. It is through such processes, and often through combinations of them, that the enzymatic processes in a living cell are very effectively regulated.

3.6 Structure and Function of Nucleic Acids

The Gene. Although DNA was known in the nineteenth century, its presence in the chromosomes of a cell was interpreted as a structural necessity. The idea of the *gene* as the functional unit of heredity already existed, and the distinction between a *genotype* and a *phenotype*, the latter being the manifestation of the function of genes (the so-called genetic expression), was recognized as useful. In the beginning of the twentieth century it was found that the gene itself was composed of DNA and not of protein, as was formerly believed. An important refinement of the concept of genetic

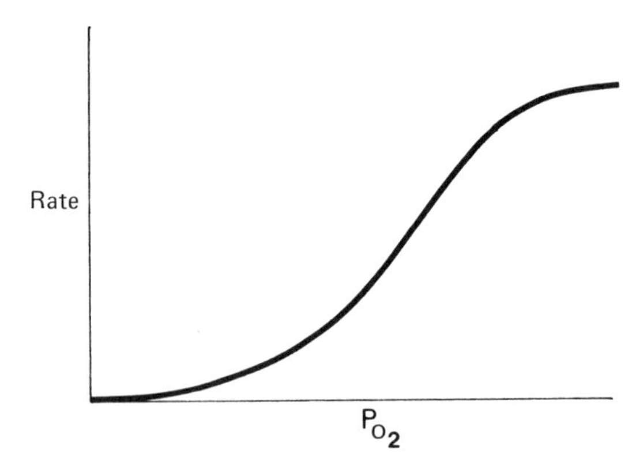

Fig. 3.21 A plot of the oxygenation of hemoglobin as a function of the partial pressure of oxygen. The sigmoid shape of the curve shows that the reaction goes faster with the more oxygen bound. This is due to the allosteric action of the enzyme.

expression was the realization that this expression occurs in a very simple way, namely through the production of a unique characteristic protein species. The problem at that stage was the question of how the information contained in a particular DNA molecule could manifest itself in a particular protein molecule. With the realization that DNA as well as protein molecules are linear polymers came the idea that the primary structure (the sequence of monomeric units) in each is all important and that the problem may well turn out to be a *coding* problem. A major factor for the confirmation of this was, of course, the 1953 determination of a satisfactory steric model of DNA by Watson and Crick and by Wilkins. For their interpretation, they made use of the techniques of X-ray diffraction described in Section 3.3.

Structure of DNA. Watson and Crick showed that the structure of DNA was made up of two antiparallel intertwisted helices with an overall diameter of about 18 Å. The phosphates in the sugar–phosphate sequence are connected at one side to the fifth carbon atom of the sugar and at the other side to the third carbon atom of the sugar (see Fig. 2.10). With respect to the bases, the sequence runs $5 \rightarrow 3$ in one strand and $3 \rightarrow 5$ in the other. These strands are made up of sugar–phosphate, sugar–etc., sequences and the bases (attached to the sugars) protrude to the inside. The bases attached to one helical strand are linked to those attached to the other helical strand by hydrogen bonds. This can be accomplished *only* when there is an exact fit. Measurements based on X-ray diffraction patterns of crystals of the purine and pyrimidine bases have shown that this exact fit exists when adenine (A) is paired with thymine (T) through two hydrogen bonds, and when guanine (G) is paired with cytosine (C) through three hydrogen bonds. As a consequence the relative concentrations of the bases in DNA must satisfy the following relation

$$[A]/[T] = [G]/[C] = 1$$

This relation had indeed been verified and was one of the pieces of evidence used by Watson and Crick to construct their model. The DNA molecule is thus like a twisted stepladder with about ten rungs per turn. The rungs, about 11-Å long, are formed by the base pairs and linked to each other as depicted in Fig. 3.22. Because of the specific base pairing, each strand can be a mold or a template to form an exact complementary copy of it, just like a photographic print is made from a negative. In the process of replication (see later) the two strands come apart and each of them serves as a template for the synthesis of a new complementary strand. The new strands then condense with the old ones forming two new molecules of DNA which are exact copies of the one old molecule. In this way the sequence

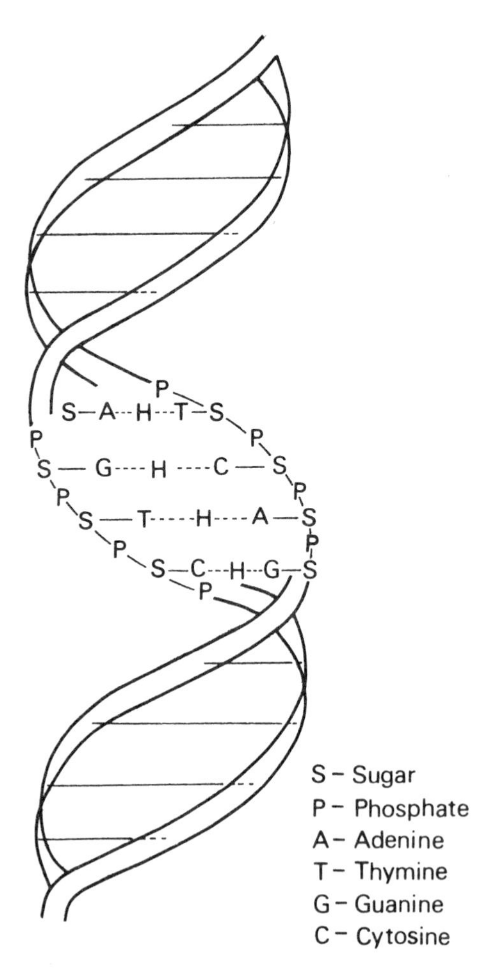

Fig. 3.22 The double helix of DNA. The two helically wound strands are bound to each other by hydrogen bonds between complementary bases. The diameter of the spiral is 18 Å and the spacing between turns is 34 Å. The "rungs" between the strands are about 11-Å long and there are about 10 "rungs" per turn. The molecules are very large. The molecular weight of bacterial DNA (*E. coli*) can be as much as 2.8×10^9. Eukaryotic cells contain a species specific number of chromosomes and each chromosome may contain one or more very large DNA molecules (the total length of all the DNA contained in the 46 chromosomes of human cells has been calculated to be about 2 *m*, which is equivalent to about 5.5×10^9 base pairs).

of the bases in the DNA is conserved completely and exactly. In the process of transcription (see later) one of the DNA strands is a template for the synthesis of a complementary copy of RNA (with the one variation that in RNA the thymine is replaced by uracyl). This RNA molecule is then used to program the synthesis of protein.

The Genetic Code. It soon became obvious that the linear sequence of the bases G, C, A, and T in DNA must contain the order for the protein programming, and a great deal of effort then went into deciphering the code. In the years following the discovery of Watson and Crick most of this effort was theoretical. If a particular sequence of two of four bases were assigned to a particular amino acid, only 16 ($=4^2$) amino acids could be accounted for. The code "word" therefore must have at least three "letters" (a "triplet"). But that gives 64 ($=4^3$) possibilities and there are only 20 amino acids in a living cell. This had been a problem in the early days because it seemed to be difficult, at that time, to think of a degenerate code (a code in which more triplets could be assigned to one amino acid). Hence, much effort went into finding groups among the 64 combinations which would turn out the magic number 20. The most brilliant among these efforts was the one of George Gamow who found "diamond-shaped" pockets in the Watson–Crick double helical model. These pockets were formed and bounded by a base on one DNA chain, the adjacent base pair (spanning both chains), and the next adjacent base beyond the pair on the other chain. The 64 possible diamond-shaped pockets could be divided into 20 categories each of which contain diamonds that do not change in character. The residues of the 20 kinds of amono acids would each be a correct fit for one of the corresponding categories of diamond-shaped pockets.

There are a number of elements in this hypothesis which have turned out to be correct. The code is indeed a "three letter" code (each of the two bases in the base pair of the boundary of the pockets is, by the complementary C–G and A–T binding, completely determined by the other so that there are only three "independent" bases in the boundaries of the pockets) and there is a certain degree of degeneracy. The most important thing which is preserved is the template idea; in Gamow's model the DNA molecule serves as a template for the peptide chain.

What turned out to be wrong in Gamow's idea is that, in his view, the DNA itself is the template for the protein synthesis. The idea of an intermediate between the DNA and the site of protein synthesis is largely owed to Jacob and Monod (1961), who defined a "messenger." Confirmed by a wealth of experimental evidence, it turned out that a special kind of RNA, the *messenger RNA* (mRNA), serves as such a messenger.

RNA. In a cell there are three kinds of RNA. First, there is *ribosomal* RNA (rRNA). This is the most abundant RNA in a cell and it is an inherent factor of the ribosomes, the sites of protein synthesis. There is very little known about the role of this large amount of RNA which, as far as the amino acids are concerned, seems to contain little information. It has a molecular weight of about 10^6 daltons. Then there is *transfer RNA* (tRNA), which serves as the *adaptor* between the mRNA and a particular amino acid. tRNA has a molecular weight of about 24,000 daltons, and there seems to be about 40 different kinds. There is evidence that both rRNA and tRNA are complementary copies of a part of the DNA of the cell. Finally, there is the *messenger RNA* (mRNA), the conveyer of the information for protein synthesis. As can be understood from its function, mRNA is heterogenous; it provides templates for some 4000 different kinds of proteins in a "typical" cell. A typical polypeptide of 40,000 daltons requires a corresponding messenger RNA of 400,000 daltons, or a "tape" length of 1200 nucleotides.

The Tape Reading Processes. The transformation of genetic information is essentially a combination of three tape reading processes. First, there is the process of *replication* in which a DNA input tape is read out by a single "tape reader" (an enzyme called replication polymerase). The outputs are two double helical molecules of DNA (Fig. 3.23a). In each of the of the new molecules one of the strands came from the original DNA molecule; the other strand is synthesized on the original one, their bases being complementary and hydrogen bonded to the bases of the original strand. The tape reading process thus turns out two new DNA tapes identical to the original one. The process must involve a complete unwinding of the DNA double helix and a simultaneous synthesis of two DNA strands, one in the opposite direction from the other. Each DNA tape is read only once during the cell cycle and the time it takes may well determine the length of

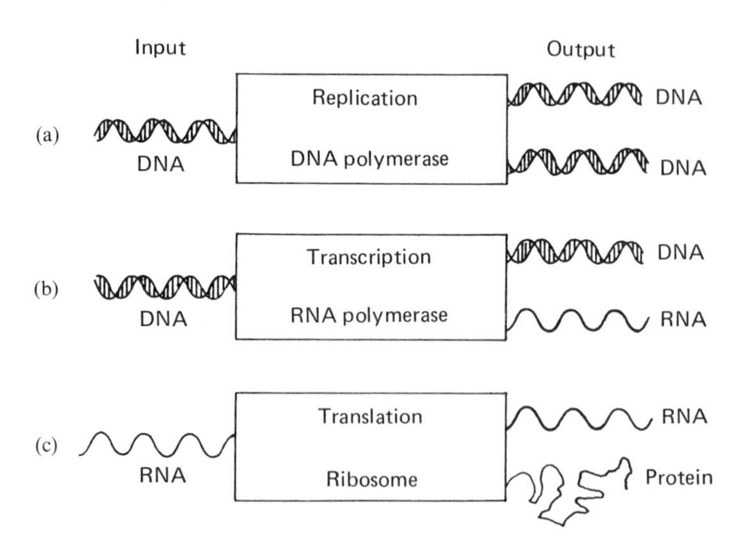

Fig. 3.23 The three tape reading processes of a cell.

the cycle. The two new DNA tapes become the complete "library" for the daughter cells after cell division.

Second, there is the process of *transcription*. In this process parts of the DNA tape are read by one of many transcription tape readers (transcription polymerases) and the output is a single stranded mRNA molecule which is the complementary copy of the readout part of the DNA molecule (Fig. 3.23b). Unlike replication, transcription copies parts of only one of the two DNA strands. It is unlikely that the DNA helix unwinds to any great extent during this process. Because many transcriptions are occurring simultaneously in a cell, many transcription tape readers must be present. Thus, the information which is incorporated in the sequence of nucleotides in the DNA molecule is transcribed in a complimentary fashion to the mRNA tapes which then carry this information to the sites of protein synthesis (the ribosomes) where it is used to form an equivalent sequence, this time of amino acids. In a chemical sense, the DNA tape is a single tape which contains a library of information sufficient for all the proteins needed by the cell. The transcription tape reading operation divides the DNA molecule into a large number of units called *cistrons* or *genes* and into contiguous coordinate groupings of cistrons called *operons*.

Finally, there is the process of *translation*. Each mRNA carries the information contained in the sequence of its nucleotides (or more precisely in the sequence of its *codons*, "words" of three nucleotide "letters" each) to the translation machines (the ribosomes) where the codons are translated into amino acids which then are incorporated into the peptide chain in the order prescribed by the order of the codons in the mRNA (Fig. 3.23c). In this process the translation machines make use of special adaptors which are the tRNAs. These relatively small RNAs consist of many nucleotides. Three of these, however, are very special; they form an anticodon (the complementary set of a codon) assigned to a specific amino acid. Due to a recognition process which still is not understood, the tRNA, through a charging process involving the energy carrier adenosine triphosphate (ATP) (see Chapter 5), binds with its assigned amino acid. The so-formed aminoacyl-tRNA can be attached to a ribosome when the latter is connected to part of the mRNA single strand (see Fig. 3.24). In this process the anticodon of the tRNA is bonded (by hydrogen bonds) to the appropriate codon of the mRNA. The adjacent codon of the mRNA can then bind the appropriate aminoacyl-tRNA through its anticodon and the two amino acids form a peptide bond when the correct factors are present (among others K^+ ion and guanosine triphosphate). The mRNA then slips one position further through the ribosome, the "uncharged" tRNA loosens itself from the ribosome, and the next aminoacyl-tRNA can attach itself, thus placing its amino acid in the right position. The exact mechanism of this process is

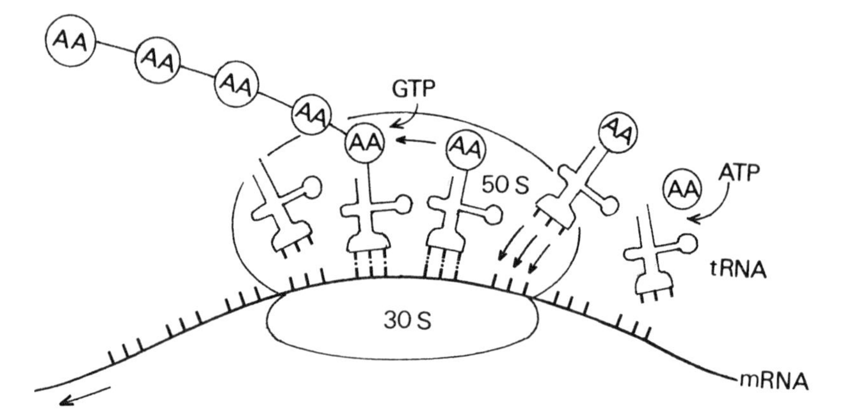

Fig. 3.24 The translating machinery of the ribosome. Specific transfer-RNA molecules form complexes each with a specific amino acid. The anticodons of these aminoacyl-tRNAs form hydrogen bonds with complementary codons of the mRNA which is attached to the ribosome. The amino acids, thus placed in a sequence dictated by the codon sequence of the mRNA, form peptide bonds after which the tRNAs detach.

not yet known but it is clearly demonstrated that the role of the tRNAs and the mRNA in the ribosome is, indeed, as described. The translation machine thus forms a polypeptide chain (a protein) exactly in the order as prescribed by the mRNA.

Regulation of Enzyme Synthesis. Most of the proteins synthesized by the tape reading processes described above, are enzymes and most of the enzymes are allosteric. As we have seen, the latter property gives the enzymes regulating power. An additional means of regulation of cell processes is the synthesis of the enzymes themselves. Such regulation makes possible economy in the synthesis of both RNA and protein, just as allosteric control results in economical use of carbon and nitrogen sources in the synthesis of small biological molecules. We could see the regulation of enzyme synthesis as a sort of coarse control, whereas the allosteric regulation provides fine control over cell metabolism. Moreover, in higher organisms the cells are differentiated; different cell types possess different sets of enzymatic activities superimposed on those enzymes necessary for the central metabolic pathways. However, all cells in such organisms (with the exception perhaps, of a few types of cells such as the red blood cell) have the same DNA library containing the same set of genes; in the course of embryonic development the synthesis of proteins must have been selectively turned on and off.

Such regulation of protein synthesis takes place not at the level of the ribosome but at the level of the DNA. Genes are turned on and off by inhibition or stimulation of the transcription process in response to the presence of an inducing or repressing substrate or product molecule. There are two kinds of regulation called *enzyme induction* (by substrate) and *enzyme repression* (by product), which were originally thought to be two entirely different processes. It now has become clear that they involve the same mechanism. In this mechanism the genes are switched off by specific proteins, named *repressors*. The existence of repressors was originally hypothesized by F. Jacob and J. Monod (1961), and has been abundantly demonstrated since then. The working of the mechanism is best explained with an example, for instance the control of the synthesis of the inductable enzyme *β-galactosidase* in the colon bacterium *E. coli*. These bacteria do not use β-galactoside (a sugar which can be hydrolyzed enzymatically and then used as an energy and carbon source) when they are supplied with glucose; they do not synthesize the enzymes, such as β-galactosidase, necessary for its metabolism. However, as soon as β-galactoside is put into the medium in which the bacteria grow, synthesis of β-galactosidase is turned on and proceeds at a high rate. The mechanism by which this "turning on" occurs is as follows (see Fig. 3.25). The part of the DNA that contains the genetic information for the synthesis of β-galactosidase has three different regions with different functions. First, there is the gene that contains the genetic code for the amino acid sequence of the enzyme. This type of gene is called a *structural gene*. Adjacent to the structural gene is a region, some 30–100 nucleotides long, which contains a specific binding site for the repressor molecule; this is the *operator*. Finally there is the gene which codes for the amino acid sequence of a specific repressor. In the absence of an inducer (the substrate, in this case β-galactoside) the repressor occurs in its free, or active, state. When it diffuses from the ribosomes where it is formed, according to the program conveyed by messenger RNA, it becomes bound to the operator. In its bound position it represses the transcription of the structural gene. But when an inducer is present it combines with the repressor protein at a specific complementary site. This binding causes a conformational change which makes the repressor incompatible with the binding site on the operator. The repressor can no longer bind to the operator and the structural gene is free to be transcribed.

Enzyme repression acts according to the same principle. Here, the repressor molecule is incompatible with the binding site on the operator when it is free. Binding with the end product (called a corepressor) changes the conformation of the repressor making it compatible to the operator binding site. The repressor then binds and the transcription of the structural gene is repressed. An example of this is repression of *histidinol dehydrogenase*,

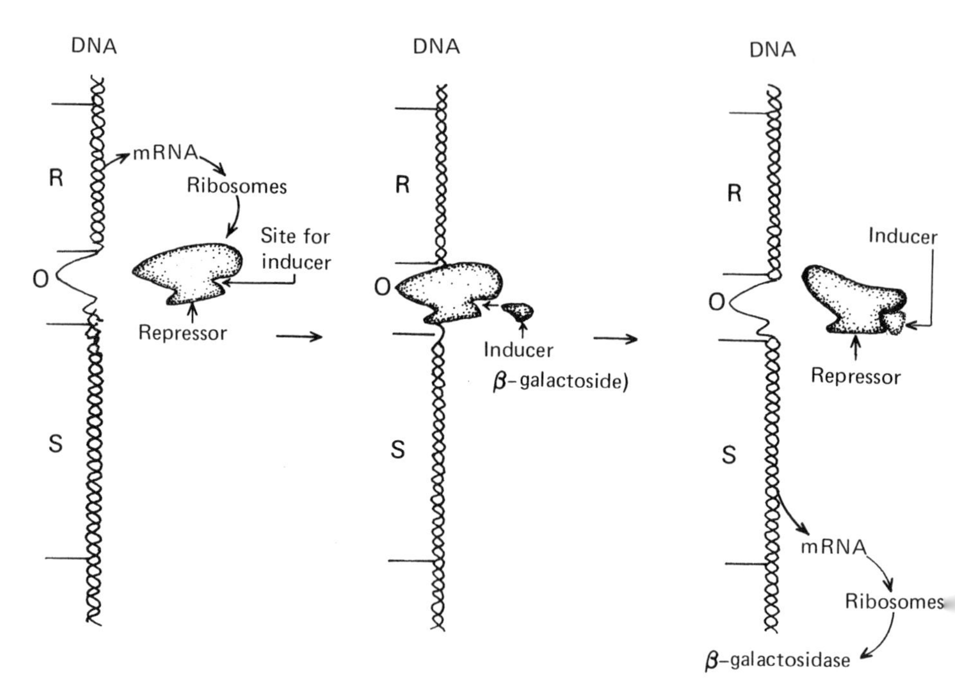

Fig. 3.25 The mechanism of induced enzyme synthesis. The regulator gene R programs for a repressor molecule which has a binding site specific for a binding site on the operator O. When the repressor is bound to O the structural gene S cannot be expressed. The inducer molecule (for instance β-galactoside) can bind to a complementary binding site on the repressor. When the inducer is bound to the repressor, the conformation of the latter is altered to such an extent that it detaches itself from the operator O. The structural gene S is then free to be expressed and enzyme (for instance β-galactosidase) synthesis can proceed.

an enzyme required for the synthesis of the amino acid histidine (Fig. 3.26).

Repression (and induction) of enzyme synthesis often occurs in coordination by a single corepressor (inducer). In many cases a product or substrate is involved in a sequence of enzymatic reactions in which the enzymes are functionally related to each other. These enzymes are coded by structural genes that are often (but not always) adjacent to each other in the DNA. Together they constitute an *operon*, and are turned on and off by a single repressor. For instance, the synthesis of histidine proceeds from phosphoribosyl pyrophosphate in ten consecutive steps involving nine different enzymes. The structural genes for these enzymes are adjacent to each other and constitute together the *his operon*. The transcription of the entire operon is repressed when histidine binds with a single repressor and the consequent binding of the repressor–histidine complex to a single operator.

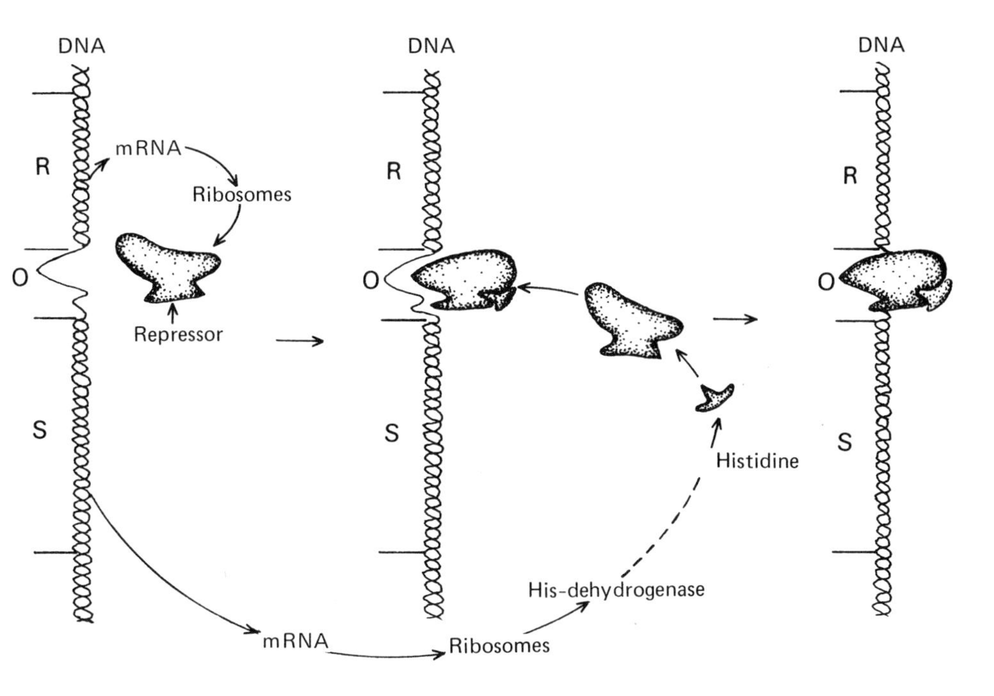

Fig. 3.26 The mechanism of repressed enzyme synthesis. The repressor molecule programmed by the regulator gene R has a conformation which is not compatible with the binding site on the operator O. When a corepressor (for instance histidine) binds to a complementary binding site on the repressor, the conformation of the latter becomes compatible with the binding site on the operator. The repressor–corepressor complex binds itself to the operator O and the expression of the structural gene S (the synthesis of, for instance, hisdehydrogenase) is blocked.

Deciphering of the Code. As mentioned before, much of the effort to assign particular codons to particular amino acids (the deciphering of the code) had, in the earlier days, been theoretical. The results were, consequently, of a rather speculative character. This changed radically when Nirenberg and Matthaei (1961) isolated an *in vitro* protein synthesizing system through which an experimental approach to the problem became possible. By letting protein synthesis be programmed by synthetic RNAs [such as the polynucleotide poly(U), or copolymers of repeating sequences, UCAUCAUCA . . . , or other variations] the assignments could be made within a few years.

The genetic dictionary is now known beyond much doubt (see Table 3.3). The codons UAA, UAG, and UGA are involved in peptide chain termination punctuation. Further, the code is highly degenerate; in a very

TABLE 3.3 Codon Assignments

	U	C	A	G	
U	phe	ser	tyr	cys	U
	phe	ser	tyr	cys	C
	leu	ser	punctuation	punctuation	A
	leu	ser	punctuation	trp	G
C	leu	pro	his	arg	U
	leu	pro	his	arg	C
	leu	pro	gln	arg	A
	leu	pro	gln	arg	G
A	ile	thr	asn	ser	U
	ile	thr	asn	ser	C
	ile	thr	lys	arg	A
	met	thr	lys	arg	G
G	val	ala	asp	gly	U
	val	ala	asp	gly	C
	val	ala	glu	gly	A
	val	ala	glu	gly	G

general way those amino acids which occur with high (or low) frequency in protein tend to have a large (or small) number of codons assigned to them. The universality of the code is remarkable. RNAs with a specific code give rise to the same polypeptide chains, irrespective of whether the protein synthesizing system originated from a bacterium or a eukaryotic cell from a higher organism.

Thus it seems that, from a cryptographic point of view, the genetic code is unraveled. There remain, however, many more questions to be answered. We have, for instance, not the slightest idea why the assignments are as they are. Or, in other words, we do not know the structural reasons for the code and what processes and forces gave rise to it. An understanding of how the code evolved to its present state may contribute answers to these questions. From this point of view, molecular genetics is still a very challenging area of biology, and very susceptible to an interdisciplinary approach, hence to biophysics.

Chapter

4 Molecular Interactions and Biology

4.1 Interaction Forces and Energy

In Chapter 3 we saw the importance of structure of a biological system. We discussed methods of measuring gross and refined molecular structures and we have seen how the structure of biological macromolecules underlies their functions in the system. In this chapter we will discuss how forces between the atoms in a molecule and between the molecules themselves determine these structures.

Interaction Energy. The physical chemistry of molecules describes many kinds of forces such as electron exchange forces, resonance forces, dipole forces, polarization forces, and van der Waals' forces. These names describe the conditions under which the forces are exerted but the kind of force is the same in all cases; it is the *electrostatic* force acting between the charged elementary particles in atoms which is described by Coulomb's law. *Electromagnetic* forces between *moving* charges are too weak to account for the formation of the structures under discussion.

A force exerted on a small object such as an atom or a molecule cannot be measured directly. We can, however, measure the *energy* by which, for instance, two atoms are held together by measuring the energy required to break the bond. If two atoms attract each other, their interactions represent a certain amount of *potential energy*. This potential energy reaches a minimum when the atoms have approached each other to such a distance that the attractive forces balance the repulsive forces. An example is the ionic bond between Na^+ and Cl^- in a molecule of NaCl vapor (in crystalline sodium chloride we cannot speak of an NaCl molecule since in the stable arrangement it is a three-dimensional crystal structure of Na^+ and Cl^- ions). The minimum of the potential energy curve occurs when the attractive Coulomb force between the two kinds of ions is balanced by the repulsive force between the nuclei (see Fig. 4.1).

The energy of the bond is the difference between the potential energy of the system with separated atoms and the potential energy of the system with bonded atoms. By *convention* the potential energy of a system of two atoms separated by an infinitely large distance is set at zero. Therefore, the bond energy is negative. Since the energy, rather than the forces, is usually the measurable quantity, it makes more sense to talk about *interaction energy* than about interaction forces. We can express the energy in ergs, or electron volts per particle or, to conform with the chemists, in calories per mole (see Table 4.1).

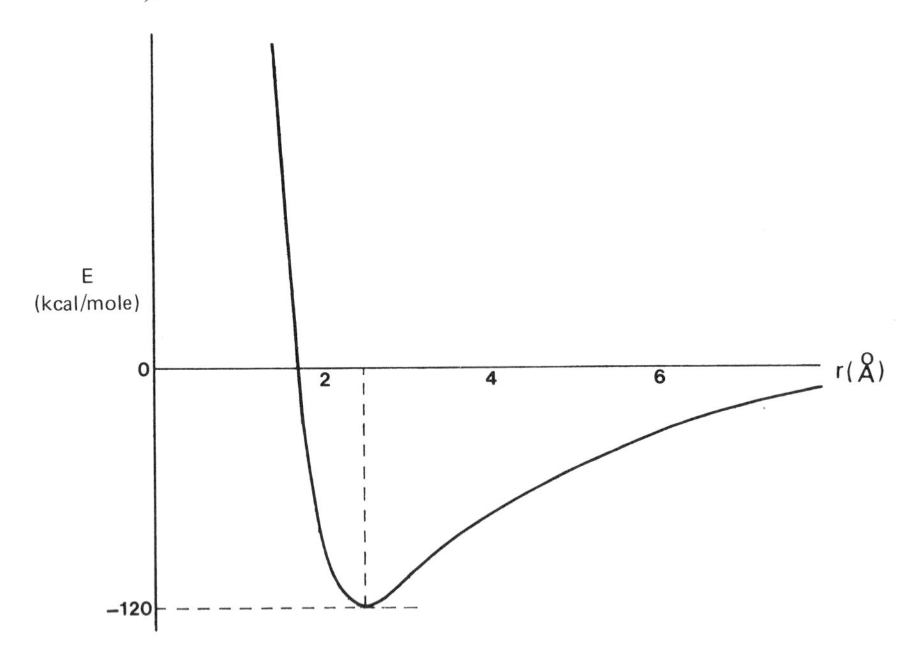

Fig. 4.1 The potential energy E of the ionic bond in NaCl as a function of the distance r between the ions Na^+ and Cl^-. The potential energy can be represented by $E = (q^2/r) + be^{-r/a}$ in which q is the ionic charge and a and b are constants.

TABLE 4.1 Energy Conversion Table

Wavelength, λ (nm)	200	400	800	1600
Wave number, $\bar{\nu} = 1/\lambda$ (cm^{-1})	50,000	25,000	12,500	6250
Frequency (sec^{-1})	15×10^{14}	7.5×10^{14}	3.75×10^{14}	1.88×10^{14}
erg	10^{-11}	5×10^{-12}	2.5×10^{-12}	1.25×10^{-12}
kcal/einstein	144	72	36	18
eV	6.3	3.12	1.57	0.79

Strong and Weak Interactions. The criterion by which we make the distinction between strong and weak interactions is the extent to which thermal motion will disrupt the interaction. The average thermal energy is kT, in which k is Boltzmann's constant and T is the absolute temperature (in degrees Kelvin). At body temperature (310 K) this is of the order of 2.5×10^{-2} eV/particle (about 0.6 kcal/mole). Strong interactions have a value many times greater than this and are, thus, unlikely to be disrupted by thermal motion. The primary structure of biological macromolecules is determined by such strong interactions. The higher-order structures are determined by weaker forces. Weak interactions are of the same order of magnitude as kT. These will be disrupted first when the molecule is heated, resulting in the loss of quaternary, tertiary, and secondary structure in that order.

4.2 Weak Interactions

Weak interactions between atoms and/or molecules cause a failure of the gas law of Boyle and Gay-Lussac, leading to the van der Waals' equation,

$$[P + (a/V^2)](V - b) = RT$$

in which a and b are (to a first approximation) constants. Weak interactions, therefore, are often called van der Waals' interactions or van der Waals' forces. These forces, electrostatic in nature, can be described as interactions in which *electric dipoles* are involved.

Dipoles. A bond between two atoms with different *electronegativities* (different affinities for negative charges or electrons) always results in a dipole moment. A typical example is the O–H bond. The electronegativity of the oxygen atom draws the electron from the hydrogen atom toward the oxygen and the result is a system in which equal charges of different sign are separated. The dipole moment of the O–H bond is 1.60 D. [1 D (= 1 debye) is equal to 10^{-18} esu cm.]

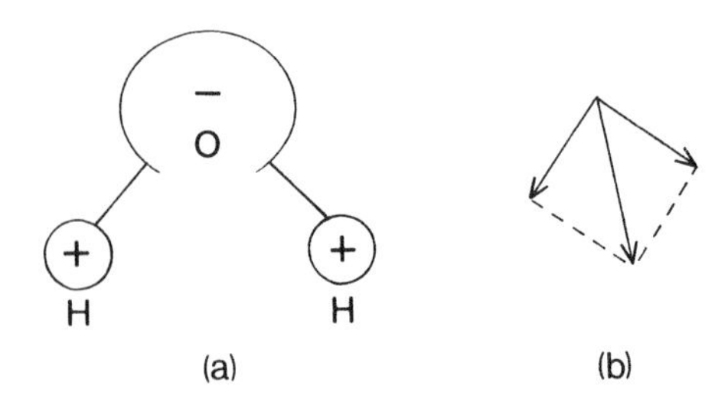

Fig. 4.2 (a) The dipole of water and (b) its vector representation.

Water consists of two such bonds in which one oxygen and two hydrogen atoms participate. The resultant dipole moment of water is 1.85 D. This means that the two O–H bonds must make an angle of 105° with each other. Only then can vector addition result in a total dipole moment as determined (Fig. 4.2).

TABLE 4.2 Dipole Moments

Molecule	Moment (Debye)
HCl	1.03
H_2O	1.85
1,3,5-trichlorobenzene	0
p-dihydroxybenzene	1.64
Glycine	15.1
Egg albumin	252

Dipole moments for some molecules are given in Table 4.2. These moments give information about (1) the extent to which a bond is permanently polarized and (2) the geometry of the atoms, especially the angle between them (as we have seen with water). The fact that the dipole moment of carbon dioxide is zero in spite of the difference in electronegativity between carbon and oxygen indicates that the molecule is linear,

$$O=C=O$$

with the two dipole moments of the C=O bonds canceling each other out. Benzene (I), *p*-dichloro benzene (II), and 1,3,5-trichloro benzene (III) must be planar to account for their zero dipole moment. The two hydroxyl groups in *p*-dihydroxybenzene (IV), however, must make an angle with the plane of the ring in order to account for the 1.64 D value of its dipole moment.

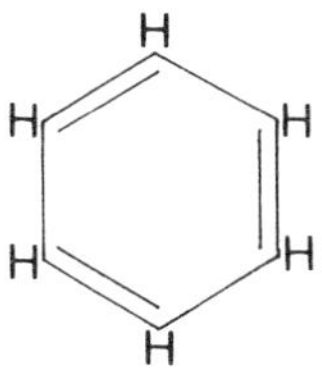

I 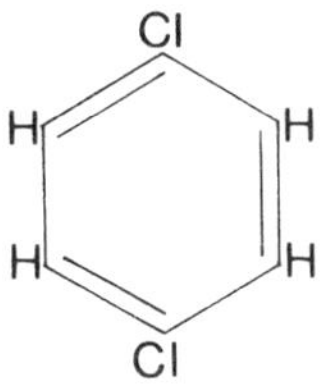 II 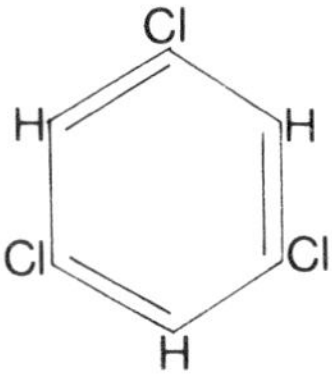 III 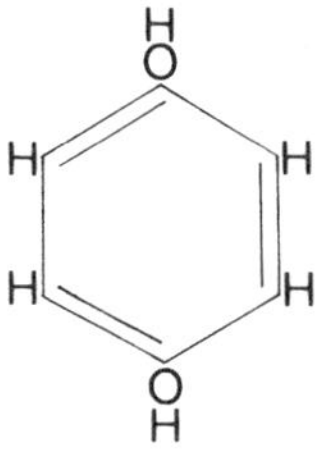 IV

Dipole Interaction. Dipoles interact with an electric field by Coulombic forces, which tend to align the dipole in the direction of the field. If the field originates at a point charge q, the force acting on the aligned dipole can be calculated by application of Coulomb's law, under the assumption that the distance between the point charge and the dipole is large as compared with the distance separating the two charges of the dipole. From such a calculation it follows that the energy of this type of ion–dipole bond is proportional to the inverse *square* of the distance (whereas the mutual energy of two ions in an ionic bond is inversely proportional to the first power of the interionic distance). Using a distance of 2.73 Å between a K^+ ion and a water molecule in the first hydration sphere and a dipole moment of 1.85 D for the water dipole, one can calculate that the binding energy of the first hydration sphere of K^+ is about 17 kcal/mole. This is about 15% of the (monovalent) anion–cation bond at the same distance. Thus, hydration of a monovalent ion such as K^+ is not much disturbed by thermal motion; at least not in the first hydration sphere. However, since the binding energy decreases with the square of the distance, the orienting force of an ion on water dipoles must become quite inefficient beyond a layer of three or four molecules. Dipole–dipole interactions contribute to the diminution of the binding energy beyond the first few hydration spheres. Dipole–dipole interaction diminishes with the third power of the dipole distance.

Permanent dipoles are characteristic of molecules in which relatively electropositive atoms are bound to relatively electronegative ones (that is when the dipoles of several bonds do not cancel each other out such as with the examples given above). All molecules acquire dipoles by relative displacement of their positive and negative charges when placed in an electric field. The magnitude of such *induced dipoles* depends on the *polarizability* α of the molecule. The interaction of induced dipoles is a second-order effect; when the dipole is induced by a field from a point charge the interaction energy is, as can be calculated easily, proportional to the fourth power of the inverse distance. Dipole-induced dipole interaction energy is proportional to the sixth power of the inverse distance.

The Hydrogen Bond. Dipole–dipole interactions (between polar molecular groups) and dipole-induced dipole interactions are important for the secondary and higher-order structure of proteins. Another type of electrostatic interaction plays a very important role in biology; this is the already mentioned *hydrogen bond.* Hydrogen, since it has only one 1s electron, can form only one covalent bond. In some situations, however, hydrogen can bond with two other atoms, instead of just one. The additional bond, which is much weaker than a normal covalent bond, is due to electrostatic attraction between the proton and some electronegative element of small atomic volume, such as N, O, or F. The proton, which is slightly isolated from the valence electron involved in the covalent bond, is extremely small; its electrostatic field, therefore, is intense and a bonding can occur due to the attraction of the positive proton for the electrons of the bonded atom.

The distances involved in this case (2–3 Å) are in the same order of magnitude as the dipole distances (about 1 Å or slightly smaller). The dipole–dipole approximation (in which the distance between the dipoles is assumed large compared to the dipole distance), therefore, cannot be applied; more detailed calculation shows that the interaction energy is some 5–10 times greater than the "close range" dipole–dipole interaction. The bonds are indicated by dashed lines in order to distinguish them from the covalent bond in which the hydrogen is engaged. The O–H---O hydrogen bond is about 0.2 eV (5 kcal/mole), the O–H----N is somewhat weaker.

Hydrogen bonds are responsible for the structure of water and ice (Fig. 4.3); the dimer of formic acid (V) is a result of hydrogen bonds. In biology

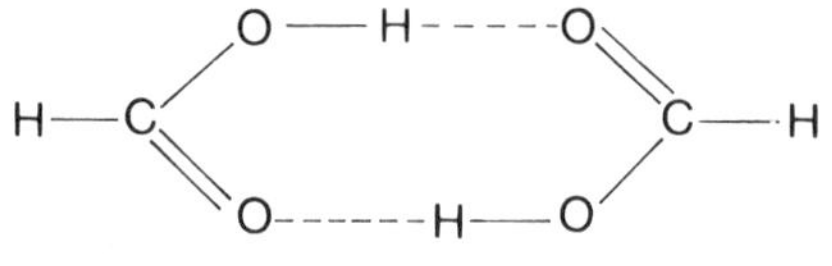

V

hydrogen bonds are, to a major degree, responsible for the tertiary structure of proteins and nucleic acids. We have met them already in the structure of DNA and have seen how they are instrumental in the processes of replication, transcription, and translation.

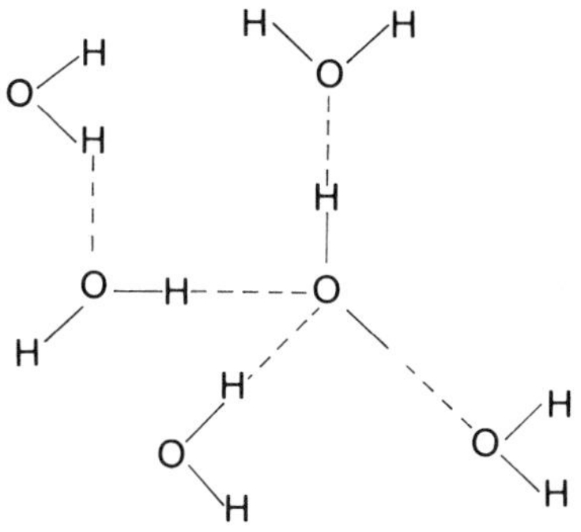

Fig. 4.3 The structure of water due to hydrogen bonding.

4.3 Strong Interactions

Molecules are held together by strong interactions, or bonds, between the atoms. These interactions hold the primary structure of biological macromolecules together. The bond energies are well in excess over thermal energies, as can be seen from Table 4.3.

Ionic and Covalent Bonds. For *polar* molecules, such as in NaCl vapor, the bond can be described as the electrostatic attraction between a positive

TABLE 4.3 Dissociation Energies of Typical Bonds

Bond		Dissociation energy (eV)
H—H		4.40
$>$C—C$<$		2.55
$>$C—H		3.80
$>$C—N$<$		2.13
$>$C—N$<$	(peptide)	3.03
$>$C=C$<$		4.35
$>$C=O		6.30
—C≡C—		5.35

and a negative ion. The explanation of *nonpolar* molecules, such as CH_4, is somewhat more complicated. For both types of molecules the *valence* (the number of atoms that each atom can bond) is usually the same. Oxygen, for example, binds two potassium atoms in the polar compound K_2O and two ethyl groups in the nonpolar compound $(C_2H_5)_2O$.

Molecules are made up of atoms. They can, therefore, be considered as systems of charged particles which assume a certain equilibrium configuration resulting from mutual attraction and repulsion of these particles. Therefore, the only thing we have to do in order to calculate the properties of the system is to set up an appropriate Hamiltonian which includes all these interactions and solve the Schrödinger wave equation,

$$\mathbf{H}\psi = E\psi \tag{4.1}$$

This equation yields, in addition to the wave functions ψ, the possible values of the energy E of the system and a number of physical quantities that can be calculated from ψ through use of the appropriate equations. The differential equation which we have to solve, however, is an extremely complex one, including all the electrons. Moreover, the terms describing the mutual interactions of the electrons bring about unsolvable mathematical problems. In practice, only systems with one electron can be rigorously solved.

Fortunately, a number of approximation procedures have been developed which result in approximate solutions of the equation. Such solutions are useful in as much as they enable us to interpret observed experimental facts in terms of the appropriate fundamental physical quantities. The *molecular orbital method* is such an approximation procedure, suitable for molecular systems. It is essentially a method in which *molecular orbitals*, wave functions for single electrons in a molecule, are constructed by the fusion of *atomic orbitals*, the wave functions for single electrons in the atoms which form the molecule. If such a fusion is carried out by linear combination of the atomic wave functions, the procedure is called the MO-LCAO (molecular orbital-linear combination of atomic orbitals) approximation.

Molecular Orbitals. The MO-LCAO procedure can be illustrated with the molecular orbital of the hydrogen molecule. In Fig. 4.4, the interactions of two hydrogen atoms A and B are shown. If the atoms are far apart the electron of each of the two atoms occupies an atomic orbital which is the single 1s orbital, $\phi_A(1s)$ and $\phi_B(1s)$. When the atoms are brought together the atomic orbitals coalesce, and if we adopt the principle that the molecular orbital can be constructed from a linear combination of atomic orbitals (LCAO) we obtain

$$\psi_{(MO)} = C_1\phi_A(1s) + C_2\phi_B(1s) \tag{4.2}$$

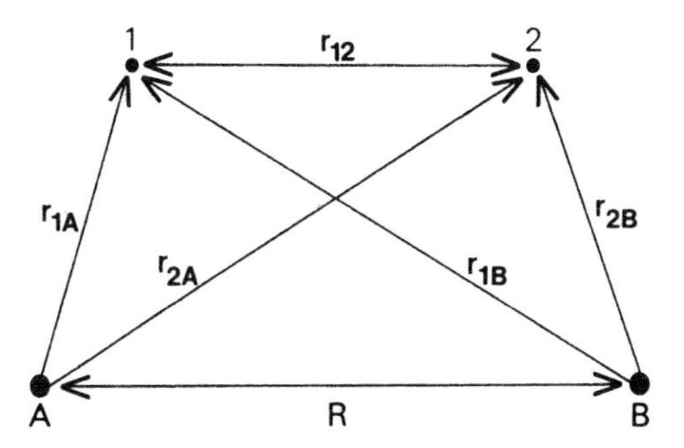

Fig. 4.4 The mutual interactions of two hydrogen atoms A and B. R is the internuclear distance, r_{ij} are the distances between nuclei and electrons.

Since the molecule is completely symmetrical C_1 must be equal to $\pm C_2$. We thus obtain two possible MOs,

$$\psi_g = \phi_A(1s) + \phi_B(1s), \qquad \psi_u = \phi_A(1s) - \phi_B(1s) \tag{4.3}$$

In Fig. 4.5 cross sections of these orbitals are sketched. According to the Pauli principle, each orbital can have two electrons provided that they have antiparallel spins.

The MO-labeled ψ_g does not change sign upon inversion through the center of symmetry in the midpoint between the two nuclei. This wave function, therefore, is called *even* (the subscript g stands for *gerade* which is German for even). The other wave function ψ_u does change sign upon inversion and is, therefore, odd (u stands for *ungerade* which means odd). The even molecular orbital ψ_g leads to a build-up of electronic density between the two nuclei, the odd molecular orbital molecular orbital has a node between the nuclei. In other words, the even orbital can be, and in the ground state is, occupied by electrons which are *shared* by the nuclei, while in the odd orbital the absence of electron density between the nuclei makes them repulse each other. Therefore, in this case the even orbitals are *bonding* and the odd orbitals are *antibonding*. Each of the orbitals can be occupied, in

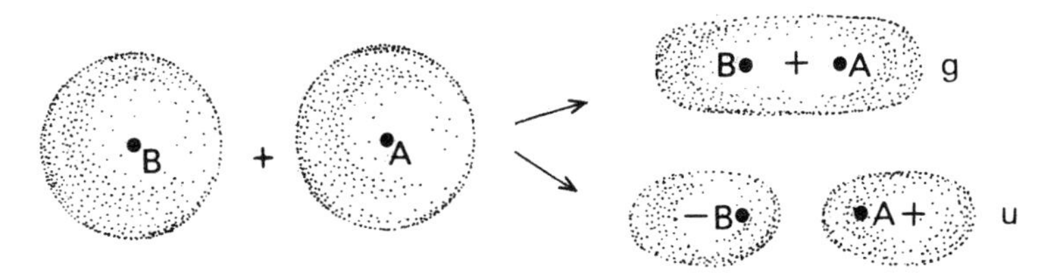

Fig. 4.5 Two molecular orbitals g and u of the hydrogen molecule formed by two atomic orbitals of hydrogen.

accordance with the Pauli principle, by two electrons. Thus, in the covalent bond of the hydrogen molecule, the bonding orbital is occupied by two electrons with antiparallel spins. Excitation of the molecule leads to the promotion of one electron to the antibonding orbital.

An elaboration of this leads to a quantum mechanical (MO-LCAO) approximation for the description of the covalent bond. It starts from the system of nuclei and adds the electrons one by one, taking into account the Pauli exclusion principle. The procedure is very similar to the procedure by which atoms can be built up by adding the s, p, d, etc., electrons in the lowest atomic orbitals, as dictated by the Pauli principle.

Orbital Quantum Numbers. As with atomic orbitals, the molecular orbitals are determined by quantum numbers. For diatomic molecules the molecular orbital quantum numbers are the principal quantum number n (with $n - 1$ being the number of nodes in the wave function) and a quantum number λ which determines the component of the angular momentum along the internuclear axis. The quantum number λ thus has an analogy with the magnetic quantum number m_l in atoms. Orbitals for which $\lambda = 0, 1, 2, \ldots$ are called σ, π, δ, ... orbitals and electrons occupying such orbitals are called σ, π, δ, ... electrons. In diatomic molecules the wave functions for σ orbitals are symmetric with respect to the axis connecting the two nuclei. π orbitals, since they have an angular momentum of unity, are antisymmetric about this axis. The higher orbitals have more complicated symmetry patterns.

In addition to σ, π, δ, ... orbitals in molecules, there are so-called *n orbitals* that resemble atomic orbitals embedded in the molecule. An n electron is confined to the neighborhood of one nucleus and interacts only weakly with the other nuclei. The wave functions of the n orbitals in a molecular atom are not very different from their counterparts in the isolated atom.

Orbitals generated by the fusion of s orbitals normally are σ orbitals, such as the one in the hydrogen molecule. p orbitals can form σ as well as π orbitals; an example is the nitrogen molecule. Nitrogen has three 2p electrons; their orbitals have a nodal point at the nucleus and have a butterflylike configuration, as sketched in Fig. 4.6. The three orbitals are at right angles to each other and can be labeled p_x, p_y, and p_z. If two N atoms come together the p_x orbitals fuse together to form a σ orbital, as sketched in Fig. 4.6a. These are the most stable molecular orbitals that can be formed from the p orbitals. The molecular orbitals formed from the p_y and p_z atomic orbitals of the two N atoms have a distinctly different form. The sides of the p orbitals coalesce to form "streamers" of charge density, one above and one below the internuclear axis (Fig. 4.6b). The angular momentum now has a component along the internuclear axis, the quantum number

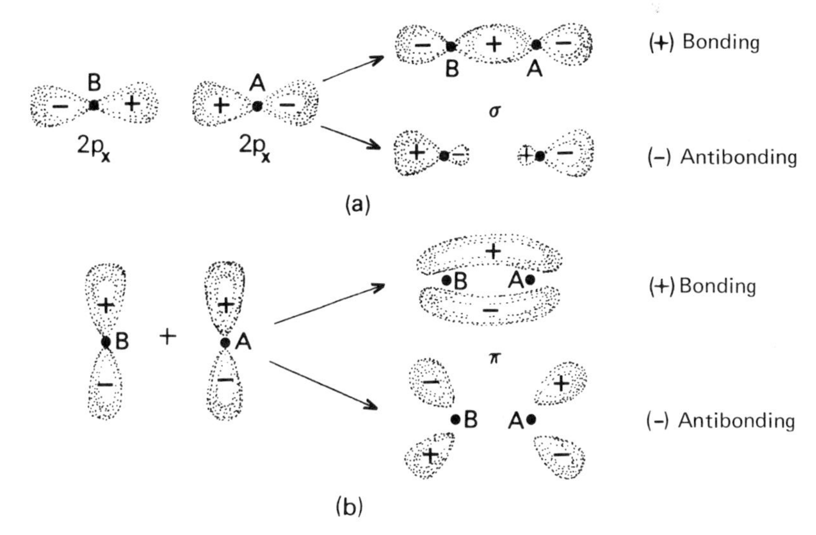

Fig. 4.6 Molecular orbitals formed by the orbitals of p electrons: (a) bonding and antibonding σ orbitals; (b) bonding and antibonding π orbitals.

$\lambda = 1$, and the orbitals are π orbitals. In this case the odd orbital (normally occupied) is bonding and the even orbital is antibonding.

Hybrid Orbitals. In multiatomic molecules the atomic orbitals often combine to form so-called *hybrid orbitals* before they form molecular orbitals. A classical and biologically important example is carbon. Carbon is bivalent in the ground state; it has two unpaired electrons. In compounds, however, C has four valences often directed to the corners of a regular tetrahedron. The hybrid orbitals explain this phenomenon. In the ground state C has two unpaired electrons in atomic orbitals p_x and p_y. In order to acquire the four valencies, one of the 2s electrons is promoted to a 2p state. The atom then obtains four unpaired electrons, a 2s electron, and three 2p electrons. This promotion requires approximately 2.8 eV, but this energy is more than compensated for by the binding energy of the four bonds. The four orbitals are now combined with each other in hybrid orbitals. An extremely stable way to do that is to combine the 2s orbital with each of the p orbitals in the following manner:

$$\begin{aligned}
\psi_1 &= \tfrac{1}{2}(\phi_s + \phi_{p_x} + \phi_{p_y} + \phi_{p_z}) \\
\psi_2 &= \tfrac{1}{2}(\phi_s + \phi_{p_x} - \phi_{p_y} - \phi_{p_z}) \\
\psi_3 &= \tfrac{1}{2}(\phi_s - \phi_{p_x} + \phi_{p_y} - \phi_{p_z}) \\
\psi_4 &= \tfrac{1}{2}(\phi_s - \phi_{p_x} - \phi_{p_y} + \phi_{p_z})
\end{aligned} \tag{4.4}$$

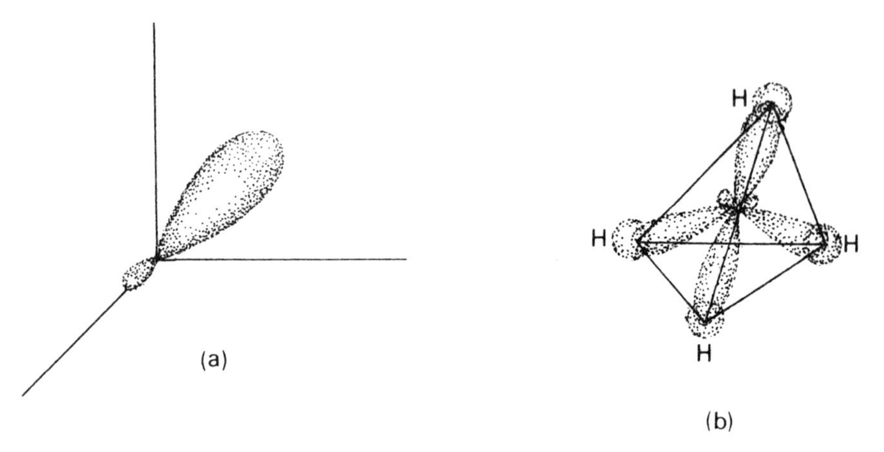

Fig. 4.7 (a) An sp^3 hybrid orbital; (b) the molecular orbitals of methane formed from sp^3 hybrid orbitals.

The resulting four *hybrid orbitals* designated as sp^3 (each having a shape shown in Fig. 4.7a), are directed to the corners of a regular tetrahedron as shown in Fig. 4.7b. In this configuration the negative charges in the orbitals avoid each other in the maximum as much as possible. If the four orbitals are now combined with the 1s orbitals of four hydrogen atoms, four σ molecular orbitals are formed in the methane molecule CH_4.

Conjugated Double Bonds (*Resonance*). Hybridization of atomic orbitals in carbon can also lead to other configurations. One example is the formation of three sp^2 hybrids (trigonal hybrids). Here the 2s, $2p_x$, and $2p_y$ orbitals are mixed, leading to three coplanar orbitals in an angle of 120° and leaving the fourth AO, the p_z, perpendicular to the plane of the others. This form of hybridization leads to the formation of *nonlocalized* molecular orbitals, an example being *benzene*. The three sp^2 hybrids are fused with each other and with the 1s orbital of the H atoms to form localized σ orbitals which are in the plane of the molecule (Fig. 4.8a). The p_z orbitals, which extend above and below the molecular plane (Fig. 4.8b), then fuse together to form π orbitals above and below the plane (Fig. 4.8c). Of the six possible orbitals three are bonding and three are antibonding. In the ground state only the three bonding (lowest energy) orbitals are occupied, each with two electrons which, in accordance with the Pauli exclusion principle, must have antiparallel spins. The six electrons in these orbitals are no longer localized to a particular atom. They can move freely in the "doughnuts." Because each electron has more space to occupy as a result of the delocalized orbitals their energy levels are lower (uncertainty principle). This extra energy (for benzene 1.8 eV or about 41 kcal/mole) contributes to the strength of the

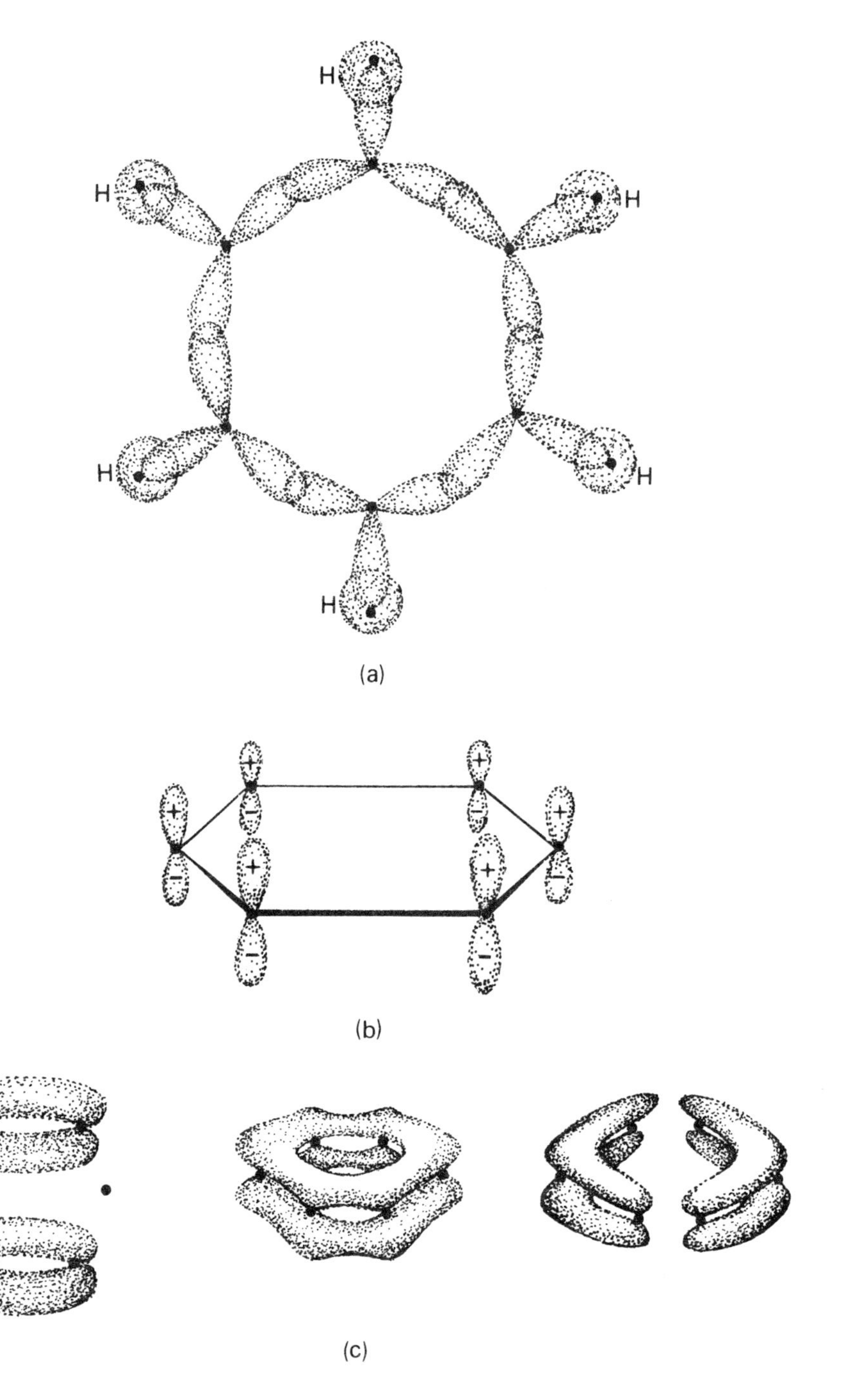

Fig. 4.8 The formation of molecular orbitals in benzene; (a) σ orbitals in the plane of the molecule; (b) the p_z atomic orbitals; (c) the three binding π orbitals formed by the p_z orbitals.

bond and is called *resonance energy*. If the structure of a molecule is seen as being made up of the superposition of various distinct valence-bond structures, we can say that the actual structure, for instance of benzene, is formed by the "resonance" of two Kekulé structures

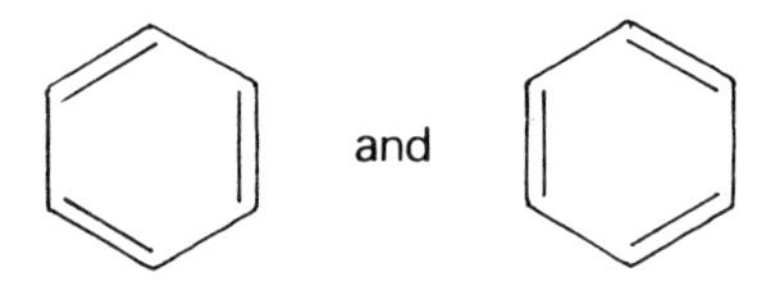

This type of bond structure is called a *conjugated double-bond structure*. Many biologically important molecules have such conjugated structures which give the molecule certain easily recognizable spectroscopic characteristics (such as the chromophores of cytochrome, hemoglobin and myoglobin, and chlorophyll).

4.4 Molecular Electronic Transitions

Proliferation of Energy Levels. If two 2p atomic orbitals coalesce into π molecular orbitals, two of these are possible; a bonding orbital (π) and an antibonding orbital (π^*). The energy levels, with respect to the energy level of the atomic orbital, are as given in Fig. 4.9. In the ground state the two electrons usually occupy the lower π orbital, having antiparallel spins according to the Pauli principle. Upon absorption of an energy quantum (light) of the right size, a transition may occur which promotes one electron from the π orbital to the π^* orbital. The energy involved in such a transition, in the visible or near uv spectral region, is far less than the energy involved in atomic electronic transitions, in the far uv; this is due to splitting of the p orbital energy level. This proliferation of energy levels is even more pronounced when a system similar to that of benzene is considered (Fig. 4.10). In this system any one of the six electrons in the ground state occupying three π orbitals can be promoted to three π^* orbitals. It is then clear that the more

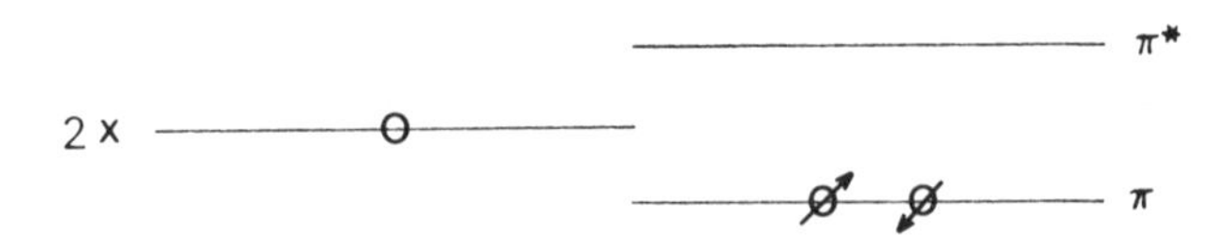

Fig. 4.9 The energy levels of the π orbitals formed by two p orbitals. In the ground state the two electrons having opposite spins occupy the lower energy orbital.

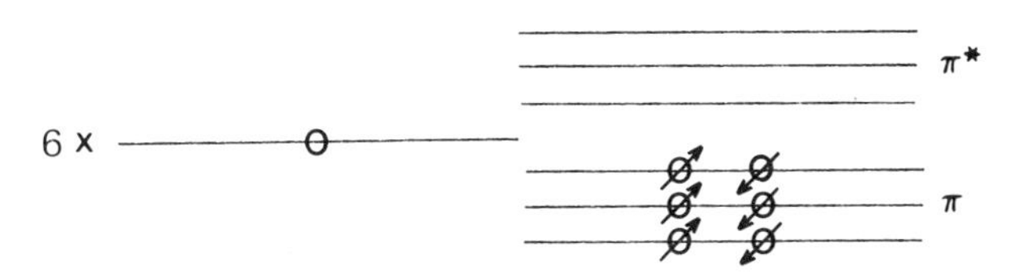

Fig. 4.10 The energy levels of the π orbitals formed by six p orbitals, as they occur in benzene. In the ground state the three lower energy levels are occupied each by two electrons having opposite spins.

extended the conjugated system the broader the absorption bands and the more they will extend toward the red part of the spectrum. This is illustrated in Fig. 4.11, which compares the spectrum of the eye pigment retinal with that of chlorophyll *a*.

Transition Diagrams. Transitions from one molecular orbital to another, e.g., from π to π^* or from n to π^*, are called $\pi\pi^*$ or nπ^* . . . transitions. In Fig. 4.12a a sequence of transitions is illustrated with an energy level diagram. In such a diagram the levels given are the energy levels of the electrons in their respective orbitals. The sequence shows

(1) the promotion of a π electron to a π^* orbital;
(2) the transfer of an electron from the n orbital to the partly vacated π orbital (the net result of these two steps is the promotion of an n electron to a π^* orbital); and
(3) the fall of the π^* electron into the vacancy in the n orbital.

The same events are more conveniently and correctly represented by the so-called transition diagram (Fig. 4.12b). In this diagram the energies of the ground state and the various excited states of the molecule as a whole are depicted. Here the nπ^* state means the state in which an electron (no matter which electron) has been promoted from the n orbital to the π^* orbital. Thus, the sequence is (1) a $\pi\pi^*$ excitation, (2) a $\pi\pi^* \rightarrow$ nπ^* interconversion, and (3) a deexcitation from nπ^* to ground state.

$\pi\pi^*$ and nπ^* Transitions. The most intense transitions are those in which a strong dipole oscillation is involved and where the orbitals have a good spatial overlap. This is the case for $\pi\pi^*$ transitions which are, therefore, highly favored and cause intense absorption bands. This can be understood if one considers that in the $\pi\pi^*$ excitation the excited π^* electron keeps a strong coupling with its partner in the π orbital, and that the π and the π^* orbital, although differing in symmetry, occupy the same region in the molecule.

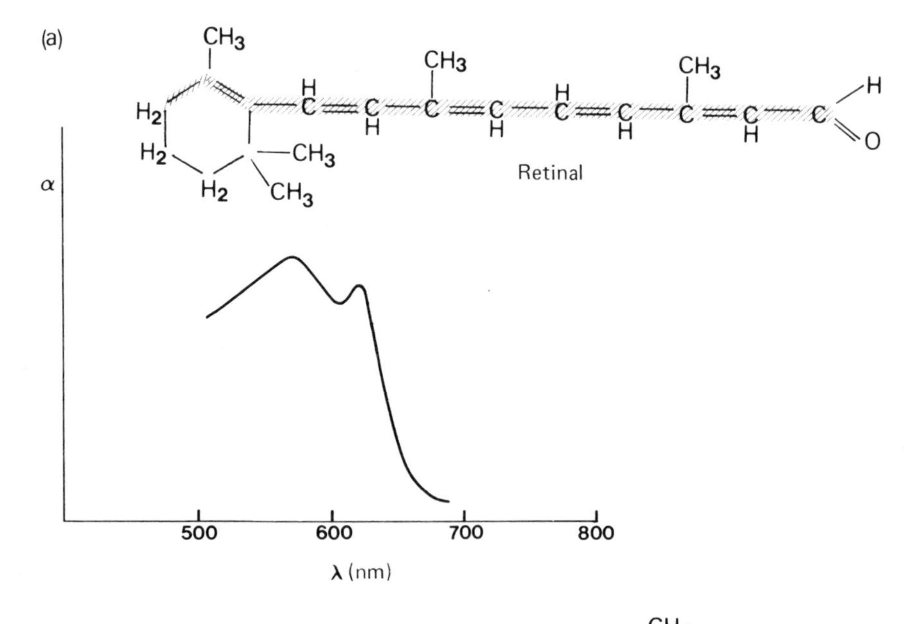

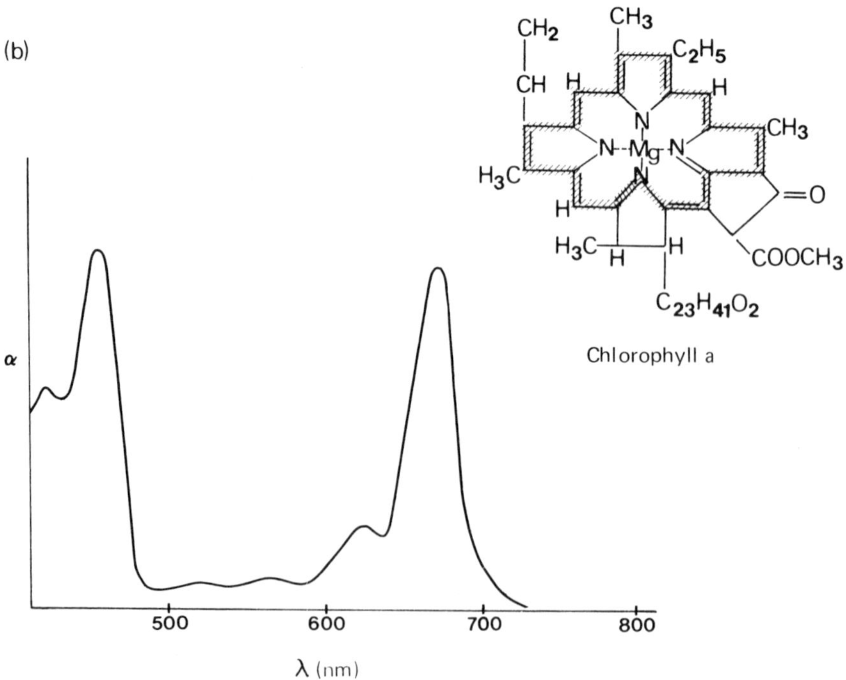

Fig. 4.11 The absorption spectra of (a) retinal and (b) chlorophyll *a*. The chemical structures of the molecules are also shown. The hatched areas designate the π systems in the molecules.

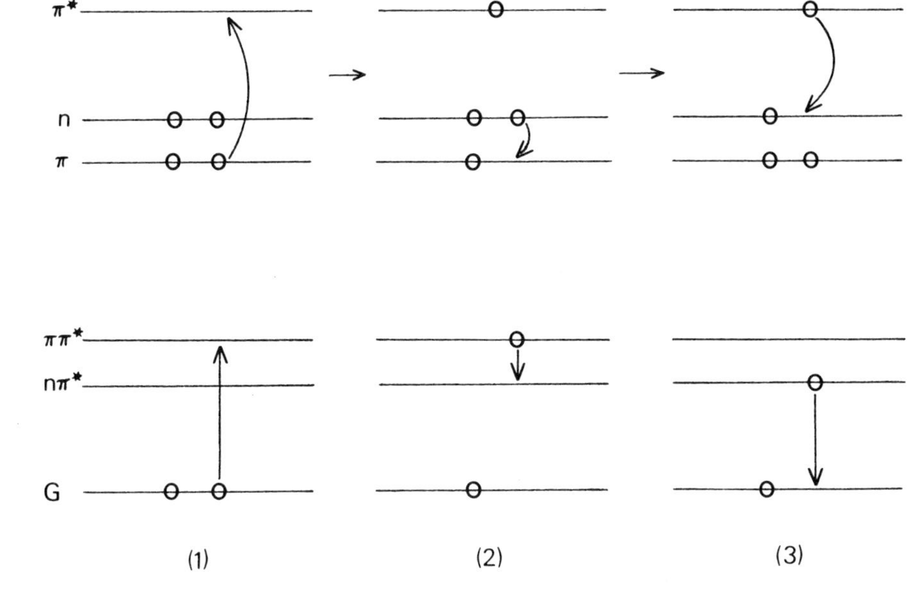

Fig. 4.12 (a) The energy level diagrams and (b) the transition diagrams of the transitions $\pi\pi^*$ (1), $\pi\pi^*$ to $n\pi^*$ interconversion (2), and $n\pi^*$ to ground transition (3).

In an $n\pi^*$ transition, an electron moves from a localized orbital to a delocalized orbital. The spatial overlap of the two orbitals, therefore, is very poor and the transition probability is correspondingly low. As a result the absorption bands of $n\pi^*$ transitions are about a hundredfold less intense than those of the $\pi\pi^*$ transitions. The intrinsic lifetimes of the $n\pi^*$ states are greater than those of the $\pi\pi^*$ states by the same factor. However, $n\pi^*$ states have a high degree of polarization due to the large electron displacement attending the transition. This may be the reason that once an $n\pi^*$ state has been attained the molecule becomes very reactive as an electron donor or an electron acceptor. Oxidation–reduction reactions are very common in biology and $n\pi^*$ states may very well be involved.

The Triplet State. When the spin of the excited electron is reversed in a transition from a singlet state a triplet state is generated. Such a spin reversal in the transition from the ground state is very improbable since the antiparallel electrons are strongly coupled in the ground state. From a first excited state a spin reversal can occur more easily since the spin–spin coupling has become looser. The principal magnetic forces that will disrupt spin–spin coupling are the orbital magnetic moments of electrons and atomic nuclei. Thus, the greater the spin–orbital coupling and the weaker the spin–spin coupling, the more likely a singlet–triplet transition can occur.

A greater spin–orbital coupling and a weaker spin–spin coupling are especially marked in nπ^* states, as contrasted with $\pi\pi^*$ states. Further, the nπ^* states are intrinsically longer lived than the $\pi\pi^*$ states. Entry into a triplet state will, therefore, more likely occur from the former than the latter state.

Triplet states have lower energy than singlet states, because the excited electron and its ground state partner have parallel spins. The more closely coupled the two electrons, the larger the energy difference between the singlet and the triplet state. Thus, the energy gap between a $\pi\pi^*$ and a $\pi\pi^T$ state is considerably larger than that between an nπ^* and an nπ^T state (see Fig. 4.13).

The lifetime of a triplet state is four to five orders of magnitude longer than that of a singlet state. Triplet states, therefore, are often implicit in photochemistry. There is, for instance, good evidence that triplet states are involved in the *in vitro* photochemistry of chlorophyll. The role of triplet states in the function of chlorophyll *in vivo* is less certain.

4.5 Absorption and Emission of Light

Electronic transitions in molecules occur with absorption or emission of light (from the uv to the near infrared spectral region). The emission of light is called *fluorescence* when it accompanies "allowed" radiative transitions such as the return from a $\pi\pi^*$ state to the ground state and short lived (in the order of 10^{-8} sec for molecules like chlorophyll). Light emitted from much longer lived, so-called metastable states (such as triplet states) is called *phosphorescence*. When light is emitted from the first excited singlet state *after* a back transition from a metastable state it is called *delayed fluorescence* or *luminescence*.

Line and Band Spectra. When the energy levels of the quantum states are well-defined and separated, as is the case in atoms, the absorption

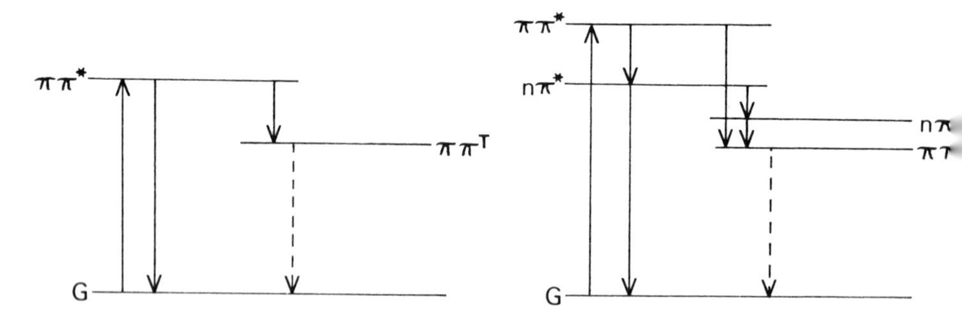

Fig. 4.13 Transition diagrams involving $\pi\pi^*$ to $\pi\pi^T$ and nπ^* to nπ^T transitions.

and emission spectra show sharp and intense lines or narrow bands. In molecules, however, there is a greater proliferation of quantum states of different energy. The electrons of atoms brought together can interact with each other and with more than one nucleus, thus becoming components of a larger system. As a result the original energy levels are split up into numerous sublevels. The relative movements, vibrational and rotational, of the nuclei with respect to each other also contribute to this proliferation of quantum states. Although one can still recognize the major electronic configuration in the ground state and the excited states (electronic states), each of these is subdivided in a manifold of substates reflecting the finer details of the interactions inside the molecule. The possibilities of transitions of different energy are thus substantially greater and the line spectra become band spectra.

Vibrational and Rotational States. The relative motions of the nuclei, vibrational and rotational, result in substates which are, although of course much closer together than the electronic states, also quantized. An electronic state thus is subdivided in a set of quantized *vibrational substates* and each of these, in turn, is subdivided in a set of rotational "*subsubstates*." The role of nuclear vibrations in molecular spectroscopy can best be explained by discussing, as an example, a model of an idealized diatomic molecule such as the one shown in Fig. 4.14a. The covalent bond determined by the orbitals of the binding electrons holds the system together while the repulsive electrostatic force between the nuclei tends to pull it apart; hence, the system vibrates. Let us assume that we can consider the system as a harmonic oscillator (which it would be when the nuclei were particles connected by a spring obeying Hooke's law). Then the potential energy of the vibration is a parabolic function, given by the bold line in Fig. 4.14b, of the distance r separating the nuclei. Classically, such a system could exist anywhere within the domain bordered by the parabola but not outside this domain. Vibration changes the internuclear distance along line segments in the parabola which, because of the conservation of energy, are parallel to the r axis. All line segments between the two arms of the parabola are possible, or in other words, the total energy of the system can assume a continuum of values. On the molecular level, however, we have to use the total energy of the system to construct a Hamiltonian operator which, acting on the wave function ψ, determines the Schrödinger wave equation. The only solutions (wave functions) of this wave equation, and hence the only possible states of the system, are those in which the energy E_{vib} is given by

$$E_{vib} = h\nu_0(v + \tfrac{1}{2}) \qquad (4.5)$$

in which h is Planck's constant, ν_0 the characteristic frequency of the

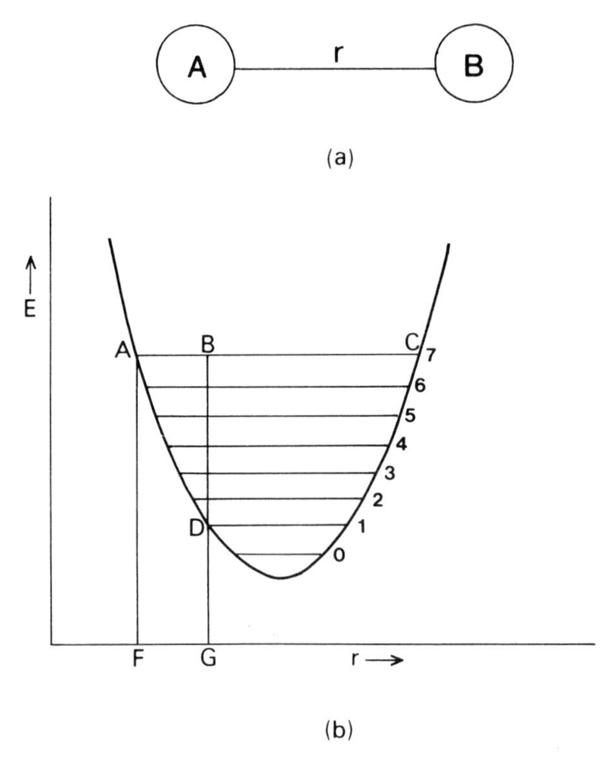

Fig. 4.14 (a) A representation of an idealized diatomic molecule. Two masses A and B are held together by a weightless spring. (b) The potential energy curve of the system shown in (a). When the system vibrates the energy can have values within the area bounded by the curve (for instance between A and C). In a quantum mechanical system only discrete levels 0,1,2, . . . are possible. At points on the curve (for instance point A) the system is at rest and the energy is all potential energy (AF). When the system moves (for instance in point B) the energy comprises kinetic energy (BD) and potential energy (DG).

oscillator, and v the *vibrational quantum number* (which can have the integer values 0, 1, 2, . . .). In Fig. 4.14b the energy levels for a number of values of v are drawn as thin lines. In classical terms we could describe the state of the molecule in the seventh vibrational level ($v = 7$) by the line segment AC. At point A or C, the velocity of vibration would be zero and the vibrational energy would only be potential. In B, however, the system moves and BD represents the kinetic part of the energy. Any point on the diagram of the horizontal line segments would then represent an instant at which the nuclei have a certain position and momentum. This statement needs to be modified, of course, owing to Heisenberg's uncertainty principle.

Franck–Condon Principle. Potential curves can be drawn for the ground states and the excited states. An electronic transition always starts from a vibrational level in one state and terminates in a vibrational level in another state. Such transitions are shown in Fig. 4.15 for a diatomic molecule. Transitions from the ground state to an excited state follow from absorption

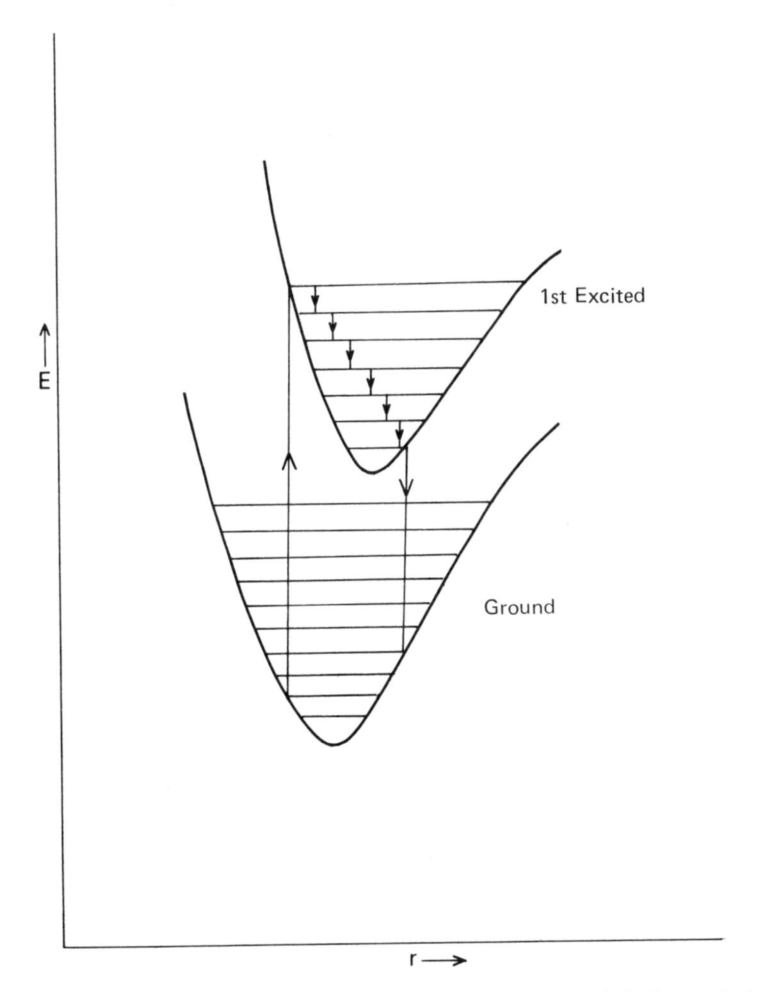

Fig. 4.15 Potential energy curves for the electronic ground state and the first excited state of a diatomic molecule. Transitions must follow the Franck–Condon principle. A transition from a lower vibrational level of the ground state to a higher vibrational level in the first excited state is followed by a rapid vibrational relaxation to lower levels after which emission can take place.

of the appropriate quanta. Since the time in which a transition occurs is much shorter than the time of a nuclear vibration, during a transition neither the momentum of the nuclei nor their relative positions will change. This is a verbal statement of the *Franck–Condon principle*. Graphically, this principle can be expressed by the statement that, in plots like the one given in Fig. 4.15, only vertical arrows represent possible transitions. Transitions from an excited state to the ground state can occur through the *emission* of photons which satisfy the following relation:

$$\Delta E = (hc)/\lambda$$

The Stokes Shift. The emission spectra, however, do not coincide with the absorption spectra; the peak of the absorption band has a higher energy than the peak of the emission band and therefore, occurs at a lower wavelength. This shift, known as the *Stokes shift*, is the result of the possibility that the excited molecule exchange (thermal) energy with its surroundings by a succession of "downward" transitions between the vibrational substates (Fig. 4.15) before it returns to the ground state. At room temperature the most probable absorption transition (having an energy span corresponding with the peak of the absorption band) originates from the lowest substates in the ground state. Owing to the Franck–Condon principle and the fact that the equilibrium position of the excited state is at a slightly larger internuclear distance, the transition terminates somewhere in the middle of the range of substates in the excited state. This event is then followed by the relaxation of the substates through subsequent intervibrational transitions (giving off heat). The most probable *emission* is, therefore, lower than that of the transitions involving absorption. This can be seen from Fig. 4.16, which shows the relative position of an (hypothetical) absorption spectrum and an emission spectrum. The amount of the Stokes shift, since it is related to the time available for the energy relaxation in the excited state, can yield information about the average time spent by a molecule in that state, the *lifetime* of the excited state.

Internal Conversion and Intersystem Crossing. Potential curves from excited states higher than the first one usually overlap with the potential curve of the first excited state. A number of vibrational states thus are "shared" by two subsequent electronic states and thermal relaxation can (and usually does) occur from higher excited states all the way down to the lower vibrational levels of the first excited state (see Fig. 4.17). This process is known as *internal conversion*. A result of this is that emission always originates from the first excited state, even when the absorption causes a primary excitation into higher states. Even though the absorption spectrum can

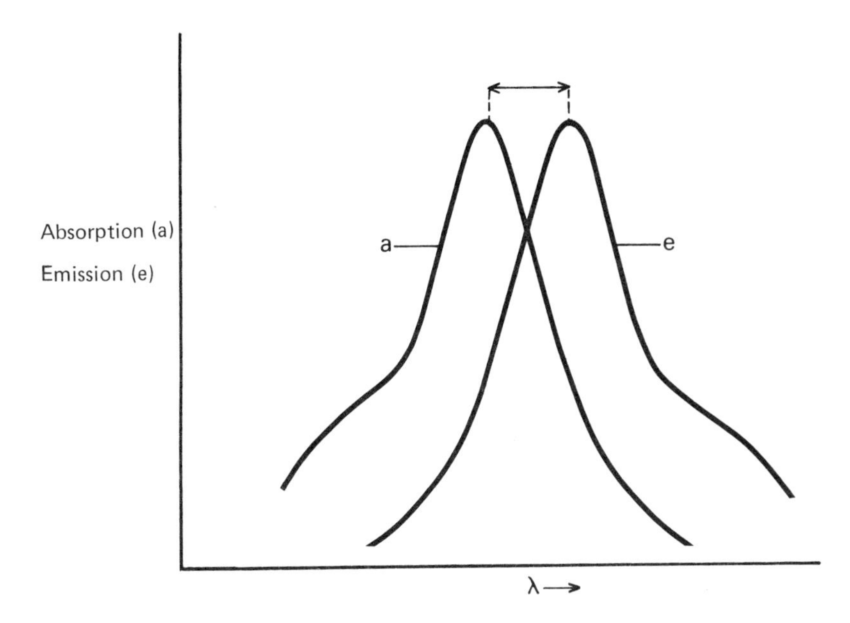

Fig. 4.16 A hypothetical absorption spectrum (a) and a hypothetical emission spectrum (e) showing the Stokes shift.

have more than one band the emission is nearly always correlative with the long wavelength band and represents transitions from the first excited states to the ground state (see Fig. 4.17).

Ordinarily the potential curves for the ground state and the first excited state do not overlap. During a molecular collision, however, the potential curves may become distorted in such a way that the ground state curve temporarily overlaps the first excited state curve. Then internal conversion from the first excited state to the ground state is possible and deexcitation occurs without the emission of photons. This is one of the factors which make the *fluorescence yield* (the ratio between the number of emitted photons and absorbed photons per unit time) smaller than 1.

Other pathways for radiationless deexcitation are transitions to metastable (e.g., triplet) states (a process often called *intersystem crossing*); the radiationless transfer of the excitation energy to neighboring molecules and/or the use of the excitation energy for a chemical reaction. Many biological reactions involve one or more of these processes. The alternative pathways for excitation and deexcitation are depicted in Fig. 4.18.

Absorption Coefficient and Intrinsic Lifetime. The intrinsic probabilities for absorption and emission (Einstein transition probabilities) are

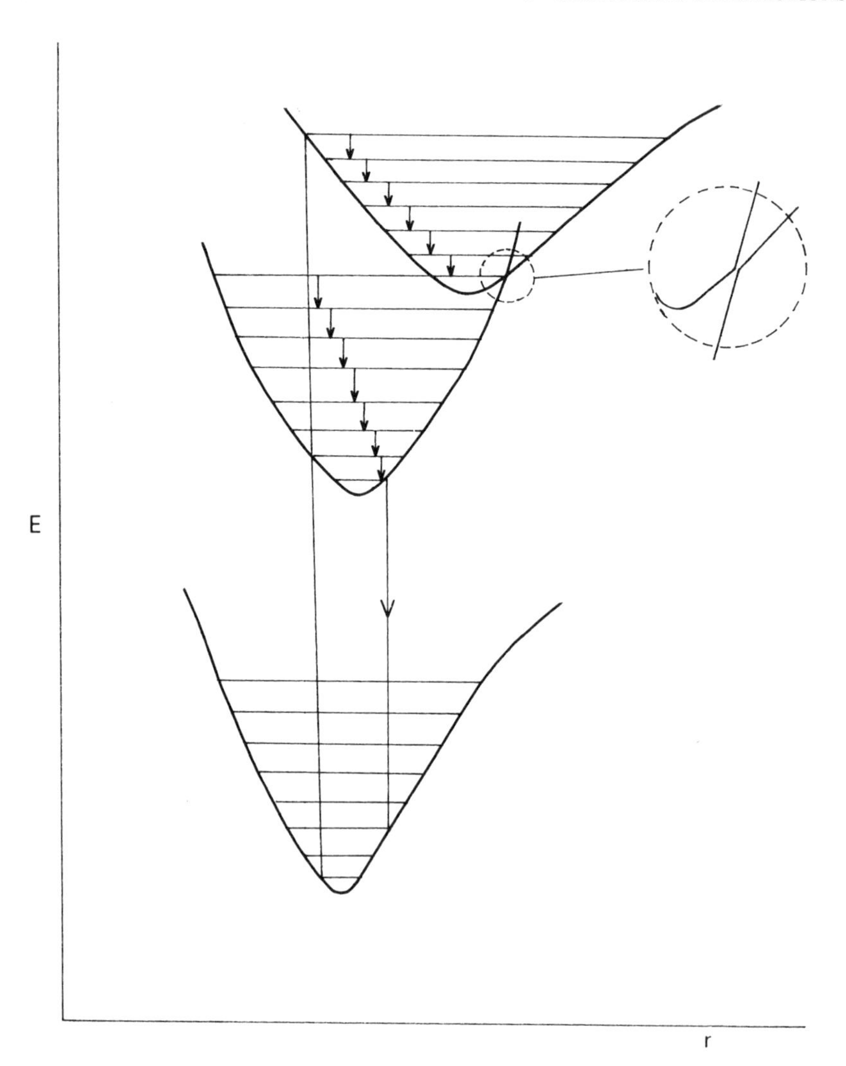

Fig. 4.17 Potential energy curves for the electronic ground state and two excited states of a diatomic molecule. Excitation of an excited state higher than the first excited state is followed by a rapid vibrational relaxation which extends through commonly shared vibrational levels. Emission thus starts always from the first excited state. The insert shows that the potential curves actually do not overlap but seem to avoid each other.

proportional. The lifetime of an excited state, therefore, varies inversely as the probability of absorption. Based on this, one can derive a relation between the intrinsic lifetime τ_0 of an excited singlet state and the integrated absorption coefficient. Since the shapes of most absorption bands are such that the area $\int \alpha \, d\bar{\nu}$ (with $\bar{\nu}$ being the wave number in cm^{-1}) is equal to

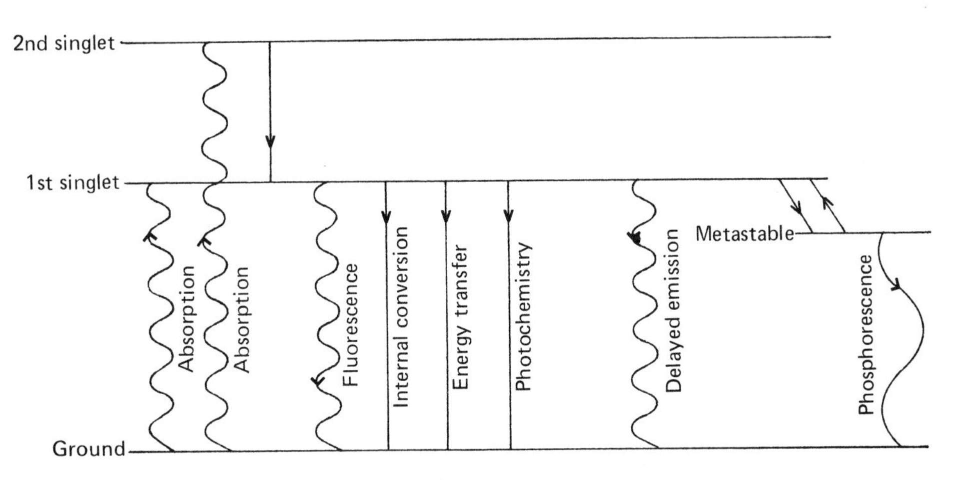

Fig. 4.18 Diagram showing the different modes of deexcitation.

the halfwidth of the band ($\Delta\bar{\nu}$) times the value of α at the peak of absorption ($\alpha_{\max}$), the approximate form of

$$1/\tau_0 = 3 \times 10^{-9} \bar{\nu}^2 \, \Delta\bar{\nu} \, \alpha_{\max} \tag{4.6}$$

can be used.

The absorption coefficient in this form is the molar extinction coefficient as defined by the optical density,

$$\mathrm{OD} = \log I_0/I = \alpha C d \tag{4.7}$$

in which C is the concentration in moles per liter, d the optical path length in centimeters, I_0 the intensity of the incident light, and I the intensity of the transmitted light.

Fluorescence Yield. The actual lifetime τ of the excited state is less than the intrinsic lifetime τ_0 by a factor ϕ_f (the *fluorescence yield*):

$$\tau = \phi_f \tau_0 \tag{4.8}$$

The actual lifetime, and hence the fluorescence yield, is determined by the rate k of deexcitation; if all modes of deexcitation are independent first-order processes, the rate is

$$k = 1/\tau = k_0 + \sum_i k_i \tag{4.9}$$

in which the k_i are the rate constants for all deexcitation processes other than fluorescence. Since $\tau_0 = 1/k_0$ and $\phi_f = k_0/k$, relation (4.8) follows immediately from (4.9).

4.6 Transfer and Trapping of Excitation Energy

In a number of biological reactions excited states are involved. An obvious example is photosynthesis, the process by which plants (in particular their leaves) and some microorganisms convert sunlight into chemical potential energy (for a more extensive description of the process, see Chapter 5). In photosynthesis light is absorbed by pigment molecules (chlorophylls, carotenoids, and phycobillins) but is not utilized for energy conversion immediately upon absorption. The apparatus is organized in so-called *photosynthetic units*, ensembles of pigment molecules within which excitation energy is transferred until it is trapped somewhere within the unit at a site specialized for photochemical conversion (a reaction center). This causes a more efficient utilization of the absorbed energy; the unit functions as a kind of funnel and the traps can turn over much faster than they would if each absorbing molecule was its own trap. A typical size of such a photosynthetic unit (in plants) is some 300 chlorophyll *a* molecule per trap.

Sensitized Fluorescence. Transfer of excitation energy is possible because of the forces involved in the redistribution of electric charge in an electronic transition; the electric dipole field of one molecule of an ensemble induces excitation in another. We can see this as a coupled event in which the deexcitation of a sensitizer molecule S is accompanied by the excitation of a different acceptor molecule A. If the latter is fluorescent, its fluorescence can be induced by exciting the sensitizer molecule:

$$\begin{aligned} \mathrm{S} + h\nu_{\mathrm{a}} &\to \mathrm{S}^* \\ \mathrm{S}^* + \mathrm{A} &\to \mathrm{S} + \mathrm{A}^* \\ \mathrm{A}^* &\to \mathrm{A} + h\nu_{\mathrm{f}} \end{aligned} \tag{4.10}$$

in which ν_a and ν_f are the frequencies of the absorbed and the emitted light, respectively.

This process of *sensitized fluorescence* can be (and has been) easily demonstrated with pairs of dyes which do not have greatly overlapping absorption spectra. The mechanism can be likened to the transfer of energy by resonance in coupled pendulums and tuning forks. The process is, therefore, often called *inductive resonance transfer*. L. Duysens (1952) has demonstrated it in *in vivo* systems; when a photosynthetic organism is illuminated with light that is absorbed by carotenoids but not by chlorophyll, fluorescence of chlorophyll *a* occurs, thus demonstrating the transfer of excitation energy from carotenoid to chlorophyll *a*.

Excitation Energy Transfer. Resonance transfer is a molecular interaction which takes place on the quantum mechanical level. Therefore, it is formally incorrect to consider the excitation as localized in a particular molecule at any one time. One has to consider the excitation as a property of the whole ensemble. Wave functions describing the system are solutions of the Schrödinger equation

$$\mathbf{H}\psi_j = E_j\psi_j$$

in which the Hamiltonian has a term describing the interaction between the molecules. Solutions are feasible only when certain simplifications are made. One can, for example, ignore contributions due to intermolecular electron orbitals (as is done in the molecular exciton model) or retain only the electric dipole portion of the radiative interaction.

One such simplification concerns vibrational interactions. Application of this leads to three distinguishable cases, depending on the magnitude of the interaction energy. Although formally incorrect, one can use "localized language" to describe these cases. In such language, one can speak of a transfer time τ_t which is the average time of residence of an excitation in any one molecule. If this τ_t is small as compared to the period of nuclear oscillations (about 3×10^{-14} sec) and, hence, small compared to the period of the intermolecular (lattice) vibrations (about 3×10^{-12} sec) as well, vibrational states do not come into play at all. The transfer is between identical electronic states of the interacting molecules and the rate of transfer is proportional to the interaction energy (hence to the inverse of the third power of the distance between the molecules in the case of dipole interaction). This is often called *fast transfer*.

One speaks of *intermediate transfer* when the transfer time τ_t is between the period of nuclear vibration and that of lattice vibrations. The resonance then is among vibrational levels in the interacting molecules. Also in this case the rate is proportional to the interaction energy (third power of the inverse distance between the dipoles). In both cases, fast and intermediate transfer, a "localized treatment" leads to incorrect results.

Molecular Exciton Model. M. Kasha and coworkers have developed the *molecular exciton model* to describe the fast transfer (ignoring vibrational interactions) in polymers. In this model the wave functions for the excited state of the polymer are linear combinations of all possible localized conditions in the polymer, thus leading to a splitting of the monomeric level with a multiplicity equal to the number of coupled monomers. Consider, for example, a dimer made up of monomers A_1 and A_2. Delocalized treatment involves a ground state $(A_1 \cdot A_2)$ and an excited state $(A_1 \cdot A_2)^*$ that is split into two levels. The wave functions for the two localized conditions $(A_1{}^* + A_2)$ and $(A_1 + A_2{}^*)$ are $\psi_1{}^* \cdot \psi_2$ and $\psi_1 \cdot \psi_2{}^*$. Linear combinations of these yield

$$(1/\sqrt{2})(\psi_1{}^* \cdot \psi_2 \pm \psi_1 \cdot \psi_2{}^*) \tag{4.11}$$

This is illustrated in Fig. 4.19 for a dimer whose transition dipole moments are perpendicular to the axis through the center of each dipole. In this case the allowed state is the upper antibonding one [positive sign in (4.11)] where the transition dipoles are parallel to each other (in phase). The lower state is forbidden. That this must be so, can be seen from the following: since the dimensions of the dipole are small compared to the wavelength of the exciting light, the molecules must be in the same region of the radiation field. The phase of the electromagnetic wave is, therefore, the same throughout that region, and the generated dipoles must be in phase with each other. If the transition dipoles are aligned with the dimer axis the lower (bonding) state would be the one in which the transition dipoles are in phase and would, hence, be the allowed one. In an oblique arrangement both dipoles have perpendicular and parallel components and both levels are allowed. The energy gap between the levels is twice the dipole interaction energy.

The above treatment can be extended to polymers with N coupled molecules. In such a case we would have N levels which are linear combinations of wave functions of the type $\psi_1 \psi_2 \psi_3 \cdots \psi_s{}^* \cdots \psi_N$. Depending on the symmetry of the array, all kinds of spectral phenomena can be predicted: blue shifts, red shifts, narrowed bands, broadened bands, band splitting, etc.

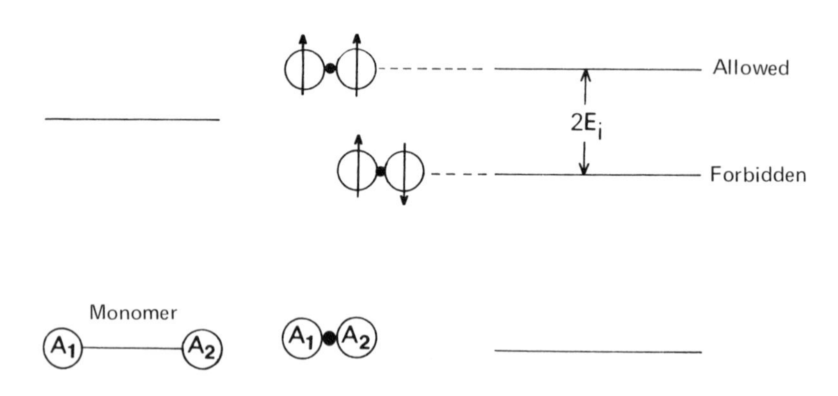

Fig. 4.19 Energy levels of an excited dimer. The arrows point to the direction of the transition dipole. In the illustrated case the transition dipole is perpendicular to the dimer axis. E_i is the dipole interaction energy.

Inductive Resonance Transfer. In the case of *slow transfer*, when τ_t is large compared to the lattice vibrations, a "localized treatment" gives an acceptable approximation. In this case the excitation resides for sufficiently long periods in a particular molecule to allow thermal equilibrium among the vibrational levels. The sensitizer molecule will settle into the lower vibrational states before transferring its energy, and the amount of energy transferred correspond to a transition from these lower vibrational states to the ground state. The same amount of energy is then gained by the acceptor molecule (see Fig. 4.20).

Since deexcitation of the sensitizer S follows the same pathway as fluorescence, the rate of transfer must be proportional to the overlap of the fluorescence spectrum of S and the absorption spectrum of the acceptor A. In Fig. 4.21a this is shown for the case in which S and A are identical molecules; Fig. 4.21b shows the case in which S and A are dissimilar. One can see that heterogeneous slow transfer can be much more efficient than homogeneous slow transfer.

T. Förster (1951) has derived a form for the ratio between the rate of slow transfer ($k_t = 1/\tau_t$) and the rate of fluorescence ($k_0 = 1/\tau_0$). This ratio can be represented by

$$k_t/k_0 = (R_0/R)^6 \tag{4.12}$$

in which R_0 is a function of the overlap integral of fluorescence and absorption spectra and the mutual orientation of the transition dipoles, and R is the distance between the interacting molecules. R_0 has the dimension of length and can be defined as the distance between two molecules at which

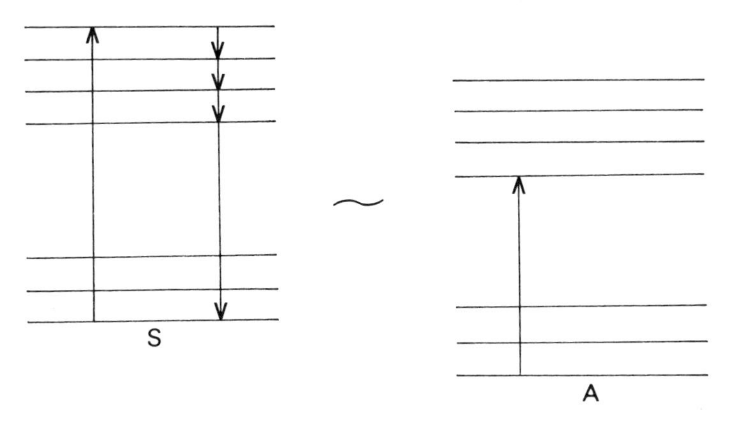

Fig. 4.20 Energy diagram showing the "slow" transfer of excitation energy from a sensitizer molecule S to an acceptor molecule A. Transfer by inductive resonance occurs when the energy lost by S in its deexcitation is gained by A in its excitation.

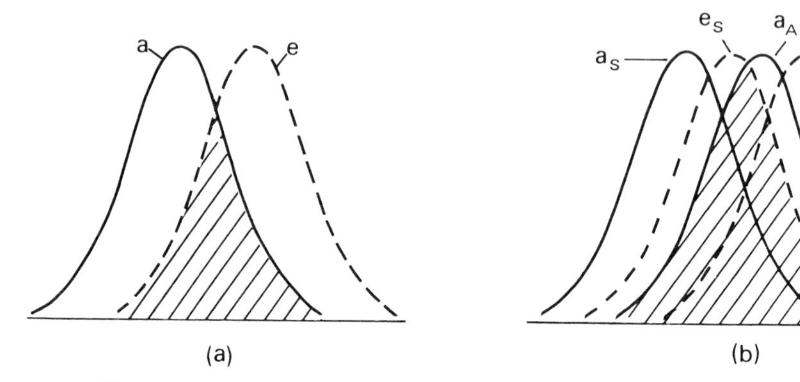

Fig. 4.21 The overlap of the emission spectrum e of S and the absorption spectrum a of A which determines the efficiency of energy transfer by inductive resonance: (a) when S and A are similar; (b) when S and A are different.

the rate of slow transfer equals the rate of fluorescence. This value is a measure of the transfer efficiency. L. Duysens (1952) has calculated an R_0 of 69 Å for *in vivo* homogeneous transfer in chlorophyll *a* and 70 Å for the transfer from chlorophyll *b* to chlorophyll *a*. Since the average distance between the molecules in a photosynthetic unit containing chlorophyll *a* and chlorophyll *b* molecules is 17 Å, this amounts to a very high transfer efficiency. The main physical reason for this is the unusually good overlap of the absorption and the fluorescence spectra of chlorophyll.

Trapping. In a photosynthetic unit the excitation energy is trapped at a specialized site. Trapping requires that the excitation become fixed in one molecule through entry into a localized excited state. Two situations can be visualized in which trapping can occur (see Fig. 4.22). One case is when the trapping molecule T has an excited singlet state which is lower than the excited singlet state of the transferring molecules M. If we consider only slow transfer we can see immediately that the T must act as an energy sink because the overlap of its absorption spectrum and the fluorescence spectrum of the M is greater than the absorption–fluorescence overlap of the M. The trapping efficiency is at least as high as the probability of excitation reaching T (Fig. 4.22a). This situation may prevail in the system I photosynthetic unit of higher plant photosynthesis (see Chapter 5), where chlorophyll *a* molecules transfer excitation energy to a specialized molecule called P700. In this system the transferring chlorophyll *a* molecules have an absorption band that peaks at 680 nm, while the peak of the absorption band of the trap molecules is at 700 nm. The energy difference is, thus, about 420 cm^{-1} (= 0.05 eV = 1.15 kcal/mole).

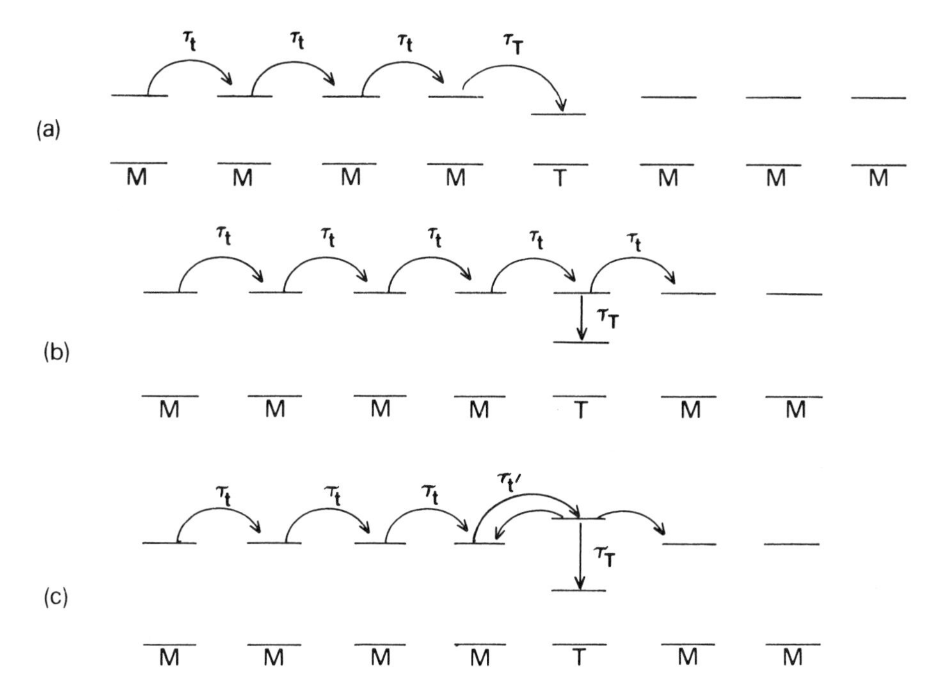

Fig. 4.22 Mechanisms for the trapping of excitation energy from an ensemble of molecules M by a trapping molecule T; (a) when the first excited state of T is lower than that of the transferring molecules M; (b) when the first excited states of M and T are equal but T has, in addition, an excited state *below* the first excited state which is absent in M; (c) when the first excited state of T is *higher* than that of M but T has in addition an excited state below the first excited state of M which is absent in M. In cases (b) and (c) efficient trapping can occur when the trapping time τ_T is far less than the transfer time τ_t.

In the second situation the singlet level of T is not different from that of the M, but another excited state (for instance nπ^* state, a triplet state, or a charge transfer state) peculiar to T can be entered through the singlet state (Fig. 4.22b). Efficient trapping can occur if the trapping time τ_T is much smaller than the transfer time τ_t. This situation may occur in most photosynthetic bacteria in which the peak of the absorption bands of the trapping chlorophyll is the same as that of the transferring chlorophyll. One species of photosynthetic bacteria is known (*Rhodopseudomonas viridis*) for which the wavelength of the peak of the trapping chlorophyll is even lower than that of the transferring chlorophyll. In this case the transfer to the traps is less efficient than the back transfer from the traps to the transferring chlorophyll but once the trapping molecule is excited trapping will be the result when $\tau_T \ll \tau_t$ (Fig. 4.22c).

The total time needed for the migration of excitation energy and trapping $\langle t \rangle$ is equal to τ_t multiplied by the number of M → M transfers required for the excitation to reach T (if $\tau_T \ll \tau_t$). If the trapped state is nonfluorescent the fluorescence yield of the ensemble cannot be greater than $\langle t \rangle/\tau_0$. Thus, the quenching of the fluorescence in the ensemble is often taken as evidence for the transfer to trapping centers (see Chapter 5).

The existence of traps can be a result of a variety of effects. An energy level in a trapping molecule could be depressed, thus forming an energy sink by environmental factors. A lowering of a $\pi\pi^*$ level can occur through distortion of the π and the π^* orbitals by local electric fields or electron orbital overlap. An $n\pi^*$ level could be lowered by an electric attraction that encourages the charge displacement involved in the transition. The polarity of the environment (aqueous or lipid phase) can have an influence on the relative position of the lowest $\pi\pi^*$ and $n\pi^*$ levels. It is known that in simple molecules the relative position of these levels can be inverted by transferring the molecule from a nonpolar to a polar solvent.

4.7 Infrared Spectroscopy

Vibrational Energy Levels. Transitions between vibrational levels in an electronic state are possible by absorption or emission of the appropriate quanta of energy. For the strict harmonic oscillator such transitions are restricted by the selection rule $\Delta v = \pm 1$, which means that only transitions between adjacent levels are possible. Therefore, if we use (4.5), the transition energy ΔE is given by

$$\Delta E = h\nu_0[(v + 1) + \tfrac{1}{2}] - h\nu_0(v + \tfrac{1}{2}) = h\nu_0 \tag{4.13}$$

and the frequency of the absorbed or emitted quantum is equal to the characteristic frequency of the oscillator.

A diatomic molecule is not a simple harmonic oscillator. A more realistic potential energy curve is given in Fig. 4.23. The curve approaches an asymptotic level which represents the energy at which the system breaks apart, the so-called dissociation energy. The vibrational energy is given by an equation slightly different from (4.5), but is also determined by the vibrational quantum number v. The levels are not, as in the harmonic oscillator, equidistant from each other, although in the lower levels the deviations from equidistancy are very small. The selection rule $\Delta v = \pm 1$ is not as strict as it is for the harmonic oscillator; although less probable transitions which involve $\Delta v = \pm 2, \pm 3, \ldots$ are possible and will, according to (4.13), result in the "overtones" approximately $2\nu_0, 3\nu_0, \ldots$.

A big polyatomic molecule can be seen as being made up of many diatomic oscillators. These molecules may, therefore, execute very complex

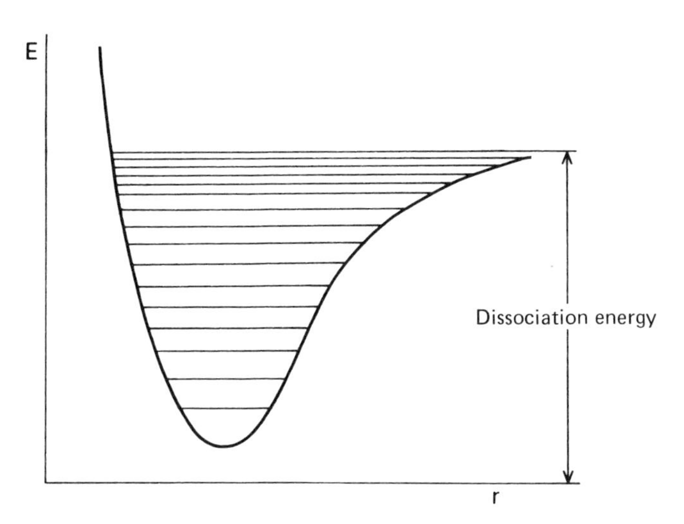

Fig. 4.23 Potential energy curve of a diatomic molecule.

vibrations which cannot be resolved into recognizable single oscillations. A protein molecule with a molecular weight of 10,000, for example, has about 1000 atoms, and the number of coordinates needed to specify the vibrational motion is about 3000. Even the simplest amino acid, glycine, exhibits 24 "diatomic" oscillations. Infrared spectroscopy, by which we can look at vibrational characteristics, therefore, is not very suitable when we want to identify a molecule as a whole. It is very useful, however, as a means of characterizing particular bonds. We can distinguish, in an infrared spectrum, frequencies characteristic of certain bonds if the groups in question are sufficiently isolated from the rest of the molecule and if the frequencies are not too near those of other bonds. Such criteria are satisfied by the end groups of molecules in which the forces holding two atoms together are roughly independent of other atom groups bonded to these groups. If hydrogen is the terminal atom, it can be seen as vibrating against a massive wall.

Stretching and Bending Vibrations. The vibrations in such end groups can be of the *stretching type* (VI) or the bending type (VII). Bending fre-

≡C—H (VI)   ≡C—H (VII)

quencies are of the order of 10^{13}/sec; stretching frequencies are usually an order of magnitude higher. Characteristic frequencies (in the wave number unit cm^{-1}) of some end groups are given in Table 4.4.

TABLE 4.4 Some Characteristic Bond Frequencies in cm^{-1} of Gases and Liquids

Group	Stretching vibration ($cm^{-1} \pm 100\ cm^{-1}$)	Group	Bending vibration ($cm^{-1} \pm 100\ cm^{-1}$)
≡C—H	3300	≡C—H	700
=C—H	3020	=C<(H)(H)	1100
>C—H	2960		
—O—H	3680 (3400)[a]	C—C≡C	300
		>N—H	1600
—S—H	2570		
>N—H	3500 (3300)[a]		
>C=O	1700		
>C=N—	1650		
—C≡C—	2050		
>C=C<	1650		
>C—C<	900		
P=O	1250–1300		

[a] Liquid hydrogen bonded.

Hydrogen Bonds. Infrared absorption spectra of a protein (keratin) and a polypeptide (made of the amino acids phenylalanine and leucine) are shown in Fig. 4.24. We can recognize bands in these spectra as belonging to stretching and bending vibrations of several groups. Upon closer examination we may discover, however, that the stretching frequencies of the N–H groups at 3300 and 3200 cm^{-1} are somewhat lower than the value of 3500 cm^{-1} for this group in the gas phase (Table 4.4). This *red shift* (a shift toward a longer wavelength) of the band is due to the hydrogen bond in which the group is engaged. The effect of hydrogen bonding will be a reduction of the "stiffness of the oscillator" as is indicated in the example of the springs in Fig. 4.25a. For the diatomic group the result will then be a widened potential curve with the energy levels depressed as shown in Fig. 4.25b. The infrared spectral data represent strong experimental evidence for the presence of hydrogen bonds in proteins and polypeptides.

4.8 Quantum Mechanics and Biology

The application of quantum mechanics in biology is obviously useful if one, for instance, looks at the biological implications of molecular spec-

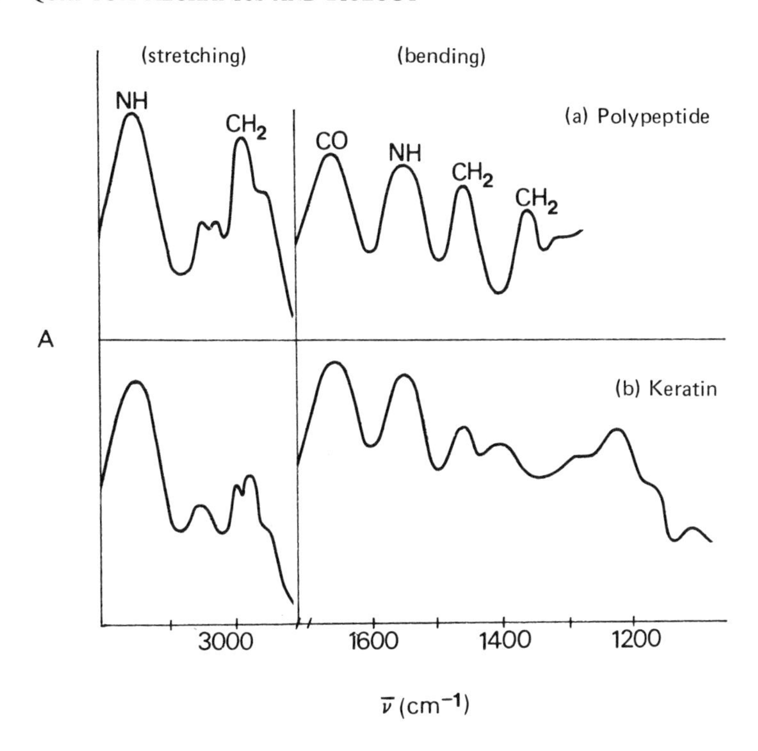

Fig. 4.24 Infrared absorption spectra of (a) a synthetic polypeptide, *polyphenylanaline-leucine* and (b) a protein *keratin*. The peaks due to the NH_2, NH, CH_2, and CO groups are easily detectable. From S. E. Darmon and G. B. B. M. Sutherland, *J. Amer. Chem. Soc.* **69,** 2074 (1947). Copyright by the American Chemical Society. Reprinted by permission.

troscopy. Spectra of biological molecules are measured and analyzed not only for identification of these molecules and their state, but also because the quantum mechanical interpretation of the spectra can lead to interpretations of biological phenomena. The development of approximation methods for the calculation of the electronic structure of biological molecules could lead to a further application, thus demonstrating much wider implications of quantum theory for biology.

Quantum Mechanical Calculations of Electronic Structures. A principal advantage of quantum theory over the usual methods of biological investigation is its universality. Experimental methods of physics and chemistry applied to biology are, in general, intended for studying one, or at best a small number, of the specific properties of a biological system. Each of these, therefore, can give only a partial understanding of the system. In quantum theory, however, a single calculation, the solution of the wave equation, can yield complete information about all the structural properties of the atomic,

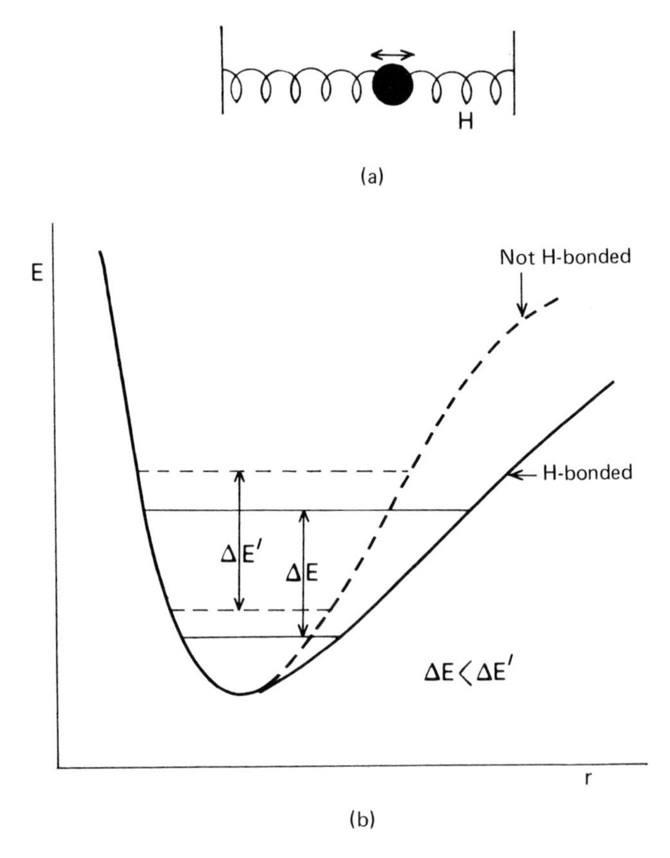

Fig. 4.25 (a) A model showing the modification of the vibration due to hydrogen bonding. (b) The modification of the potential energy curve due to hydrogen bonding. The curve is stretched and the differences between the vibrational energy levels have become smaller.

molecular, and more extended systems and, thus, offer a much more complete understanding of the behavior of such systems. The calculation, however, would involve interactions of a tremendous number of electrons with each other and with the nuclei; the differential equation which must be solved is an extremely complex one and in most cases the mathematics are inadequate for such a task. The only hope for some degree of success is the development of methods of approximation which, although reducing the equations and yielding only partial information, should still cover a variety of aspects wide enough to be practical. The exploration of such approximation methods recently got underway and some success appears to have come from the LCAO (linear combination of atomic orbitals) approximation of molecular orbital theory. We have discussed this method briefly in Section 4.3.

When such approximate calculations are carried out it is possible to determine (at least on a relative scale) a number of *electronic indices* which allow conclusions about quite a number of biological characteristics, Table 4.5 summarizes this. The LCAO–MO calculation gives two sets of result. One is derived from the energy of the molecular orbital and the other from the coefficients of the atomic orbitals which, in the linear combination, make up the molecular orbital. From the first set of results, the *resonance energy* (the additional stabilizing energy in conjugated double-bond systems), for example, yields information about oxidation–reduction potentials in reversible systems, thermodynamic stability in chemical transformation, and

TABLE 4.5 Principal Applications of the Electronic Indices[a]

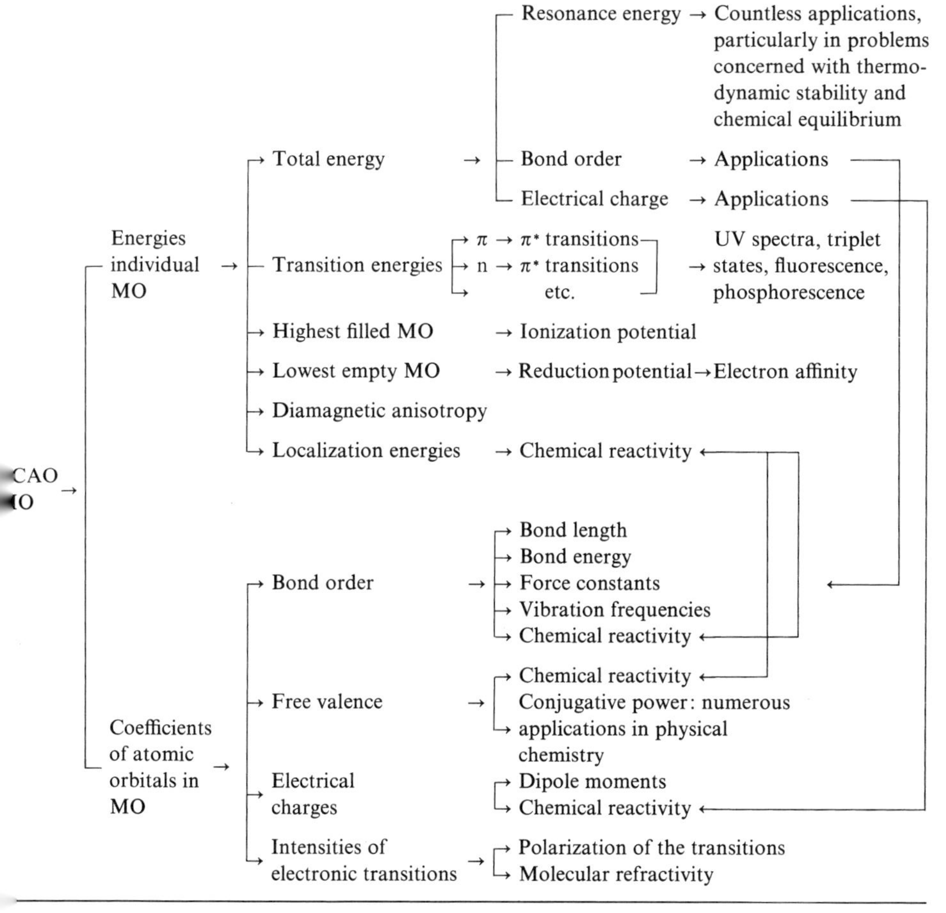

[a] From Pullman (1965).

many other related problems. Such data are important for the study of bioenergetic processes in cells (see Chapter 5). Another example is the *energy of the highest filled and the lowest empty molecular orbital.* These energies tell us something about the electron donor and electron acceptor properties of the system in question. *Transition energies* give spectroscopic information, and chemical reactivity can be derived from a number of other electronic indices. The atomic orbital coefficients, in addition to information related to the energy indices, can yield information about things like electric charges and *dipole moments*, which are of eminent importance for secondary and higher-order structures of the molecules.

Calculations based on the LCAO–MO method, in which the approximations are made on a more or less empirical basis, have only recently been initiated. A number of simple organic and biological molecules were used; some conclusions based on these calculations have confirmed experimental evidence about a number of properties while other conclusions were experimentally confirmed later. The finding by B. Pullman and A. Pullman (1963) that the most important stabilizing factor in the double helix structure of DNA is not the hydrogen bonding between the complementary bases in the two strands, but the interaction resulting from the vertical stacking of the bases in an example of such a conclusion. Calculations have thus far included a number of coenzyme molecules such as quinones, pyridine proteins, and flavoproteins. The results of all these calculations have substantiated the suspicion that delocalized mobile electrons in the conjugated (π) systems are of essential importance for the phenomenon of life. Indeed, not only the basic molecules, nucleic acids and proteins, but also the essential coenzymes such as the pyridine nucleotides and flavonucleotides, quinones, and the prosthetic groups of the cytochromes, are all conjugated compounds. It may very well be that an essential feature that make these molecules the bearers of life and life processes is the electronic delocalization in their conjugated systems.

This application of quantum mechanics is one which allows, at least in principle, a description of the electronic structure of the atoms, molecules, and molecular aggregates making up a cell. According to some workers a complete and detailed description along these lines, if at all feasible, could lead to a prediction of a cell's behavior and the emergence of "biological" laws from pure quantum mechanics and hence, physical principles. Others, however, have expressed the view that there may be a more fundamental distinction between biology and physics and that a broader application of quantum mechanics (a formulation of a biological complementarity principle, for example) would lead to a better understanding of the phenomenon of life (see Chapter 7).

Chapter

5 Bioenergetics

5.1 The Biological Energy Flow

Biological Cycle. Work is continuously being performed by a living organism. Living cells do work either within themselves or on their environment. This work may be of different kinds; mechanical work of muscle contraction, electrical work when charges are transported, osmotic work when material is transported across semipermeable barriers, or chemical work when new material is synthesized. In order to perform all this work the cell needs mechanisms by which energy can be transformed (converted) in the proper way. At the constant temperature prevailing in most cells a net output of work can be obtained only when energy is *dissipated*, that is, when it is converted to a less useful form. The living cell has complex and very efficient devices at its disposal to accomplish this.

The ultimate source of energy for life is the sun. Green plants, algae, and a few types of bacteria are able to capture energy from sunlight and convert it into a form suitable for sustaining their own life and that of the rest of the living world. The process by which this conversion occurs is known as *photosynthesis*. The product of photosynthesis—a large amount of chemical potential energy (food)—is then used in a "reverse" process, yielding the form of energy suitable for the performance of work (Fig. 5.1).

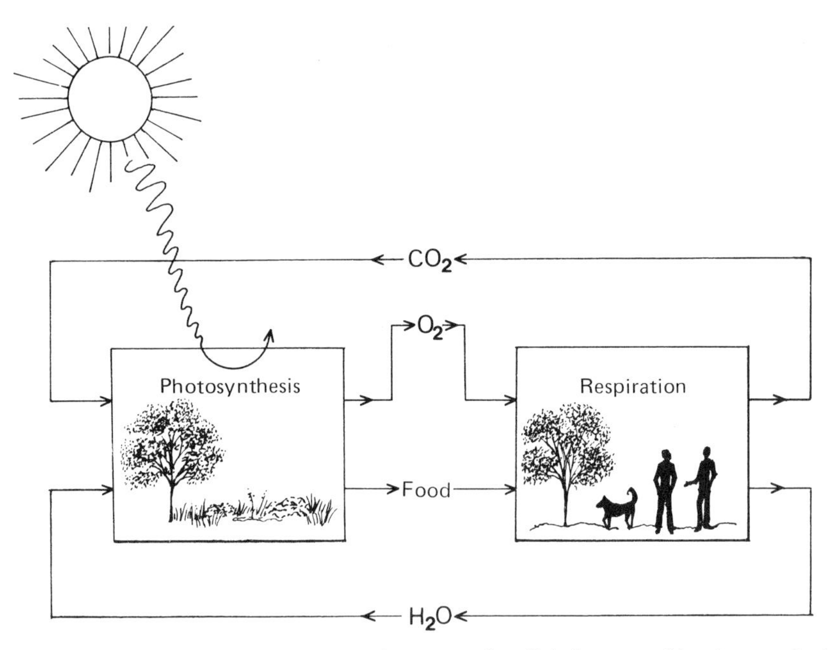

Fig. 5.1 The biological energy cycle. The energy of sunlight is captured by photosynthesis in the form of the chemical potential energy of food and oxygen. The respiratory processes recombine the two to form the lower energy products carbon dioxide and water. The energy liberated in this way is used to preserve and reproduce life. Carbon dioxide and water are again used to carry out the photosynthetic reactions, capturing sunlight, and the cycle is repeated.

Photosynthesis. In photosynthesis, the excitation of a specialized kind of molecules in the so-called *reaction centers* (see Section 5.3) leads to a primary oxidation–reduction reaction which sets in motion a sequence of reactions of the same type (see Section 5.3). The final result of this is the oxidation of a hydrogen donor H_2A and the concomitant production of a relatively strong reductant H_2X:

$$H_2A + X + \text{light} \rightarrow H_2X + A \tag{5.1}$$

The reductant is then used to reduce carbon dioxide to a sugar,

$$12H_2X + 6CO_2 \rightarrow (HCOH)_6 + 6H_2O + 12X \tag{5.2}$$

in a reaction which does not need light. The overall reaction of photosynthesis can, thus, be written as

$$12H_2A + 6CO_2 + \text{light} \rightarrow (HCOH)_6 + 6H_2O + 12A \tag{5.3}$$

In photosynthetic bacteria, A can be a variety of substances ranging from sulfur to organic groups. In higher plants and algae, however, A is always oxygen so that the donor is always water. The overall reaction for these organisms thus becomes

$$6H_2O + 6CO_2 + \text{light} \longrightarrow (HCOH)_6 + 6O_2 \tag{5.4}$$

(the form in which it is generally known). The energy term in the left side of this equation (light) is the electromagnetic energy of absorbed light. Energy is represented in the right-hand side of the equation by the *chemical potential energy* in the substances $(HCOH)_6$ and O_2. One can visualize this by realizing that the compounds on the left-hand side of the equation, water and carbon dioxide, are more stable than the sugar and oxygen on the right-hand side; the total binding energy on the left-hand side is lower (more negative) than that of the right-hand side. Part of this difference is necessarily lost because the reaction is not reversible. The rest is the energy stored as chemical potential energy in the sugar and oxygen.

Respiration. Recombination of the sugar and the oxygen can lead to the liberation of this stored energy. If this recombination occurs by combustion in a test tube all the energy is liberated in the form of heat which as such cannot be used to do work at the constant temperature prevailing in most organisms. When this release of energy occurs, however, by a balanced sequence of oxidation–reduction reactions such as those occurring in the mitochondria of living cells, the energy is liberated in a stepwise manner and caught in a form (again as chemical potential energy) which makes it possible to perform work at the necessary time and place. This form of energy capture occurs in the compound *adenosine triphosphate* (universally known by its initials ATP) which is produced during the stepwise oxidation of food. The hydrolysis reaction

$$ATP + H_2O \longrightarrow ADP + P_i \tag{5.5}$$

in which ADP stands for adenosine diphosphate and P_i for inorganic phosphate, occurs with a release of energy which, when the reaction proceeds under controlled conditions, can be used for work. The controlled progress of the biological energy cycle can be depicted by the running of a series of watermills driving generators which charge batteries (Fig. 5.2). When photosynthesis is compared to a pump used to bring "water" to an elevated level "using electromagnetic energy," respiration can be represented as the stepwise downfall of the "water" which drives the "watermills" charging the "ATP batteries." The batteries then can be transported to sites where work has to be done; when properly connected, they can be "discharged" by the hydrolysis reaction when work is performed.

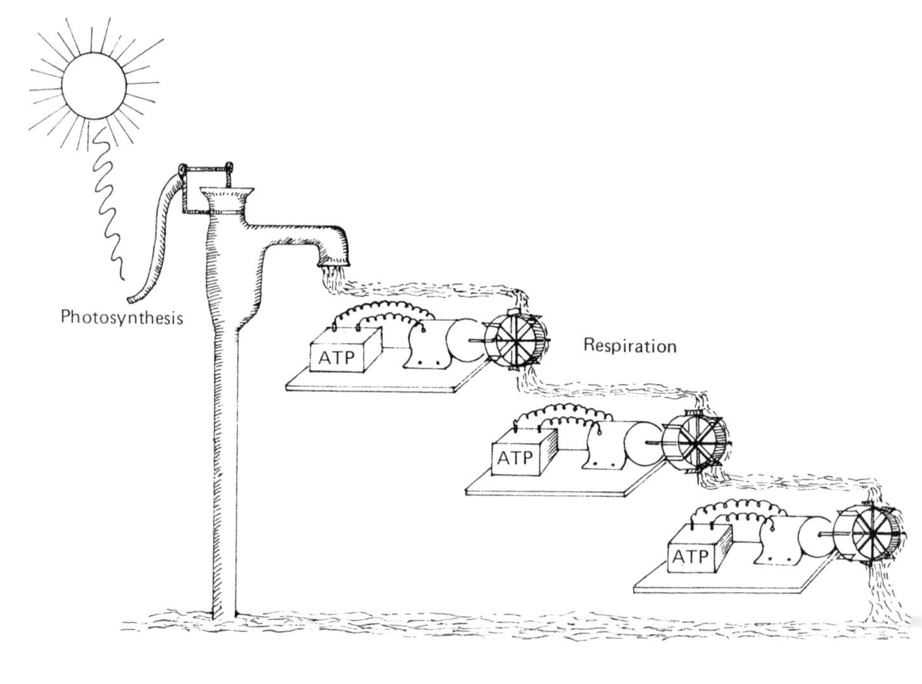

Fig. 5.2 The "ATP-charging water mills" described in the text.

Although the coupling of energy-releasing oxidation reactions and ATP formation, as well as the hydrolysis of ATP during the performance of chemical, electrical, osmotic, or mechanical work, may be thermodynamically simple, its molecular mechanism (or mechanisms) is far from known. This constitutes one of the most central and challenging problems of present research in bioenergetics. We will discuss it in more detail in the following sections of this chapter.

5.2 Adenosine Triphosphate in Coupled Reactions: Pyridine Nucleotides

High-Energy Phosphates. ATP, the universal energy currency in living systems, is a nucleotide. It consists of a base, adenine, of the purine group which is linked by a glycosidic linkage to a molecule of D-ribose. A row of three phosphate groups is attached to the 5′-position of the ribose (see Fig. 5.3). If the terminal phosphate group is detached the nucleotide becomes

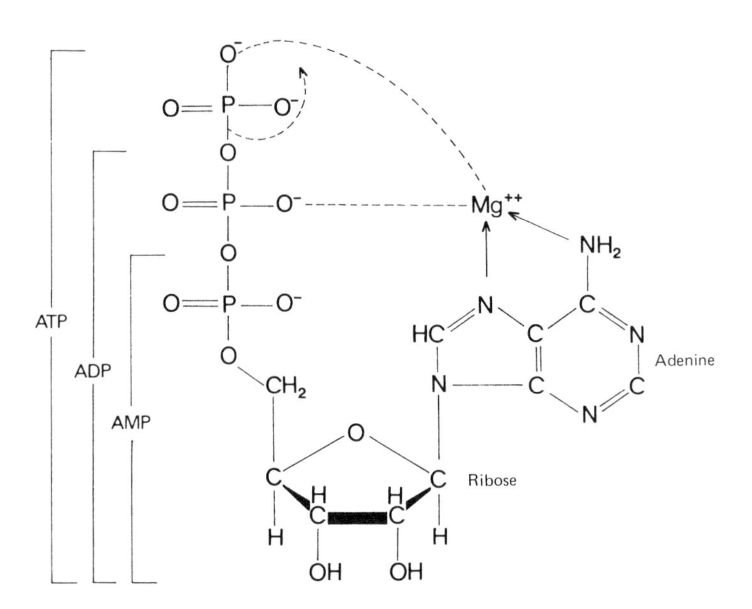

Fig. 5.3 The chemical structures of adenosine triphosphate (ATP), adenosine diphosphate (ADP), and adenosine monophosphate (AMP). ATP can easily form a complex with Mg^{2+} as shown. The molecule then is twisted and folded somewhat.

adenosine diphosphate (ADP). With only one phosphate group we have *adenosine monophosphate* (AMP).

In an intact cell, at a pH of about 7, the ATP molecule is highly charged; each of the three phosphate groups is ionized and the molecule, therefore, has four negative charges. The molecule can easily form complexes with divalent cations such as Mg^{2+} and Ca^{2+}. The result of this is that living cells have little ATP present as a free anion; most of it is bound to Mg. This feature may have something to do with the specific enzymatic hydrolysis of ATP by which the chemical potential energy is converted into work.

Free Energy of Hydrolysis. The energy-carrying function of ATP is not connected to the chemical bonds of the phosphate groups, as may be suggested by the inaccurate term "high-energy bond." It is due to the strongly negative free energy of the hydrolysis reaction

$$ATP^{4-} + H_2O \longrightarrow ADP^{3-} + HPO_4^{2-} + H^+ \tag{5.5a}$$

which, at equilibrium under standard condition, is about -7 kcal/mole. The equilibrium of the reaction is far to the right because the reaction products are stabilized as a result of their negative charges and the formation

TABLE 5.1 Standard Free Energy of Hydrolysis of Some Phosphorylated Compounds[a]

Compound	G° (kcal)
Phosphoenolpyruvate	−14.80
1,3-Diphosphoglycerate	−11.80
Phosphocreatine	−10.30
Acetyl phosphate	−10.10
Phosphoarginine	−7.70
ATP	−7.30
Glucose 1-phosphate	−5.00
Fructose 6-phosphate	−3.80
Glucose 6-phosphate	−3.30
Glycerol 1-phosphate	−2.20

[a] From A. L. Lehninger, "Biochemistry," Worth Publishers, New York, 1970, p. 302.

of new hybrid molecular orbitals. ATP is not the only phosphate compound in biological systems having this feature. In fact, there are many other phosphate compounds which have higher (and many more which have lower) standard free energies of hydrolysis. Table 5.1 summarizes the standard free energy of hydrolysis of a number of compounds in biological systems. We can see from this table that the value for ATP is actually somewhere in the middle of the range of standard free energies of hydrolysis. This makes it very suitable for the function of energy transfer by coupled reactions having *common intermediates.*

Coupled Reactions. In coupled reactions a reaction having a negative free energy change can be used to "drive" another reaction having a positive free energy change. Consider, for example, the reactions

$$\mathrm{A} \rightleftharpoons \mathrm{P} + \mathrm{Q} \tag{5.6}$$

and

$$\mathrm{P} + \mathrm{B} \rightleftharpoons \mathrm{R} \tag{5.7}$$

The free energy change of reaction (5.6) is

$$\Delta G_{\mathrm{a}} = -\mu_{\mathrm{A}} + \mu_{\mathrm{P}} + \mu_{\mathrm{Q}} = -\mu_{\mathrm{A}}^{\circ} + \mu_{\mathrm{P}}^{\circ} + \mu_{\mathrm{Q}}^{\circ} + RT \ln C_{\mathrm{P}} C_{\mathrm{Q}} / C_{\mathrm{A}} \tag{5.8}$$

and that of reaction (5.7) is

$$\Delta G_{\mathrm{b}} = -\mu_{\mathrm{P}} - \mu_{\mathrm{B}} + \mu_{\mathrm{R}} = -\mu_{\mathrm{P}}^{\circ} - \mu_{\mathrm{B}}^{\circ} + \mu_{\mathrm{R}}^{\circ} + RT \ln C_{\mathrm{R}} / C_{\mathrm{P}} C_{\mathrm{B}} \tag{5.9}$$

in which the μ are the appropriate chemical potentials and the C designate concentrations (or rather, activities). We can assume conditions under which ΔG_a is negative and ΔG_b is positive. Such conditions could exist when, for instance, the concentration of P (which is common to both reactions) is very low, thus giving μ_P a large negative value. Such conditions would make reaction (5.6) go to the right and reaction (5.7) to the left until the concentrations of all reactants reached values which make both ΔG_a and ΔG_b zero. The free energy liberated would be dissipated in the form of heat and the only result would be an increase in the temperature of the solution. If, however, the substance P formed in reaction (5.6) is not allowed to enter the solution but is used directly in reaction (5.7) to form R, the two reactions are *coupled* and the total free energy change will be the algebraic sum of ΔG_a and ΔG_b. If this sum is negative the overall reaction

$$A + B \rightleftharpoons Q + R \tag{5.10}$$

proceeds to the right, thus ensuring that the free energy change of reaction (5.6) is used to do useful work [synthesizing R from B in reaction (5.7)].

Many coupled reactions of this sort take place in biological systems and ATP is involved in most of them. An example is the reaction in which the energy liberated during enzymatic oxidation of 3-phosphoglyceraldehyde is stored in ATP. This reaction is a part of *glycolysis*, the anaerobic breakdown of sugar in cells (see Section 5.4). In this reaction the aldehyde is not oxidized directly into carboxylic acid, but rather is first oxidized (in the presence of phosphate) into an intermediate called 1,3-diphosphoglycerate. If we denote the aldehyde by

$$\text{Ⓟ}-R\langle^{O}_{H}$$

(in which Ⓟ stands for phosphate), the reaction can be represented by

$$\text{Ⓟ}-R\langle^{O}_{H} + \text{Ⓟ}^{2-} \rightleftharpoons \text{Ⓟ}-R^{\nearrow O}-\text{Ⓟ}^{2-} + 2H \tag{5.11}$$

The standard free energy change of this reaction is about -7 kcal/mole but since the diphosphoglycerate concentration may initially assumed to be low, the actual free energy change has a larger negative value. Reaction (5.11) is coupled to the phosphorylation of ADP by the diphosphoglycerate and the formation of phosphoglycerate and ATP:

$$\text{Ⓟ}-R^{\nearrow O}-\text{Ⓟ}^{2-} + ADP^{3-} \rightleftharpoons \text{Ⓟ}-R\langle^{O}_{O^-} + ATP^{4-} \tag{5.12}$$

Since this reaction has a standard free energy change of some 7 kcal/mole, it can proceed to the right and the coupling of both reactions ensures that ATP is synthesized by expending the free energy change of the oxidation of the phosphoglyceraldehyde. The synthesis of sucrose from glucose and fructose is a process in which the hydrolysis of ATP is coupled to a synthetic reaction. This reaction requires energy; the standard free energy is $+5.5$ kcal/mole. ATP provides this energy by first phosphorylating the glucose into glucose 1-phosphate which then reacts with fructose to form sucrose and inorganic phosphate. The function of the ATP/ADP couple as energy and phosphate "conveyers" is clearly demonstrated by these reactions.

The molecular mechanism of the coupling of reactions as described is largely unknown. Enzymes are required in all known cases, many of which are found in or on membranes. Enzymes that are involved in reactions coupled to the *hydrolysis* of ATP are called *ATPases*. Examples are the Na–K–ATPases in membranes which mediate the active transport of the cations Na^+ and K^+ across membranes using the energy of the hydrolysis of ATP, the actomyosin system in muscles which causes contraction and, under special conditions, the *coupling factor* in mitochondria, chloroplasts, and bacterial systems. These systems will be discussed in more detail in later sections of this chapter.

Pyridine Nucleotides. As has been stated before, the formation of ATP is always coupled to oxidation reactions having a relatively large negative free energy change. These reactions are invariably enzymatic reactions and often also require the presence of coenzymes. The two pyridine nucleotides, *nicotinamide adenine dinucleotide* (NAD) and *nicotinamide adenine dinucleotide phosphate* (NADP), are among the most common coenzymes. These cofactors mediate the oxidation–reduction reactions by acting as *electron carriers*. The oxidation of the phosphoglyceraldehyde in glycolysis, for instance, is mediated by NAD which itself, becomes reduced in the process. The structures of NAD and NADP are given in Fig. 5.4. Both are dinucleotides, consisting of two sugar molecules (D-ribose) connected to each other by two phosphate groups. Adenine is attached to one of the riboses, which in the case of NADP contains an additional phosphate, and the other ribose holds the base nicotinamide. The oxidation–reduction reactions take place at the nicotinamide. In the oxidized form the nitrogen in the nicotinamide ring bears a positive charge. Reduction, which requires two electrons and a proton, neutralizes the charge and adds a hydrogen to the ring, as indicated. Thus, the oxidation of phosphoglyceraldehyde occurs concomitantly with the reduction of NAD^+ and the overall equation becomes

$$ⓟ\!-\!R\!\begin{smallmatrix}\nearrow O \\ \searrow H\end{smallmatrix} + P_i^{2-} + NAD^+ + ADP^{3-} \longrightarrow ⓟ\!-\!R\!\begin{smallmatrix}\nearrow O \\ \searrow O^-\end{smallmatrix} + H^+ + NADH + ATP^{4-} \tag{5.13}$$

or, written in a way in which it is most often presented,

$$ⓟ\!-\!R\!\begin{smallmatrix}\nearrow O \\ \searrow H\end{smallmatrix} \xrightarrow[NAD^+ \;\curvearrowright\; NADH]{ADP^{3-} + P_i^{2-} \;\curvearrowright\; ATP^{4-}} ⓟ\!-\!R\!\begin{smallmatrix}\nearrow O \\ \searrow O^-\end{smallmatrix} + H^+ \tag{5.13a}$$

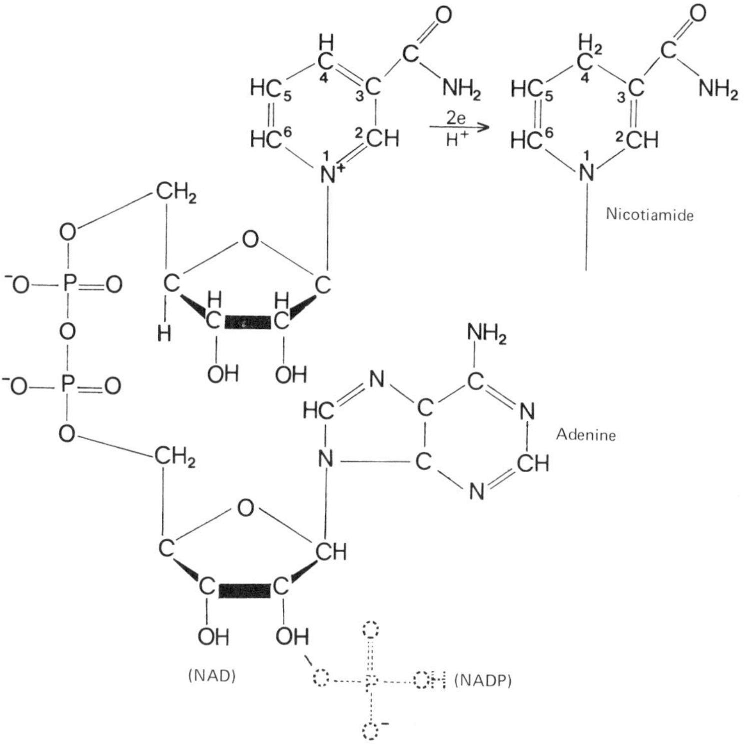

Fig. 5.4 The chemical structure of nicotinamide adenine dinucleotide (phosphate). In NADP the hydroxyl on the 2nd C atom in the ribose of the adenine nucleotide is replaced by a phosphate (dashed structure). In the oxidized form the nucleotides carry a positive charge on the nitrogen in the 1 position of the nicotinamide. Two electrons and a proton (in fact a hydride ion, $:H^-$) from a reducing substrate cause the reduction of the 1 and the 4 positions of the base.

Transhydrogenase. Although the two pyridine nucleotides NAD and NADP have near structural identity and both are oxidation–reduction mediators, their biological functions seem to differ somewhat. In general, NADP is used for reductive syntheses while NAD is more functional in energy metabolism. The cell, however, has an available mechanism by which NADPH can be converted to NADH and vice versa. This is an enzymatic reaction and the enzymes are called *transhydrogenases*. Measurements of the levels of NADPH and NADH in intact mitochondria have shown that under conditions in which energy is available the level of NADPH is in excess of that of NADH. This suggests that hydrogen is transferred from NADH to NADPH when it is not needed for further conservation of energy and can be used for reductive syntheses. The cell thus would have means of coordinating the synthesizing and energy delivering processes.

5.3 Photosynthesis

Photosynthetic Structures. Photosynthesis (as mentioned above) is the process by which the energy of sunlight is captured for use in sustaining life. The organisms which are capable of performing this process are plants and algae (multi-, as well as unicellular), and certain kinds of bacteria. In eukaryotic photosynthetic organisms (all plants and most algae) the process occurs within an organelle called the *chloroplast* (see Chapter 2). The photosynthetic apparatus is embedded in lamellae which are densily packed in some regions (grana) and in other areas extend into the stroma (intergranal lamellae). The lamellae are closed double membranes forming the flattened closes vesicles called *thylakoids* (see Fig. 5.5). The photosynthetic apparatus of prokaryotic photosynthetic organisms (blue-green algae and photosynthetic bacteria) is also embedded in membraneous structures which in these organisms extend throughout the cell. Upon rupture of the cells one often finds membrane fragments which are formed into closed vesicles from 300 to 500 Å in diameter. These vesicles are called *chromatophores* or *chromatophore fractions.* (Fig. 5.6.)

The Photosynthetic Unit. Photosynthesis begins with the absorption of light. In all photosynthetic organisms light which is absorbed by the photosynthetic pigments causes the excitation of a pigment aggregate. The transfer of excitation energy to the so-called *reaction center*, the site at which the conversion into chemical energy actually takes place, occurs by the

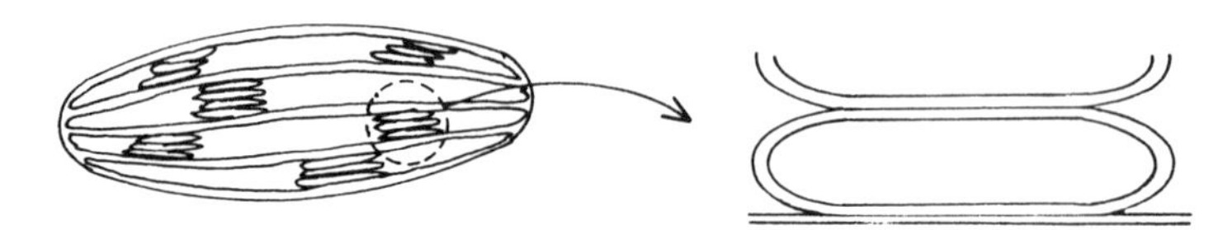

Fig. 5.5 A schematic representation of a chloroplast, showing the grana consisting of the flattened vesicles called thylakoids.

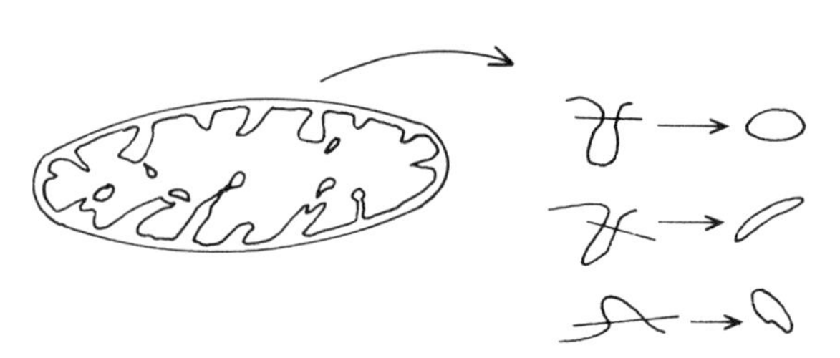

Fig. 5.6 A schematic representation of a photosynthetic bacterium showing the inward folded cytoplasmic membrane containing the photosynthetic apparatus which, upon disruption, form the closed vesicles called chromatophores.

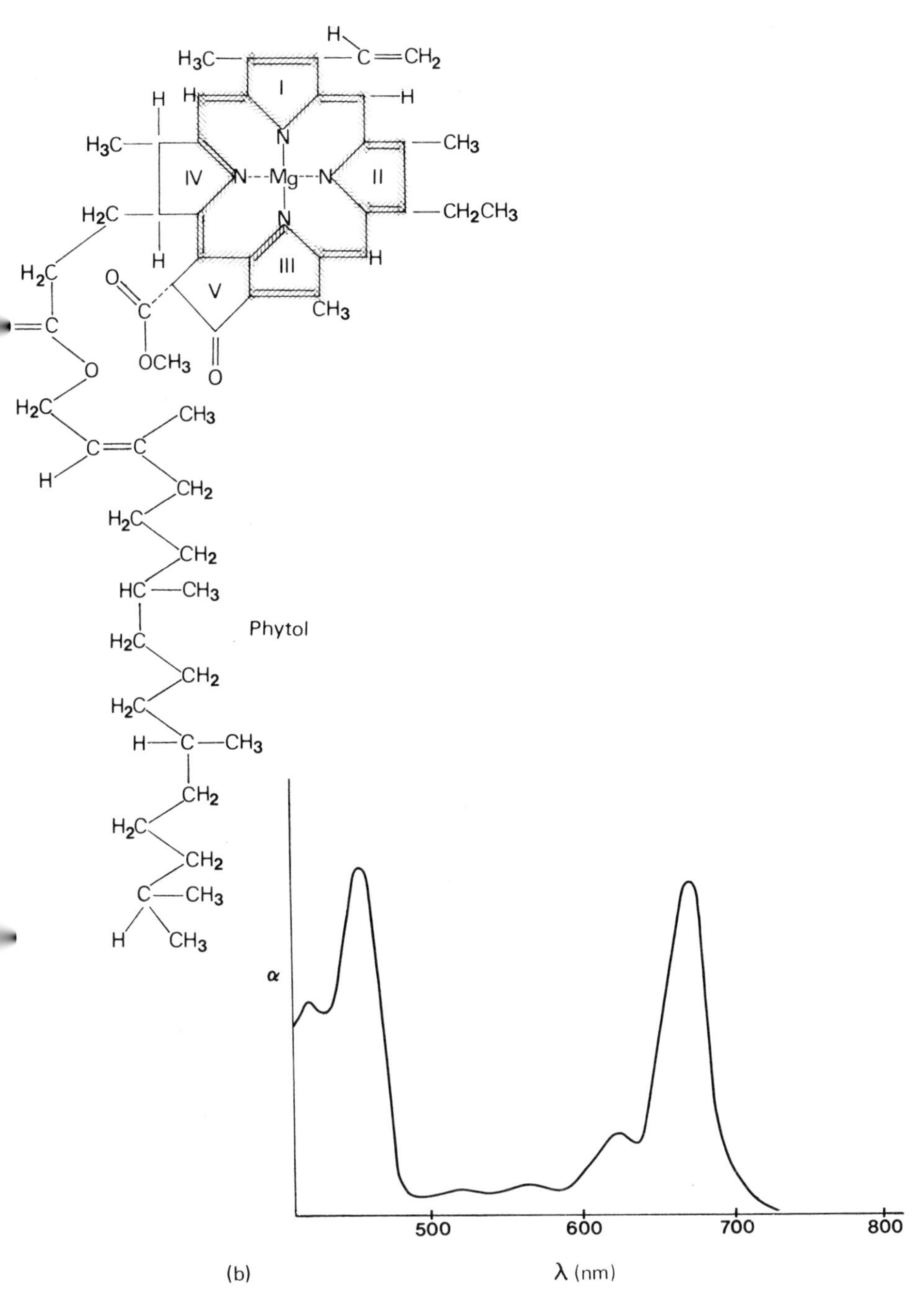

Fig. 5.7 (a) The chemical structure of chlorophyll *a*; the hatched parts designate the π-system of the molecule. (b) An absorption spectrum of chlorophyll *a* in organic solvent.

physical processes described in Section 4.6. All photosynthetic organisms are organized this way; the aggregate of the photosynthetic pigments and the reaction center together is called a *photosynthetic unit*. In higher plants the size of a photosynthetic unit is somewhere between 200 and 400 chlorophyll molecules per reaction center, in photosynthetic bacteria the size is about 50 bacteriochlorophyll molecules per reaction center.

Chlorophyll. Absorbing pigments are found in a wide variety in photosynthetic organisms. The most important of these are the chlorophylls, a group of highly conjugated structures, which differ among each other in only minor aspects. The conjugated structure is a closed tetrapyrole, called *porphyrin*, enclosing a magnesium atom. A long hydrocarbon chain called the *phytol tail* is attached at ring IV. Figure 5.7a shows the structure of chlorophyll *a*, the major chlorophyll found in all plants and algae. Figure 5.7b shows the absorption spectrum of chlorophyll *a* dissolved in ether. The major chlorophyll in photosynthetic bacteria is called *bacteriochlorophyll.*

Assessory Pigments. Most photosynthetic organisms also contain the so-called *assessory pigments*. Green plants and green algae have chlorophyll *b*, red and blue-green algae have *phycobilins* (open tetrapyroles), and all photosynthetic organisms have one or more variations of the group of *carotenoids* (unsaturated hydrocarbon chains with aromatic end groups at both sides).

The absorption spectrum of a unicellular alga, *Chlorella pyrenoidosa* is given in Fig. 5.8. The spectrum is more or less representative, also for higher plants. The peak at 675 nm is the *in vivo* absorption maximum of chlorophyll *a*. The shoulder at 650 nm in the spectrum is due to chlorophyll *b*; the absorption bands in the blue-green region are largely due to carotenoids.

Comparison with the spectrum of chlorophyll *a* dissolved in ether (Fig. 5.7b) shows that the *in vivo* absorption maximum is shifted toward the red end. This is due to the fact that the chlorophyll molecules form complexes with proteins and with each other. This complex formation results in even more profound spectral changes in some photosynthetic bacteria. A summary of the different kinds of pigments in photosynthetic organisms and their absorption characteristics *in vivo* and *in vitro* is given in Table 5.2.

The Primary Reaction. In Section 5.1 we suggested that photosynthesis essentially occurs in two phases. The first phase is the production, in the presence of light, of a reductant, and the second phase is the utilization, in the absence of light, of this reductant to synthesize sugar by the reduction

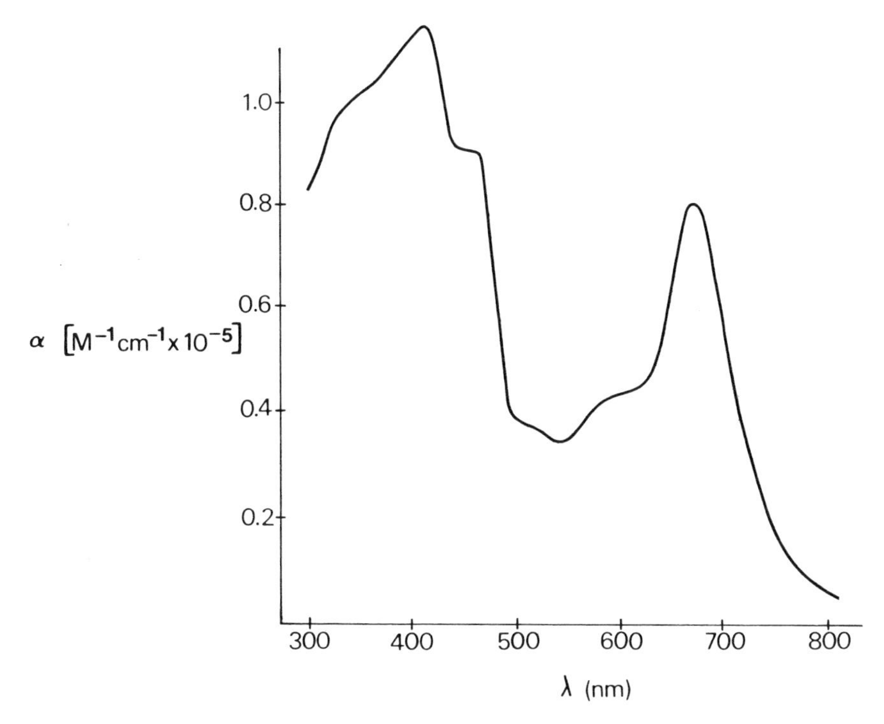

Fig. 5.8 An absorption spectrum of the green alga *Chlorella pyrenoidosa*. The maximum at 675 nm and the shoulder at about 650 nm are due to chlorophyll *a* and chlorophyll *b*, respectively.

of CO_2. The light-produced reductant is a pyridine nucleotide, more specifically (for higher plants and algae) NADPH. The light-induced reduction of $NADP^+$ is initiated by so-called *primary reactions* of the type

$$(P \cdot A) \xrightarrow{h\nu} (P \cdot A)^* \longrightarrow (P^+ \cdot A^-) \qquad (5.14)$$

These reactions occur in the reaction centers, within a time shorter than 10^{-8} sec after excitation, through the antenna pigment aggregate. Thus, the primary reaction amounts to the extremely rapid production of a *primary oxidant* (P^+) and a *primary reductant* (A^-).

Higher Plant and Algal Photosynthesis. In higher plants and all algae (including the prokaryotic blue-green algae) *two* such primary reactions are needed to complete the sequence of oxidation–reduction reactions leading to the production of NADPH [cf. Eq. (5.1)]. The sequence, including

TABLE 5.2 Photosynthetic Pigments

Pigment	Absorption maxima (nm) In organic solvents	Absorption maxima (nm) *In vivo*	Occurrence
Chlorophyll *a*	420, 660	435, 670–680	Higher plants, all algae
Chlorophyll *b*	453–643	480, 650	Higher plants, green algae
Chlorophyll *c*	445–625	645	Diatoms, brown algae
Chlorophyll *d*	450–690	740	Some red algae
Bacteriochlorophyll *a*	365, 605, 770	585, 800–890	Purple bacteria
Bacteriochlorophyll *b*	368, 582, 795	1 017	Some purple bacteria, instead of bacteriochlorophyll *a*
Bacteriochlorophyll *c* (chlorobium chlorophyll 1)	425, 650	750	Green bacteria
Bacteriochlorophyll *d* (chlorobium chlorophyll 2)	432, 660	760	Some green bacteria, instead of bacteriochlorophyll *c*
Phycoerythrin	490, 546, 576	Same	Red algae (small amounts in blue-green algae)
Phycocyanin	618	Same	Blue-green algae (small amounts in red algae)
Carotenoids	420–525	Same	All photosynthetic organisms; some seven different kinds in different organisms

the two primary reactions, is shown in Fig. 5.9 in which the vertical axis measures the redox potentials (relative to the standard hydrogen electrode at pH = 7) of the electron transport mediators involved. The arrows indicate the transfer of electrons from one intermediate to the other. It can be seen from this figure that the primary reaction, labeled *photosystem II*, produces a strong oxidant Z^+, capable of oxidizing water and a relatively weak reductant Q^-. The other primary reaction, labeled *photosystem I*, produces a strong reductant X^- which, through a number of intermediates such as the nonheme iron protein *ferredoxin* and an enzyme called pyridine nucleotide reductase (a flavoprotein), can reduce $NADP^+$. A relatively weak oxidant $P700^+$ is concomitantly produced in this primary reaction. The weak reductant Q^- of photosystem II then reacts with the weak oxidant $P700^+$ of photosystem I through a number of intermediates, such as plastiquinone, cytochromes, and a copper protein, plastocyanin, thus closing the redox chain. The electron transport from Q^- to $P700^+$ is coupled to the formation of ATP. Based on the difference in redox potential between these two reactants, the coupling is thermodynamically in favor of the production of one

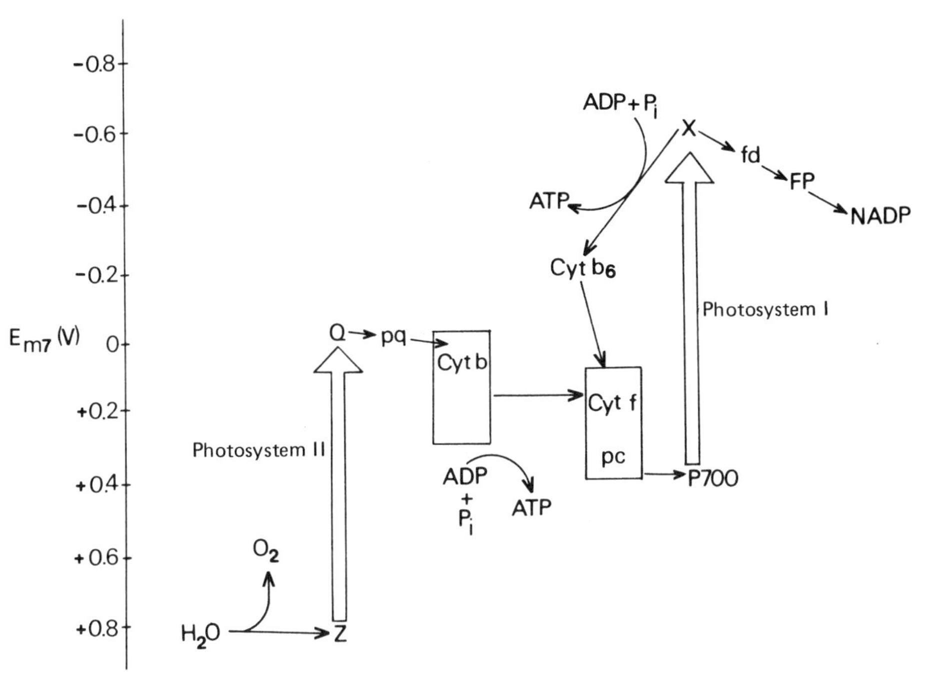

Fig. 5.9 The "Z-scheme" of higher plant and algal photosynthesis. The fat arrows represent the reaction center associated electron transport from the primary donor to the primary acceptor. For system II these are Z (presumably a specialized chlorophyll *a* called P680) and Q (perhaps specialized plastoquinone), respectively; for system I the primary donor is P700, a specialized chlorophyll *a* and the primary acceptor is unknown. pq designates plastoquinone, cyt *b* and cyt *f*, respectively, are cytochrome *b* and cytochrome *f*, pc is plastocyanin, fd is ferredoxin, (an nonheme iron protein), and FP is a flavoprotein.

ATP molecule for every two electrons transported from Q^- to $P700^+$. One has to be careful, though, in drawing conclusions from redox potentials which are defined for, and often measured under, equilibrium conditions. The actual potentials may be very different since the reactions may occur far from equilibrium. In fact, actual measurements of the ATP/electron ratio give conflicting results.

Apparently, the photoreduction of $NADP^+$ is not the only photoreaction which can result from the primary reaction of photosystem I: Electrons can also cycle back from the primary reductant X^- to the main chain, thus performing a cyclic electron transport. This cyclic electron transport is coupled to the formation of ATP, and a cytochrome *b* seems to be a physiological intermediate. In isolated chloroplasts the cycle can be stimulated to a great extent by nonphysiological intermediates such as phenazine methosulfate, or vitamin K. The purpose of this light-induced cyclic electron transport may be to provide an additional means for the production of ATP; as we will see later, the production of ATP in the main chain (one ATP molecule per two electrons transported through the chain) is not sufficient for the fixation of carbon dioxide in the cyclic series of dark reactions occuring in most higher plants and algae.

Bacterial Photosynthesis. To date there seems to be no sufficient evidence for more than one type of primary reaction in photosynthetic bacteria. The electron transport reactions set in motion by the primary reaction in these organisms are more primitive, but also more flexible than those in higher plants. Most probably the main electron transport chain is cyclic, like the cyclic electron transport system described for photosystem I in higher plants. During this cyclic electron transport ATP, or some other "high energy intermediate," is produced and can then be used to drive reactions against the thermodynamic gradient in coupled reactions. It has been demonstrated, for example, that preparations of photosynthetic bacteria can catalyze the reduction of NAD^+ by succinate, a reaction with a positive free energy change. The energy for such a reaction, occurring in darkness, can be obtained from the coupled hydrolysis of ATP or even of pyrophosphate. Such reactions most probably do occur under *in vivo* physiological conditions, so that the main task of the light reaction in photosynthetic bacteria (Fig. 5.10) would be to provide energy intermediates in suitable form (ATP or others). The occurence, under physiological conditions, of other direct light-driven electron transport reactions still cannot be ruled out.

Reaction Centers. There is, as yet, no certainty about the identity of most of the participating components of the primary reactions in plants, algae, and photosynthetic bacteria. Only the identity of the primary donor of photosystem I in plants and algae and of the primary donor of the bacterial system are known; they are, respectively, chlorophyll *a* and bacteriochlorophyll *a* in a specialized environment. In the case of photosystem I in plants and algae, the special environment results in a shift of the main red absorption band to about 700 nm. The pigment is, therefore, called P700; it traps the excitation energy transferred to it from the antenna pigment system and becomes excited as a result. A transfer of charge then takes place in a very short time ($<10^{-10}$ sec) and the pigment subsequently becomes oxidized. Its absorption band at 700 nm disappears and this change in the absorption spectrum can be measured without much difficulty. Figure 5.11 shows a light-minus-dark difference spectrum (the difference between a

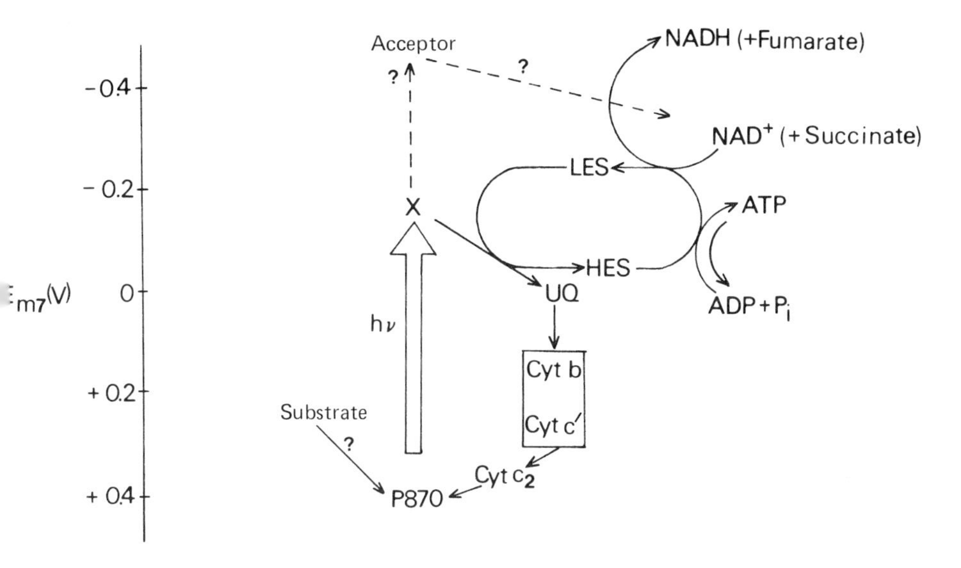

Fig. 5.10 A diagram of the light-induced electron transport in photosynthetic bacteria. The fat arrow represents the reaction center associated electron transport from the primary donor P (a specialized form of bacteriochlorophyll *a*) to a primary acceptor (with an unknown identity). This transport drives a cyclic reaction in which participate ubiquinone (UQ), and the cytochromes *b*, *c′*, and c_2. The cyclic electron transport creates a high-energy state (HES) which can be discharged to a low-energy state (LES) either by causing the reduction of NAD by succinate or by the phosphorylation of ADP to ATP. A direct light-induced electron transport, perhaps through an alternative acceptor to NAD is still questioned.

spectrum measured when the sample is illuminated and one measured when the sample is kept in the dark), which is identical to an oxidized-minus-reduced difference spectrum, of a preparation of spinach chloroplasts. The bleaching of P700 is clearly seen. Also, the primary donor in photosynthetic bacteria is an excitation energy trapping pigment which undergoes spectral changes when it is oxidized by light or by chemical means. Its absorption maximum (or rather, one of its absorption maxima) is at a wavelength between 835 and 890 nm, depending upon the species. Denominations vary, therefore, from P840 (green bacteria), P870 and P880 (*Rhodopseudomonas spheroides* and *Rhodospirillum rubrum*) to P890 (*Chromatium*). The light-minus-dark difference spectrum in the far-red spectral region of a preparation of *Rhodospirillum rubrum* is shown Fig. 5.12. The spectra of the oxidized and reduced forms of the preparation are also shown on the same wavelength scale. The light causes an oxidation which, in the spectrum, results not only in a bleaching at about 870 nm but also in a shift toward the blue region of the spectrum of an absorption band at about 800 nm. This feature

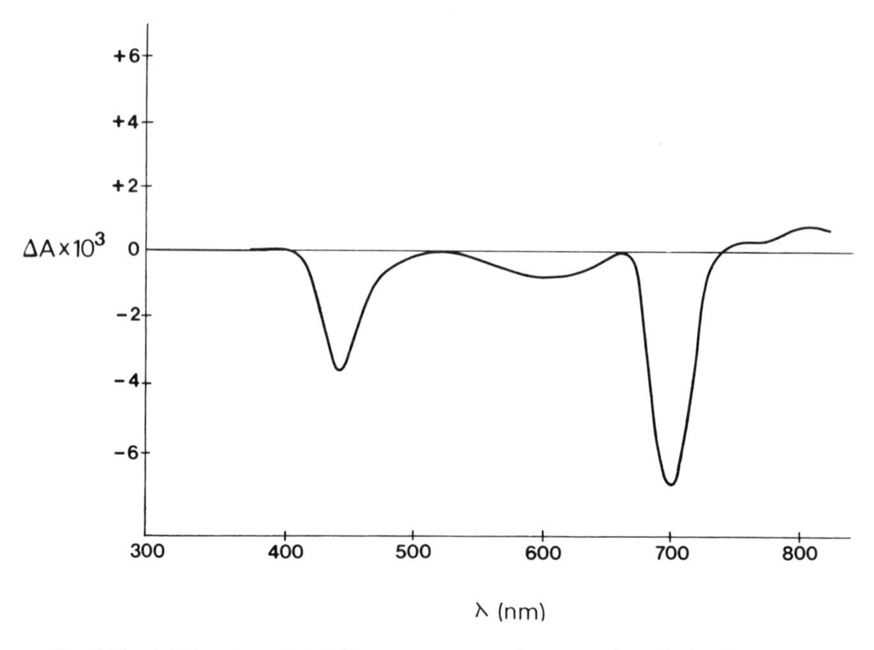

Fig. 5.11 A light-minus-dark difference spectrum of a suspension of spinach chloroplasts. The bleaching of P700 is demonstrated by the troughs at 700 and 438 nm.

demonstrates the fact that the bacteriochlorophyll *a* in the reaction center is organized in a rather complicated form.

Neither the primary donor nor the primary acceptor of the reaction center of photosystem II in higher plants and algae are known with certainty. The mechanism of the charge transfer (from primary donor to primary acceptor) is also still partly a "black box" reaction. From experiments on the kinetics of flash-induced oxygen evolution by chloroplast and algae, one can conclude that the primary donor of photosystem II has to accumulate four positive charges (by donating four electrons to the primary acceptor Q) before it can react with water to give an oxygen molecule. Manganese seems to be an essential factor in this process.

Chlorophyll Fluorescence Yield Changes. The transfer of excitation energy from the antenna pigments to the reaction center can be followed by observing the changes of the fluorescence yield of the antenna pigments occuring during the transfer. When a reaction center is excited by a transfer of excitation energy from the antenna pigment aggregate (usually followed by a chemical change, such as an oxidation of the primary donor and a reduction of the primary acceptor), the reaction center is inaccessible to following

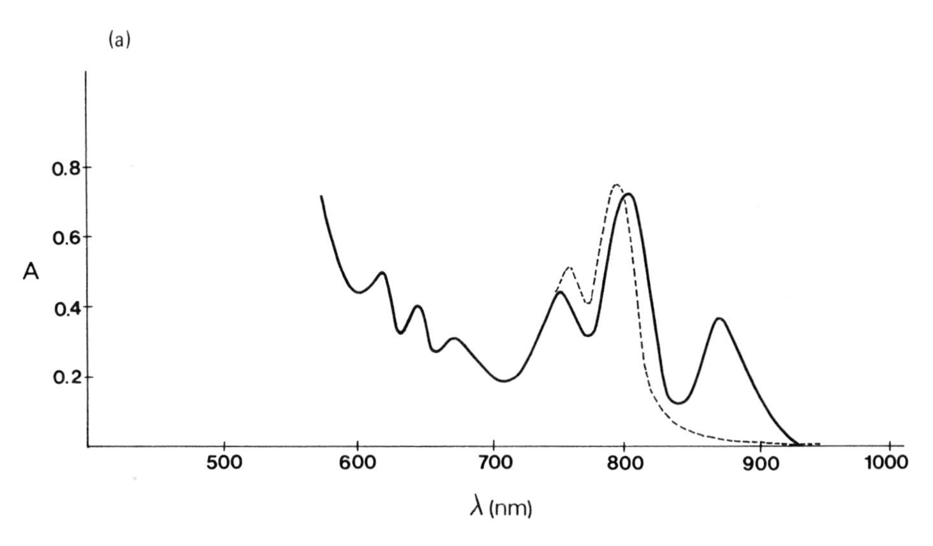

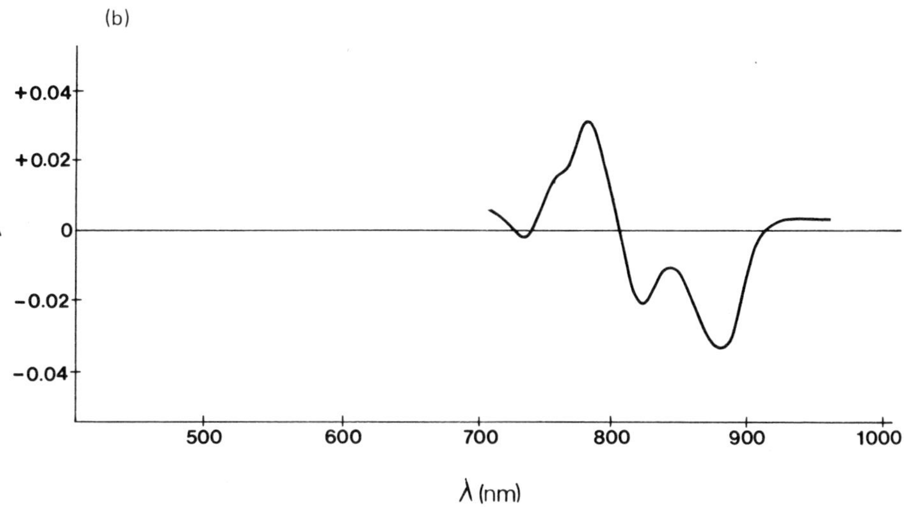

Fig. 5.12 (a) Spectra of the reduced (——) and the oxidized (– – –) form of a reaction center preparation from the photosynthetic bacterium *Rhodospirillum rubrum*. (b) The light-minus-dark difference spectrum of the same preparation.

excitation quanta until the original (ground) state of the reaction center is restored. From Eqs. (4.8) and (4.9) it immediately follows that at high intensities (when many reaction centers are excited at the same time) the fluorescence yield ϕ_f must be high, and at low light intensities (when many reaction centers are still "open" for excitations) the fluorescence yield ϕ_f must be low (one of the k_i can be seen as the rate constant for energy transfer, ultimately to the reaction centers). The variations of the fluorescence yield of an antenna pigment can, thus, give information about the oxidation–reduction processes in the reaction center. This, of course, is only possible when there is an antenna pigment that does fluoresce. This is the case for the chlorophyll *a* of the pigment aggregate of photosystem II in higher plants and algae. Thus, even though the redox reactions in the reaction center of photosystem II cannot be followed directly, the kinetics of the changes in the fluorescence yield of chlorophyll *a* reflect those redox reactions.

There is no fluorescent pigment in the antenna pigment aggregate of photosystem I (consisting of chlorophyll a_1). In this system we can only follow the light-induced changes of the absorbance of P700, reflecting its oxidation. In photosynthetic bacteria, however, the redox reactions of the reaction center can be followed directly by observing the spectral changes due to the oxidation of the reaction center bacteriochlorophyll and indirectly by looking at the changes of the fluorescence yield of the antenna bacteriochlorophyll *a*. This fortunate situation has allowed the establishment of a relation between the absorbance changes and the changes of the fluorescence yield, at least for the steady-state situation at different intensity levels of exciting light. From Eqs. (4.8) and (4.9) it follows that

$$\phi_{\mathrm{f}} = \frac{k_0}{k_0 + k_i + k_{\mathrm{t}} C_{\mathrm{p}}} \tag{5.15}$$

if it is assumed that the rate constants for the deexcitation processes are of first order and that the trapping rate of the reaction center is proportional to the concentration C_{p} of "open" reaction centers. In this equation the rate constants k_0, k_i, and k_{t} are those for fluorescence, radiationless transitions, and trapping, respectively. The concentration of the "open" traps C_{p} is related to the absorption change ΔA according to

$$\Delta A = K_1 + K_2 C_{\mathrm{p}}$$

where K_1 and K_2 are constants related to the rate constants and the molar absorption coefficients of the reaction center bacteriochlorophyll in reduced and oxidized form. It is now easy to demonstrate that the inverse of the fluorescence yield is linearly related to the absorption change:

$$\frac{1}{\phi_{\mathrm{f}}} = \alpha + \beta\, \Delta A \tag{5.16}$$

in which α and β are constants derived from the constants K_1 and K_2. This linear relationship has been experimentally verified for steady-state conditions at different intensities of light. Figure 5.13 shows the result of such an experiment. A great deal of the biophysical research in photosyn-

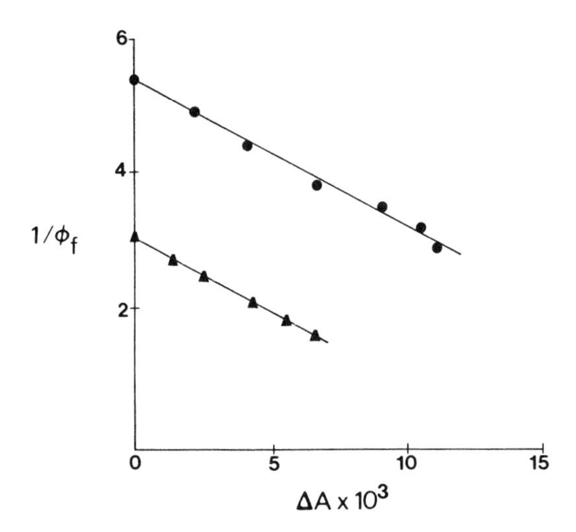

Fig. 5.13 Plots of the inverse fluorescence yield ϕ_f of bacteriochlorophyll as a function of the extent of the light-induced change of the absorption of the reaction-center bacteriochlorophyll in a suspension of the purple bacterium *Rhodospirillum rubrum* in two different media (from Vredenberg and Duysens, 1963).

thesis makes use of this technique. It is based, as we have shown, on the fact that changes in the fluorescence yield always reflect changes in the photochemistry as long as the latter starts from the fluorescent state, irrespective of whether or not the chemistry can be measured.

Photophosphorylation and Carbon Fixation. One of the two products of the light reaction of photosynthesis, in photosynthetic bacteria as well as in higher plants and algae, is ATP. ATP formation is coupled to the light-induced transport of electrons by a mechanism which is most probably not essentially different from the coupling mechanism of ATP synthesis during respiratory electron transport (see Section 5.4). When coupled to photosynthetic electron transport, the process is known as *photophosphorylation;* the ATP synthesis coupled to respiratory electron transport is called *oxidative phosphorylation.* Since much of the biophysical research is directed to the coupling mechanism, we will devote a separate section to it (Section 5.6).

The second phase of photosynthesis is the fixation of CO_2, a process which occurs in the dark. The formation of sugar, and eventually of other cell constituents, results from a cycle of chemical reactions in which CO_2 is incorporated into a five-carbon compound and the resulting unstable six-carbon compound is split into two three-carbon compounds which are

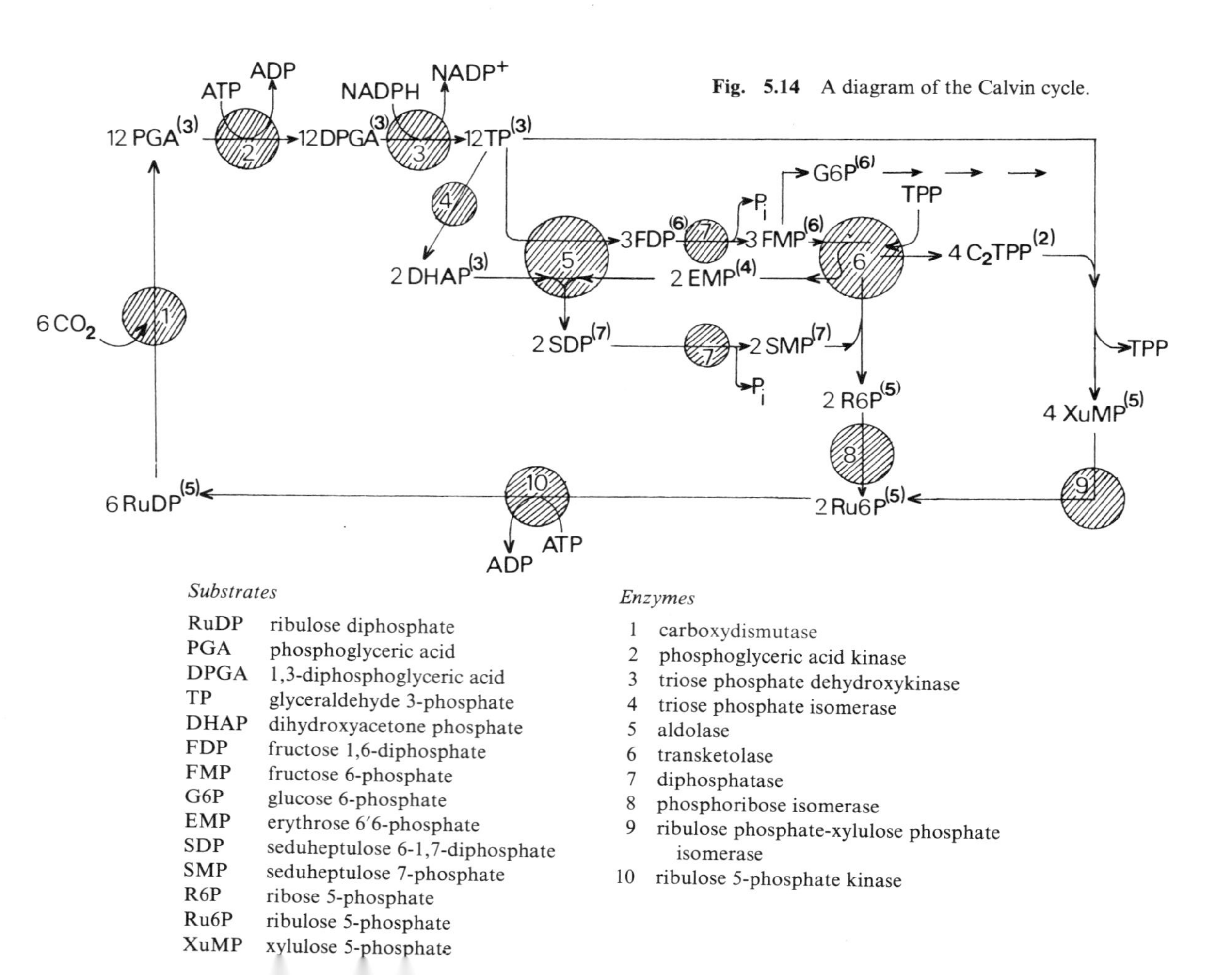

Fig. 5.14 A diagram of the Calvin cycle.

Substrates

RuDP	ribulose diphosphate
PGA	phosphoglyceric acid
DPGA	1,3-diphosphoglyceric acid
TP	glyceraldehyde 3-phosphate
DHAP	dihydroxyacetone phosphate
FDP	fructose 1,6-diphosphate
FMP	fructose 6-phosphate
G6P	glucose 6-phosphate
EMP	erythrose 6′6-phosphate
SDP	seduheptulose 6-1,7-diphosphate
SMP	seduheptulose 7-phosphate
R6P	ribose 5-phosphate
Ru6P	ribulose 5-phosphate
XuMP	xylulose 5-phosphate

Enzymes

1	carboxydismutase
2	phosphoglyceric acid kinase
3	triose phosphate dehydroxykinase
4	triose phosphate isomerase
5	aldolase
6	transketolase
7	diphosphatase
8	phosphoribose isomerase
9	ribulose phosphate-xylulose phosphate isomerase
10	ribulose 5-phosphate kinase

subsequently reduced. The Calvin cycle, as it is often called after its discoverer Melvin Calvin, is given in a schematic and somewhat simplified form in Fig. 5.14. The five-carbon compound *ribulose diphosphate* picks up a molecule of carbon dioxide in a reaction that is catalyzed by the enzyme *carboxydismutase* and the resulting six-carbon compound immediately breaks down into two molecules of the three-carbon compound 3-phosphoglyceric acid. The phosphoglyceric acid is then phosphorylated (by ATP) and reduced (by NADPH) into glyceraldehyde 3-phosphate (which is a triose phosphate). A sixth of the glyceraldehyde is then converted into glucose through a series of enzymatic reactions. The remaining part goes through a cycle of reactions which includes one more ATP-mediated phosphorylation step, to form again the five-carbon compound ribulose diphosphate. Thus, to "grind out" one molecule of glucose the cyclic reaction "mill" has to carboxylate and dismutate six molecules of ribulose diphosphate through the formation of twelve molecules of phosphoglyceric acid. These twelve molecules need twelve ATP molecules and twelve NADPH molecules to form twelve glyceraldehyde phosphate molecules. Two of the glyceraldehyde phosphate molecules then join to form ultimately one glucose molecule. The other ten glyceraldehyde phosphate molecules go through the cycle to form six new molecules of ribulose diphosphate, utilizing six more ATP molecules. The total process, thus, needs twelve molecules of NADPH and *eighteen* molecules of ATP; in other words, for each *two* molecules of $NADP^+$ reduced in the light reaction, *three* ATP molecules are needed to perform the Calvin cycle as described. An extra photophosphorylation site in a light-induced cyclic electron pathway would fulfill this need.

5.4 Fermentation and Respiration

In the previous section we have seen how photosynthesis stores the energy from sunlight in a large number of different compounds together known as food. The way in which the nonphotosynthetic part of the living world recovers the energy from food in order to perform "work for living" is the subject of this section.

Anaerobic Oxidation. As we have stated before the energy is always recovered in the form of ATP, the synthesis of which is coupled to a stepwise oxidation of the food. Such an oxidation can occur even in the complete absence of oxygen. In some organisms such an anaerobic oxidation process is the only form of energy conversion. In most cells, however, anaerobic as well as aerobic energy conversion can take place and in the cells of all higher organisms an obligatory anaerobic step precedes the aerobic step.

The anaerobic oxidation of sugar (or amino acids or fatty acids) is called *fermentation*. There are several forms of fermentation, with different initial substrates and different final products. Some yeast cells live anaerobically by fermenting glucose into ethanol. Some bacteria can ferment glucose into acetone or butanol, while other bacteria ferment glucose into lactic acid. The latter process is the most widespread and best understood. It is called *glycolysis* and is the type of glucose breakdown that precedes the further sequence of oxidations in higher organisms.

Glycolysis. Glycolysis occurs in the cytoplasm of the cell. It is a series of enzymatic reactions in which the six-carbon glucose is phosphorylated, isomerized, and again phosphorylated after which it breaks down into two three-carbon fragments which are readily interconverted. One of these fragments, 3-phosphoglyceraldehyde, is then oxidized with the formation of ATP, and after another isomerization and an elimination of a water molecule another phosphorylation step occurs which is coupled to an intramolecular oxidation–reduction reaction and the formation of ATP. Figure 5.15 gives the details of this sequence. Glucose is phosphorylated into glucose 6-phosphate by an enzyme called hexokinase. This phosphorylation goes on at the expense of the third phosphate of an ATP molecule. After the isomerization into fructose 6-phosphate (by the enzyme phosphoglucomutase) another phosphorylation, at the expense of another ATP molecule, takes place. The enzyme catalyzing this process is phosphofructokinase. The product at this stage is fructose diphosphate which can now be split, by the enzyme aldolase, into the two triose phosphates, 3-phosphoglyceraldehyde and dihydroxyacetone phosphate. The process, thus, needs to be "primed"; in order to be able to synthesize ATP it must use ATP. Of course, more ATP has to be produced than expended, otherwise the situation would not make any sense. In fact, if we follow the sequence further, we see that the 3-phosphoglyceraldehyde is oxidized by NAD^+ into 3-phosphoglycerate, *after* first being phosphorylated, and that the phosphorylated oxidized 1,3-diphosphoglycerate in turn phosphorylates ADP into ATP. The enzyme catalyzing this step is glyceraldehyde 3-phosphate dehydrogenase (this reaction was used as an example of coupled reactions in Section 5.2). The 3-phosphoglycerate is then isomerized into 2-phosphoglycerate (the enzyme is phosphoglyceromutase) and converted into phosphoenolpyruvate (with enolase) eliminating a water molecule. Finally, phosphoenolpyruvate, phosphorylates another ADP molecule into ATP (with pyruvate phosphokinase), producing pyruvate. Two molecules of ATP, thus, are synthesized in one sequence from 3-phosphoglyceraldehyde to pyruvate. But since the enzyme triose phosphate isomerase converts the other triose phosphate into 3-phosphoglyceraldehyde, four molecules of ATP are synthesized and two mole-

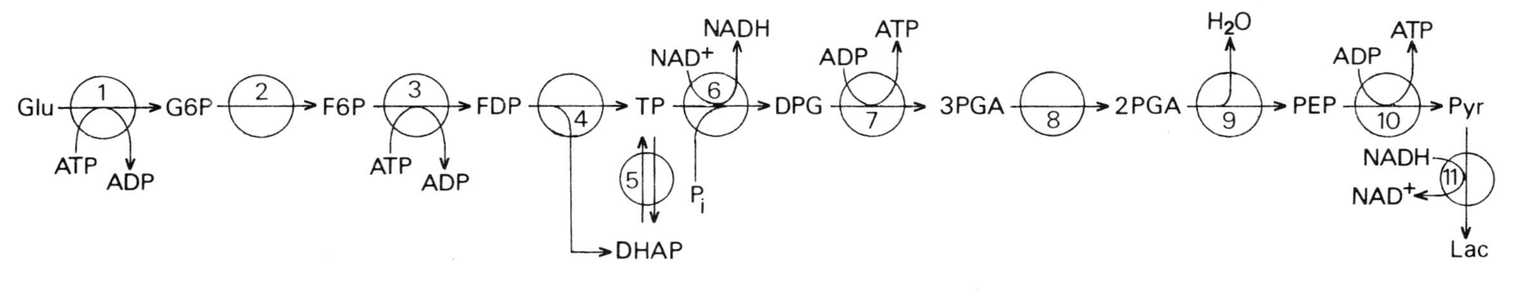

Fig. 5.15 The glycolitic sequence.

Substrates

Glu	glucose
G6P	glucose 6-phosphate
F6P	fructose 6-phosphate
FDP	fructose 1,6-diphosphate
TP	3-phosphoglyceraldehyde
DHAP	dihydroxyacetone
P_i	inorganic phosphate
DPG	1,3-diphosphoglyceric acid
3PGA	3-phosphoglyceric acid
2PGA	2-phosphoglyceric acid
PEP	phosphoenolpyruvic acid
Pyr	pyruvic acid
Lac	lactic acid

Enzymes

1	hexokinase
2	phosphoglucomutase
3	phosphofructokinase
4	aldolase
5	triose phosphate isomerase
6	glyceraldehyde phosphate dehydrogenase
7	diphosphoglycerate kinase
8	phosphoglyceromutase
9	enolase
10	pyruvate phosphokinase
11	lactate dehydrogenase

cules of ATP are expended for each sequence glucose–pyruvate. The net yield, thus, is two molecules of ATP per glucose molecule.

The aldolase reaction, which results in the cleavage of fructose diphosphate, has an equilibrium of about 90% in the direction of the fructose diphosphate. Essentially, we have seen the reaction occurring in the reverse direction in the Calvin cycle. The reason why the reaction goes forward in glycolysis is that one of its reaction products, 3-phosphoglyceraldehyde, is removed by oxidation, the energy of which is conserved in the production of ATP. In the Calvin cycle the *reduction* of phosphoglycerate requires energy, which is furnished by the oxidation of NADH, initially reduced by light. Thus, for the glycolytic sequence to proceed at all it is necessary that the 3-phosphoglyceraldehyde be oxidized. As we will see later, this provides a means of control.

When glycolysis is not followed by aerobic energy conversion (in glycolytic anaerobic cells or in facultative cells in the absence of oxygen) the pyruvate is reduced into lactate; in this reaction NADH is oxidized. Thus, in anaerobic metabolism NAD^+ reacts in a cyclic manner (see Fig. 5.16a).

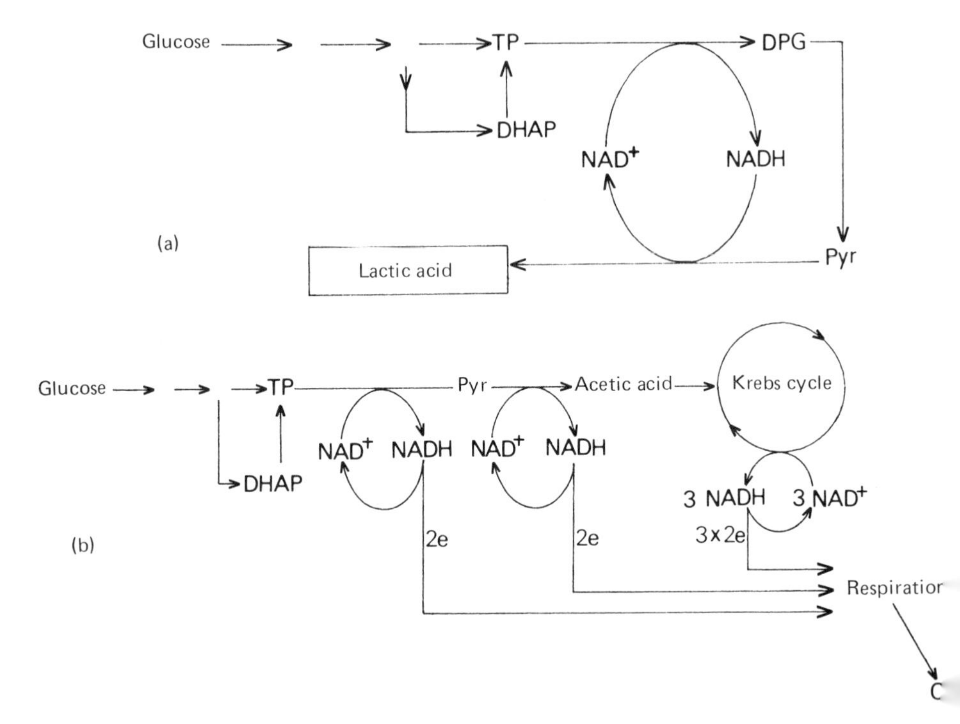

Fig. 5.16 (a) The cyclic nature of the NAD reaction in anaerobic metabolism. (b) The electron intermediate function of NAD in aerobic metabolism.

There is no net oxidation, but energy is conserved from an oxidative process all the same.

Aerobic Oxidation. In aerobic cells glycolysis is followed by the respiratory reactions which, in eukaryotic cells, occur inside the mitochondria (Fig. 5.17). This organelle consists of two membranes; the outer membrane is permeable to most smaller molecules and the inner membrane convolutes to the inside forming the many inward folds called *cristae*. The surface of this membrane is thus tremendously increased. It encloses an inner compartment, the *matrix*, and is permeable only to water and a limited number of small neutral molecules, such as urea and glycerol. The membrane contains, however, several permeases, "enzymes" which are carriers for specific metabolites such as amino acids, acetates, the reducing equivalents of NADH, ADP, ATP, phosphate, and others. The respiratory reactions occur in two groups of steps, the first of which takes place in the matrix. This cycle is known as the Krebs cycle (after its discoverer, Hans Krebs), citric acid cycle, or tricarboxylic acid cycle. A number of decarboxylations and oxidations take place in the Krebs cycle and the reduced products of these reactions (NADH and reduced flavin) enter the second set of reactions, the respiratory chain, which is bound to the cristae. The final oxidation to the level of oxygen and the membrane-coupled synthesis of ATP occur through this chain.

Krebs Cycle. Although the principal role of the Krebs cycle is the coupling of the glycolytic breakdown of sugar in the cell cytoplasm to the series of oxidation–reduction reactions leading to the production of ATP in the mitochondria, it also serves to regulate the synthesis of a number of compounds required by the cell. Several intermediates of the cycle are

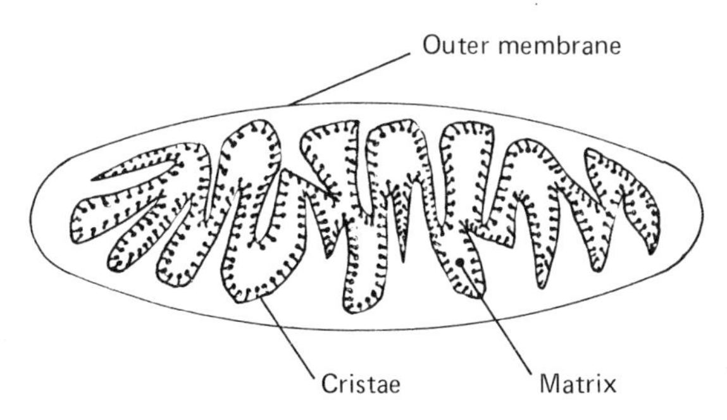

Fig. 5.17 A representation of a cross section of a mitochondrion. The inner membrane convolutes to the inside, the matrix, thus forming the many folds called cristae. The components of the respiratory chain are embedded in the cristae.

branch points from which the synthesis of, for instance, amino acids and fat can start. Many of these "branching off" reactions are reversible, so that they can also serve to generate the cycle intermediates.

In the Krebs cycle the 4-carbon compound *oxaloacetate* incorporates a 2-carbon *acetyl* group to become the 6-carbon *citric acid*. Subsequent decarboxylations and oxidations then lead to the regeneration of oxaloacetate, thus closing the cycle. The acetyl group is furnished by glycolysis. To this end the pyruvate is not reduced to lactate, as in anaerobic cells, but oxidized and decarboxylated (CO_2 is liberated) by the enzyme *pyruvate dehydrogenase*. NAD^+ is again a coenzyme and the reaction is, thus, coupled to its reduction. The oxidation product is not free acetate but an acetyl group which is fused to another coenzyme called *coenzyme A* (CoA). Coenzyme A (see Fig. 5.18) is a complex molecule consisting of a base (adenine), a sugar (ribose), phosphate groups, and a tail to which a sulfhydryl group is attached. The sulfhydryl is the active part and the molecule is usually written as CoA-SH; the reaction with pyruvate (from glycolysis) and NAD^+, thus, produces acetyl-S-CoA (Fig. 16b).

Acetyl-S-CoA then enters the Krebs cycle by reacting with oxaloacetate to form citrate and free CoA-SH (Fig. 5.19). The first oxidation of the cycle is preceded by an isomerization of the citrate into isocitrate. Then the isocitrate is oxidized into oxalosuccinate concomitant with the reduction of NAD^+. The next step is a decarboxylation; CO_2 is again removed and α-ketoglutarate is formed. This compound is the first branch point of the cycle; α-ketoglutarate is a precursor for a number of amino acids in reactions called *transamination reactions*. In the cycle itself the α-ketoglutarate is decarboxylated and oxidized in one reaction. NADH is again the reduced product and the succinyl group which is formed is attached to CoA-SH in exactly the same manner as the acetyl group. The succinyl-S-CoA, thus formed, is another branch point of the cycle; from it proceeds the synthesis of porphyrin, the backbone of heme, which is the prostethic group of cytochromes and hemoglobin. Succinyl-S-CoA, as well as acetyl-S-CoA, are intermediates in fatty acid metabolism. The cycle proceeds with the formation of succinate. This reaction is an energy-conserving step; it is coupled to the phosphorylation of guanosine diphosphate (GDP) into guanosine triphosphate (GTP). The GTP then undergoes a phosphate group transfer reaction with ADP:

$$\text{GTP} + \text{ADP} \rightleftharpoons \text{GDP} + \text{ATP} \tag{5.17}$$

The next step of the cycle is the oxidation of succinate into fumarate. This reaction differs from the other oxidation–reduction (dehydrogenase) reac-

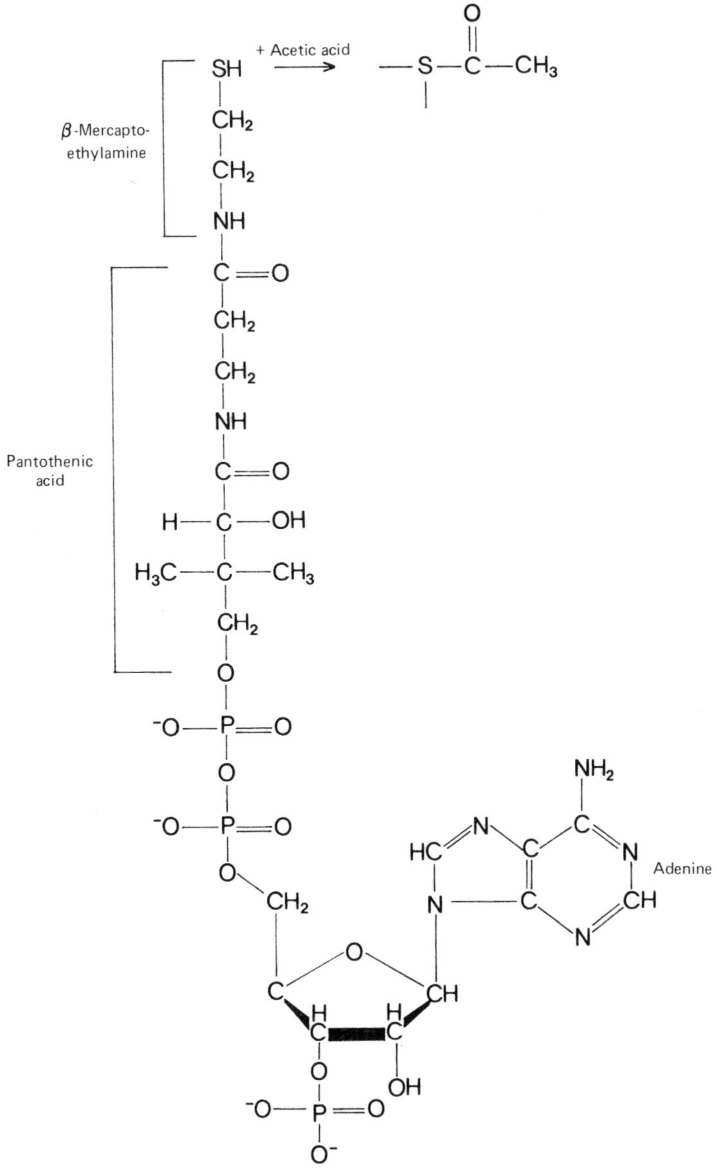

Fig. 5.18 The chemical structure of coenzyme A. The reaction with acetic acid (or with succinic acid) takes place at the SH terminal of the molecule, thus forming acetyl-S-CoA (or succinyl-S-CoA).

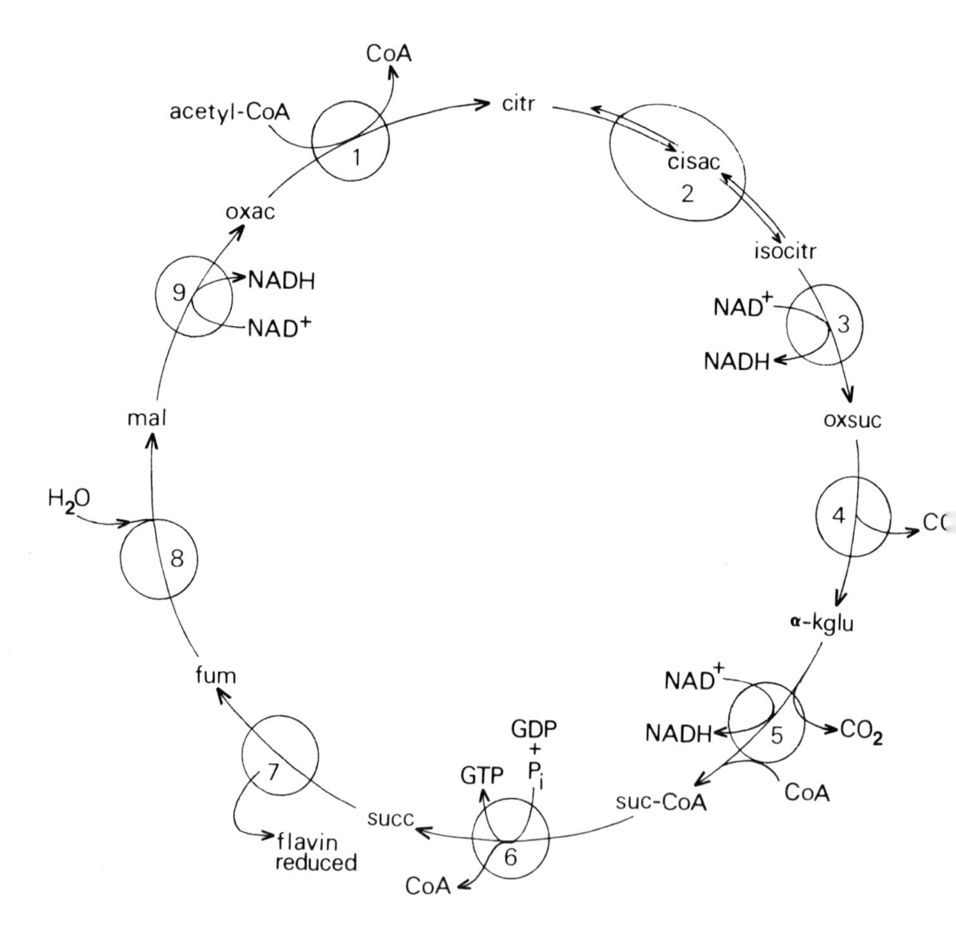

Fig. 5.19 The Krebs cycle, citric acid cycle, or tricarboxylic acid cycle.

Substrates

citr	citric acid
cisac	*cis*-aconic acid
isocitr	isocitric acid
oxsuc	oxalosuccinic acid
α-kglu	α-ketoglutaric acid
suc-CoA	succinyl coenzyme A
succ	succinic acid
fum	fumaric acid
mal	malic acid
oxac	oxaloacetic acid

Enzymes

1	citrate synthase (condensing enzyme)
2	aconitase
3	isocitrate dehydrogenase
4	oxalosuccinate decarboxylase
5	α-ketoglutarate dehydrogenase
6	succinyl-CoA synthase
7	succinate dehydrogenase
8	fumarase
9	malate dehydrogenase

tions of the cycle; its reaction partner is *not* the coenzyme NAD but rather the succinate dehydrogenase enzyme itself. Succinate dehydrogenase is a protein with *flavin adenine dinucleotide* (FAD) as a prosthetic group. When succinate is oxidized, FAD is reduced to $FADH_2$. This enzyme is tightly bound to the mitochondrial membranes and is an entrance port into the respiratory chain for electrons, as we will see later. Fumarate is involved in nitrogen metabolism. The enzyme fumarase then takes care of the hydration of fumarate into malate and a final oxidation step (again concomitant with the reduction of NAD^+) regenerates oxaloacetate, ready to react again with another acetyl-S-CoA molecule. Figure 5.19 gives a summary of the reaction cycle, including the names of the enzymes involved.

Anaplerotic Reactions. The pivotal function of the Krebs cycle, in the first place for the aerobic oxidation of sugar and in the second place for provision of intermediates for biosynthesis, makes it necessary that the cycle be kept running within close tolerances. First, the siphoning off of intermediates for biosynthesis must be compensated for lest the cycle come to a halt. Part of such a compensation is accomplished by the fact that many of the branching-off reactions are reversible, so that Krebs cycle intermediates can be synthesized from amino acids via these reversed pathways. But there are other special, so-called *anaplerotic* ("filling up") *reactions*. The most important of these is the enzymatic carboxylation of pyruvate into oxaloacetate:

$$\text{Pyruvate} + CO_2 + \text{ATP} \longrightarrow \text{oxaloacetate} + \text{ADP} + P_i \qquad (5.18)$$

The enzyme catalyzing the reaction is *pyruvate carboxylase* and is primarily found in the mitochondria of the liver cells of most species. This reaction is a very good example of the use of the allosteric properties of enzymes for control. Pyruvate carboxylase is an allosteric enzyme with acetyl-CoA as a positive modulator; thus, the higher the level of acetyl-CoA, which is the fuel for the Krebs cycle, the better the reaction rate, the more oxaloacetate produced, and the more acetyl-CoA oxidized in the cycle. In the absence of acetyl-CoA, the rate of the reaction is very low.

Another anaplerotic reaction is the synthesis of malate from pyruvate by the *malic enzyme*, a reaction which involves the oxidation of NADPH. Many plants and bacteria can also use a sort of short circuit of the Krebs cycle to produce more succinate and malate directly from isocitrate. In this so-called *glyoxylate cycle*, isocitrate is split into succinate and glyoxylate by the enzyme *isocitritase*. Another enzyme, *malate synthase*, then promotes the reaction of glyoxylate with acetyl-CoA to form malate and CoA-SH.

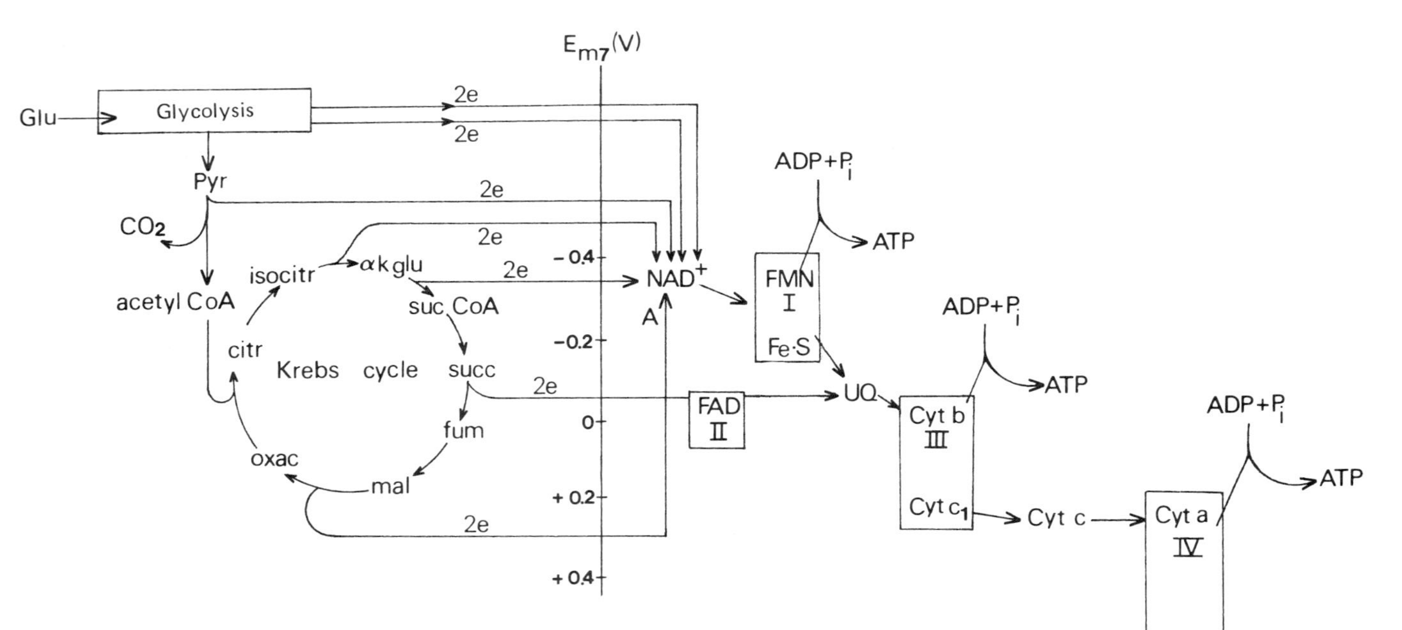

Fig. 5.20 The respiratory chain: the electron transport components forming active complexes. Complex I contains FMN (flavin mononucleotide or NAD dehydrogenase) and nonheme iron protein; complex II contains FAD (flavin adenine dinucleotide or succinate dehydrogenase); complex III contains cyt *b* (cytochrome *b*) and cyt c_1 (cytochrome c_1); complex IV contains cyt *a* and cyt a_3 (cytochromes *a* and a_3 or the respiratory enzyme). Between complex I and II on one hand and complex III on the other hand UQ (ubiquinone) reacts and between complex III and complex IV cyt *c* (cytochrome *c*) reacts. The complexes I, III, and IV are presumably the oxidative phosphorylation sites. Complex I receives electrons from NADH and complex II from succinate.

The Respiratory Chain. Most of the electrons involved in the oxidation reactions of the Krebs cycle end up in NADH. Only those from the reaction of succinate are directly bound to the succinate dehydrogenase enzyme itself. NADH is also a product of the dehydrogenation of pyruvate. Moreover, for every glucose molecule entering the glycolytic chain, two NADH molecules are produced in the dehydrogenation step. All these NADH molecules react with the enzyme NADH dehydrogenase, which is another entrance port to the respiratory chain. NADH dehydrogenase, like succinate dehydrogenase, is a flavin-linked enzyme; its prosthetic group is *flavin mononucleotide* (FMN) which is reduced when NADH is oxidized. It also is tightly bound to the mitochondrial membrane.

The respiratory chain is a series of oxidation–reduction enzymes and coenzymes which actually take part in the reactions themselves. They are more or less tightly bound to the cristae membrane and the processes, in fact, involve the transport of electrons from NADH and succinate to oxygen (Fig. 5.20). The chain components are, in addition to the two dehydrogenases mentioned above, nonheme iron proteins, an abundantly present quinone called ubiquinone (or CoQ), and a series of cytochromes, proteins with a heme as prosthetic group. A heme group is a tetrapyrole, just like the one in chlorophyll, with iron chelated in its center (Fig. 5.21). An oxidation (reduction) of a cytochrome is the removal (addition) of an electron from (to) the iron which, thus, makes a ferrous–ferric (ferric–ferrous) transition.

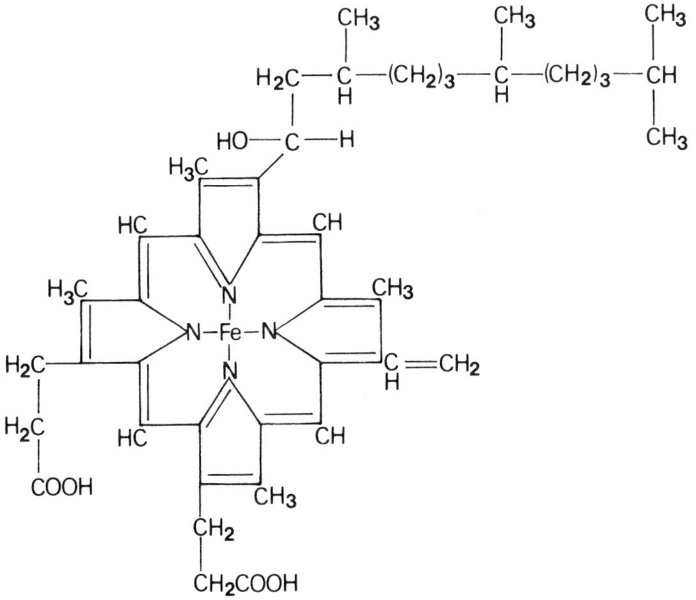

Fig. 5.21 The chemical structure of the heme group of cytochrome *c*.

The types of cytochromes found in the respiratory chain are, in order of increasing redox potential, cytochrome *b*, cytochrome c_1, cytochrome *c*, cytochrome *a*, and cytochrome a_3. Cytochrome *c* is easily extracted from the mitochondria by strong salt solutions. It can be purified and crystalized and is, consequently, the best characterized of them all. The other cytochromes are much more tightly bound to the membrane and, therefore, difficult to obtain in soluble and homogeneous form. Cytochromes *a* and a_3 form a complex which is often called *cytochrome oxidase* or the *respiratory enzyme*.

The respiratory chain is depicted in the right-hand part of Fig. 5.20. The arrows in the figure indicate the movement of the electrons through the chain, and the vertical axis is a redox potential scale. As shown, the electrons move down a gradient determined by the redox potentials of the respiratory components. From NADH the electrons are picked up by NADH dehydrogenase (FMN), probably transported to a nonheme iron protein, and from there to ubiquinone. This latter component can also receive electrons from succinate dehydrogenase (FAD). Reduced ubiquinone transmits its electrons to cytochrome *b*, from which the electrons go, via cytochromes c_1 and *c*, to the cytochrome $a + a_3$ complex. This complex, in turn, is reoxidized by molecular oxygen and, with protons from the environment, water is formed.

Oxidative Phosphorylation. The negative change in the free energy of the redox reactions in the chain is conserved in the form of ATP. We can distinguish three regions in the chain in which the difference in redox potential between consequetive components is relatively large. Thus, the difference in redox potential between NADH and ubiquinone is some 0.27 V, that between cytochrome *b* and cytochrome c_1 is about 0.22 V, and that between the cytochrome $a + a_3$ complex and oxygen is 0.53 V. Each of these three energy gaps is sufficiently large to provide for the free energy of the ADP → ATP phosphorylation reaction. This, obviously, would suggest three sites at which the phosphorylation could be coupled to electron transport. In fact, it has been demonstrated that for each couple of electrons from NADH, three molecules of ATP are formed and that a couple of electrons from succinate can provide for only two ATP molecules. This, of course, is consistent with three coupling sites indicated in Fig. 5.20. In this respect, the finding that when mitochondria are treated with specific detergents under predefined conditions the components of the cristae membranes can be separated into four complexes (numbered I–IV), each enzymatically active and containing several proteins, is suggestive. Complex I contains the NADH dehydrogenase (FMN) and there is good evidence that a nonheme iron protein is also present. Complex II contains succinate dehydrogenase (FAD) and may also have nonheme iron protein. It seems then that complexes I and II are connected through ubiquinone to complex III containing cytochromes b and c_1. Cytochrome c may be the link between complexes III and IV, the latter containing the cytochrome oxidase. This arrangement of the complexes may be essential for the coupling of ATP synthesis to electron transport (see Section 5.6 and Fig. 5.20).

The Coupling Factor. The coupling of ATP phosphorylation to electron transport is still obscure; the various conflicting hypotheses will be discussed in Section 5.6. The coupling can be broken by a number of compounds that either inhibit the whole process or only stop the synthesis of ATP while leaving the electron transport going, often even stimulating it. Compounds exhibiting the latter effect are called *uncouplers*. An effective uncoupler for mitochondrial phosphorylation is 2,4-dinitrophenol (DNP). A certain class of antibiotics are also uncouplers of phosphorylation in both respiratory and photosynthetic systems. The uncoupling effect of such compounds has led to a picture in which the synthesis or generation of a "high-energy intermediate" by electron transport precedes the phosphorylation of ADP to ATP. At present, however, no consensus exists as to the nature of such an "high-energy intermediate" (see Section 5.6). The coupling does require a proteinous factor, the so-called *coupling factor*, which can be removed from the membranes by special treatment. Particles devoid of coupling factors have lost their ability to synthesize ATP but still carry out electron transport. When purified coupling factor is added to the deficient particles, ATP synthesis is partially restored. There is good evidence that the coupling factor in mitochondria is located in little spherical particles with a diameter of 80–90 Å, and connected to the inner membrane by a narrow stalk sticking out toward the inside of the matrix space (Fig. 5.22).

Although many aspects of respiratory and photosynthetic electron transport and energy conservation (especially those of the coupling mechanism) are not yet known and still vigorously under investigation, the last two decades have seen substantial progress in our knowledge of the process. This is largely due to new (biophysical) approaches and the development of new (biophysical) techniques. Among these are novel applications of spectrophotometry. Cytochromes absorb light in the visible spectral region and their spectra undergo substantial changes when the enzyme changes its redox state. Figure 5.23 shows the spectrum of cytochrome c in the reduced and oxidized states. Using sensitive absorption difference spectrophotometry one can follow the oxidation–reduction reactions of the cytochromes and, thus, determine the sequence of the oxidation–reduction reactions of the electron carriers. A difference spectrum of respiring versus nonrespiring rat liver mitochondria is given in Fig. 5.24. These processes can be more closely

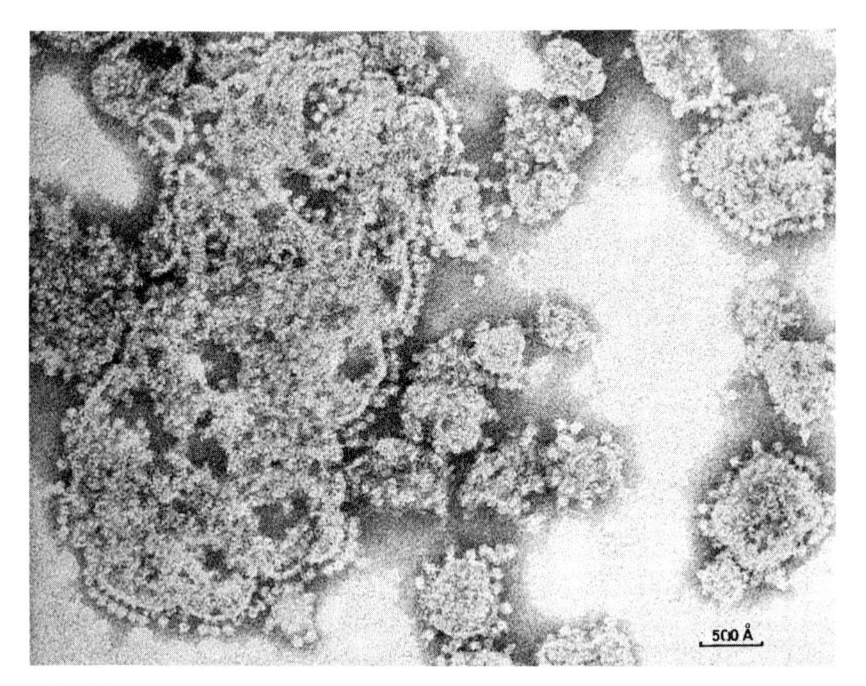

Fig. 5.22 Electronmicrograph of a preparation of a mitochondrion showing the coupling factor as little knobs on stalks sticking out into the matrix (courtesy of Dr. E. Racker, Cornell University).

examined by using specific inhibitors. The insectiside *rotenone*, for example, specifically blocks the electron transport from NADH to flavoprotein. Other inhibitors are the antibiotic *antimycin*, which blocks between the cytochromes *b* and *c* and *cyanide*, which prevents oxygen from oxidizing cytochrome oxidase. Using such blocks it is easy to demonstrate that all components before the block in the direction of oxygen are completely reduced while all components after the block remain oxidized.

Control Mechanisms. Fermentation, in particular glycolysis, and respiration in living cells must proceed at rates and levels which are adapted to specific functions and conditions. This is most obvious for the eukaryotic cells of the higher organisms; thus, in liver cells, for example, these requirements are very different from those of muscle cells. Different conditions set different requirements in prokaryotic cells as well. To provide for such flexibility, both enzyme systems must operate under tight controls. Control of glycolysis is already affected by the rate at which the product of the aldolase reaction, 3-phosphoglyceraldehyde, is removed by the oxidation step. This is dependent on the level of the NAD^+ which, under anaerobic conditions,

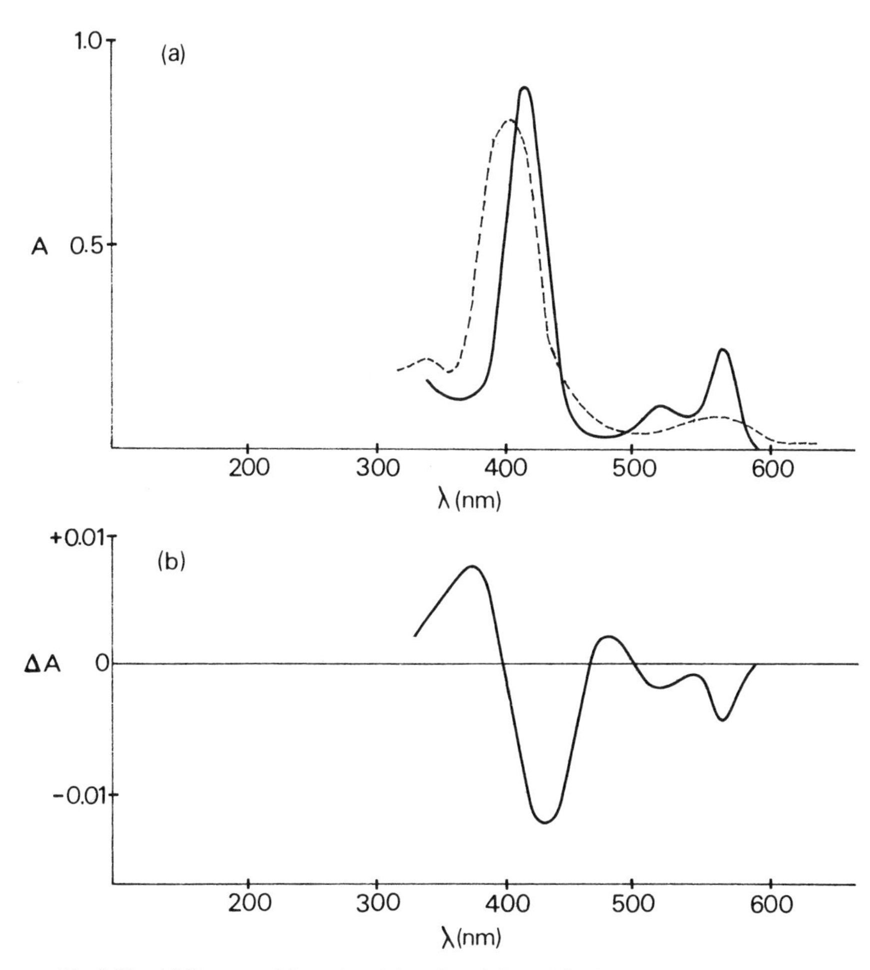

Fig. 5.23 (a) Spectra of the reduced (——) and the oxidized (– – –) form of cytochrome *c*. (b) Oxidized-minus-reduced difference spectrum of cytochrome *c*.

is regenerated from NADH by the reduction of pyruvate to lactate. Thus, in facultative cells, utilizing glucose from an ample supply under anaerobic conditions, glycolysis runs at a high rate and lactate is accumulated. But when such cells are transferred into aerobiosis the rate of glycolysis decreases dramatically and the accumulation of lactate is reduced to zero. This inhibition of glycolysis by oxygen is called the Pasteur effect after Louis Pasteur, who discovered it during his investigations of the fermentation processes of wine making. Since respiration is reoxidizing NADH, thus leading to an increased $NAD^+/NADH$ ratio, one would expect a stimulation rather than

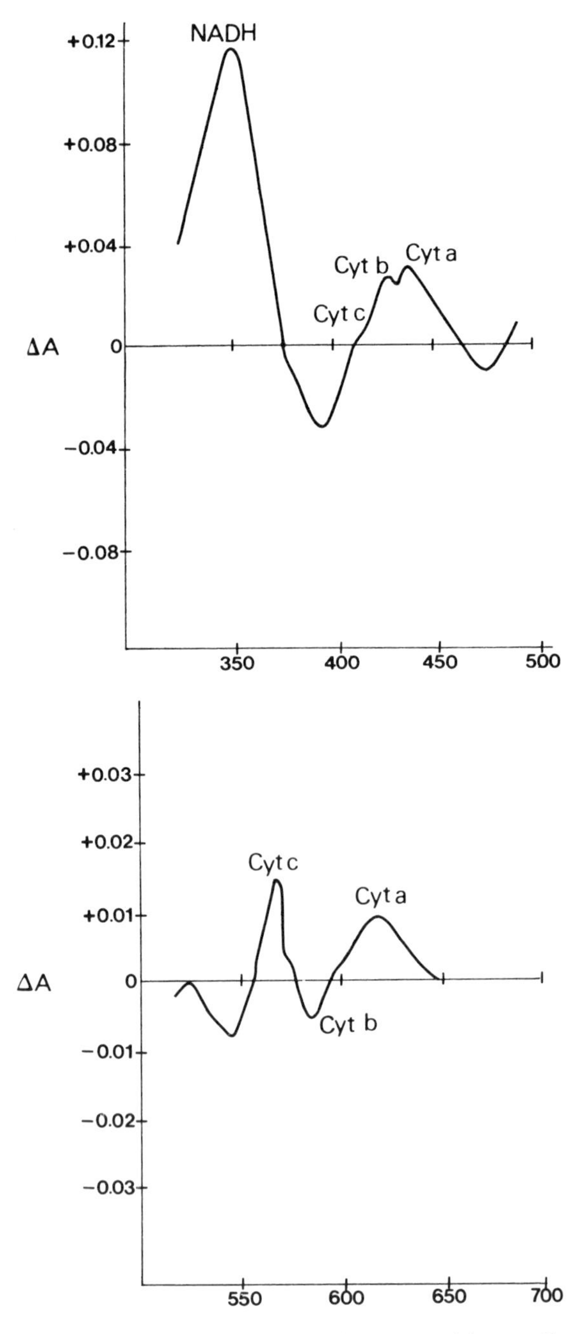

Fig. 5.24 Difference spectrum of respiring-minus-nonrespiring rat liver mitochondria (from A. L. Lehninger, "Biochemistry," Worth Publishers, New York, 1970, p. 380).

an inhibition of glycolysis by oxygen. Of course, a controlling effect is exerted by the "shuttle" mechanism that forms the communication link between the cytoplasmic pool and the intramitochondrial (matrix) pool of NAD^+ and NADH. However, a major factor of control is that the activity of one of the key enzymes of glycolysis, phosphofructokinase, is allosterically inhibited by ATP and stimulated by ADP and phosphate. Phosphofructokinase catalyzes the rate-limiting reaction in which fructose 6-phosphate is phosphorylated into fructose 1,6-diphosphate; when during respiration the ATP/ADP ratio increases the phosphofructokinase is gradually "turned off," thus slowing down the rate of glycolysis. Although phosphofructokinase seems to be the major regulating enzyme of glycolysis, there are a number of secondary control points in the glycolytic sequence which are influenced by the ATP/ADP ratio as well as by the level of intermediates such as citrate, acetyl-CoA and glucose 6-phosphate.

The ATP/ADP ratio not only controls the rate of glycolysis but also has a profound influence on the rate of respiration. When the supply of respiratory substrates is ample, a high rate of oxygen consumption occurs when the ADP and phosphate concentration is high and the concentration of ATP is low. When the concentration of ATP rises and the concentrations of ADP and phosphate are reduced to zero the respiratory rate becomes very low. This effect is called *respiratory control* or *acceptor control*. This can easily be demonstrated by an experiment such as that schematically diagrammed in Fig. 5.25. Mitochondria, in the presence of phosphate but in the absence of ADP, show a very low rate of oxygen consumption. When

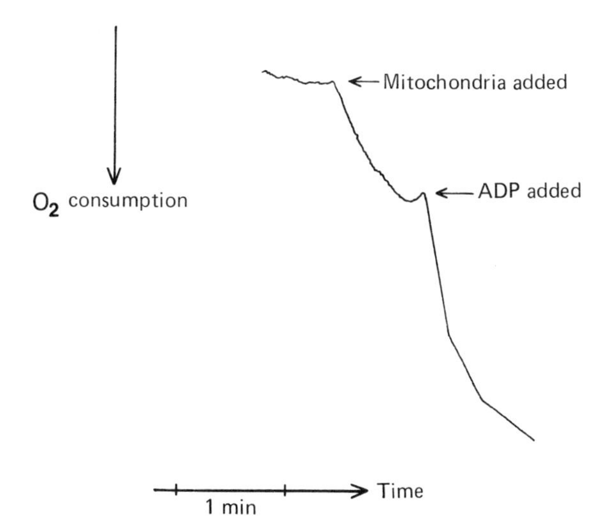

Fig. 5.25 An experiment showing respiratory control. The uptake of oxygen increases sharply when ADP is added to a suspension of respiring mitochondria.

ADP is added the rate increases by as much as a factor of 20. When all added ADP is phosphorylated the rate returns to the original low level.

The mechanism of respiratory control is not known. It is seen by some as evidence for the existence of a high-energy intermediate of phosphorylation. In their explanation the high-energy intermediate itself inhibits respiration. The phosphorylation of ADP to ATP, which uses the high-energy intermediate, thus relieves the inhibition. The fact that uncouplers of phosphorylation, such as dinitrophenol, stimulate the rate of respiration is seen as supporting evidence; uncouplers are supposed to cause a breakdown of the high-energy intermediate (see Section 5.6).

The relative concentrations of ATP and ADP in the cell are, thus, the most important controlling elements. This is true for all processes generating or utilizing the energy incorporated in the phosphate bond. Many regulatory enzymes, as well as those involved in biosynthetic pathways, are responsive to the levels of ATP and ADP (and also of AMP). Regulation, therefore, is accomplished in all these reactions by a delicate balance of the concentrations of these important nucleotides.

5.5 Passive and Active Transport: Membrane Permeability

Transport across Membranes. Transport processes are an integral part of biological function. For example, the energy converting processes which we have discussed in the previous section need a continuous supply of substrates and a continuous disposal of products and waste. It is evident that there can be no respiration when there are no means for oxygen and substrates (glucose) to penetrate the cells and the organelles; carbon dioxide has to be removed as well. Often, ATP produced at one point in the cell must be transported to another. Many other substances, neutral as well as charged, have to be transported in order to make vital processes function.

Compartmentalization seems to be the structural feature by which cells carry out their function. This is more conspicuous in the higher developed and differentiated eukaryotic cells than it is for the prokaryotes, although it seems essential for these more primitive cells as well. Compartmentalization is accomplished by membranes, and where some transport of matter occurs through channels bordered by membranes (the endoplasmic reticulum and the Golgi apparatus, for example), the *selective* transport often occurs through the membranes themselves. By *passive* and *active* transport the chemical integrity inside the compartments of the cell and the cell organelles is kept constant within narrow limits, thus providing optimal conditions for the life processes. By *passive transport* we mean diffusion *in the direction* of the thermodynamic gradient; *active transport* is the movement of solutes *against* the thermodynamic gradient. The latter requires an energy source and mechanisms to couple the energy input to the transport. The selectivity is a consequence of the permeability of the membrane itself, often determined by the particular molecular mechanism of the transport. Membrane transport and permeability are the subjects of this section. In cells or cell organelles large differences in the concentrations of (charged or uncharged) solutes between the inside and outside of the membrane-surrounded vesicle can be found, even when the membrane is permeable to such solutes. In red blood cells, for example, the cytoplasmic membrane is perfectly permeable to K^+ and Na^+. The concentration of K^+ inside the cell, however, is many times higher than the concentration of the ion in surrounding medium, while Na^+ has a lower concentration inside than it has outside. One would say that a situation like this can only be maintained by active transport of K^+ to the inside and of Na^+ to the outside. This is, indeed, true for the red blood cell; the membrane of the red blood cell has a so called K^+–Na^+ ATPase which takes care of the transport of the two ions (each in opposite directions) at the expense of the hydrolysis of ATP. There are situations, however, in which a concentration gradient of a solute, charged or uncharged, can be maintained *at equilibrium*. This occurs when the membrane has semipermeability characteristics.

Osmotic Equilibrium. Consider, for example, a membrane separating two compartments I and II (see Fig. 5.26). Compartment I contains a solvent S, while in compartment II a solute A is dissolved in the solvent S. Let the membrane only be permeable to the solvent. The molecules of the solvent tend to move from compartment I to compartment II because of the concentration difference. Net movement in equilibrium, however, is counteracted by a buildup of pressure in compartment II. This can be translated into thermodynamical terms as follows.

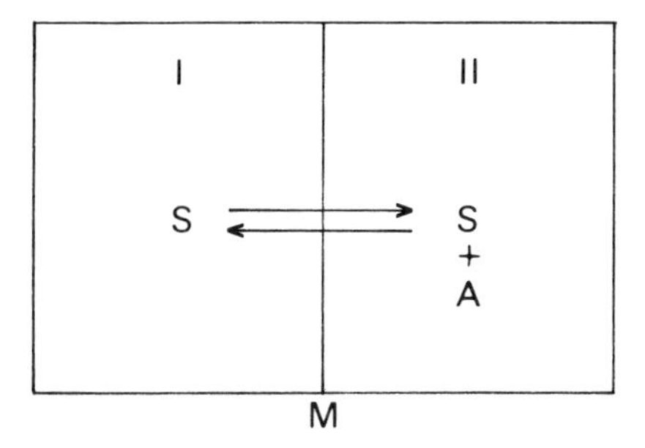

Fig. 5.26 Two compartments I and II, separated by a semipermeable membrane permeable to the solvent S but not to the solute, A.

The change of free energy when a mole of solvent moves from compartment I to compartment II is

$$\Delta G = \mu_S^{II} - \mu_S^{I} \tag{5.19}$$

in which μ_S^{I} and μ_S^{II} are the chemical potentials of the solvent in compartments I and II, respectively. At equilibrium $\Delta G = 0$. Using the expressions for the chemical potential given in terms of the mole fraction x_S (the ratio between the number of moles of the solvent and the total number of moles of solvent and solute) Eq. (5.19) becomes

$$\mu_S^{\circ II} + RT \ln x_S^{II} - \mu_S^{\circ I} - RT \ln x_S^{I} = 0 \tag{5.20}$$

This reduces to

$$\mu_S^{\circ II} - \mu_S^{\circ I} + RT \ln x_S^{II} = 0 \tag{5.21}$$

since $x_S^{I} = 1$ because there is no solute in compartment I. The μ_S° depend only on pressure. In order to find this dependence we can apply the Gibbs equation and the defining differential of the Gibbs free energy to the solvent (see Appendix II). This yields

$$d\mu_S = v_S \, dP \tag{5.22}$$

in which v_S is the molar volume of the solvent. Integrating Eq. (5.22) between the appropriate limits,

$$\int_{\mu_S^{\circ I}}^{\mu_S^{\circ II}} d\mu_S^{\circ} = \int_{P^I}^{P^{II}} v_S \, dP$$

yields

$$\mu_S^{\circ II} - \mu_S^{\circ I} = v_S(P^{II} - P^{I}) \tag{5.23}$$

if the solvent is assumed to be incompressible, which in our case is a reasonable assumption. Substituting Eq. (5.21) into Eq. (5.23), and rearranging the terms, gives

$$P^{II} - P^{I} = \pi = -\frac{RT}{v_S} \ln x_S^{II} \tag{5.24}$$

defining $P^{II} - P^{I} = \pi$ as the *osmotic pressure*. We can express the osmotic pressure in terms of the solute concentration by putting $x_S = 1 - X_A$ in which X_A is the total mole fraction of solute. For dilute solutions X_A is small with respect to x_S and we can make the approximation

$$X_A = \frac{n_A}{n_S + n_A} \approx \frac{n_A}{n_S} \tag{5.25}$$

Since $n_S v_S$ is the solvent volume V we have

$$\frac{n_A}{n_S} = v_S C_A \tag{5.26}$$

in which C_A is the concentration of the solute. Furthermore, the logarithm of Eq. (5.24) can be expanded to yield

$$\ln(1 - X_A) = -X_A - \frac{X_A^{\ 2}}{2} - \frac{X_A^{\ 3}}{3} - \cdots \tag{5.27}$$

in which, again for dilute solutions, the higher-order terms can be neglected. Using these approximations, Eq. (5.24) can be reduced to

$$\pi = RTC_A \tag{5.28}$$

which is the well-known van't Hoff equation.

Ionic Equilibrium. The osmotic pressure can thus, be seen as raising the chemical potential of the solvent in the solution to that of pure solvent as required. A similar situation exists when we have a charged solute. Suppose that the membrane separates two compartments I and II with different concentrations of an electrolyte C^+A^- and that it is permeable only to ions of one sign, for instance, the cations C^+. At equilibrium, the change in the free energy when a mole of these cations pass from compartment I to compartment II is zero:

$$\Delta G = \tilde{\mu}_C^{II} - \tilde{\mu}_C^{I} = 0 \tag{5.29}$$

in which the $\tilde{\mu}_C$ are the *electrochemical* potentials, consisting of a chemical component μ and an electrical component $z\mathscr{F}\psi$ (with z as the valence of the ionic species, $\mathscr{F}$ as the Faraday constant, and ψ as the electrical potential). Substituting the appropriate expressions for the electrochemical potentials in (5.29) we obtain

$$\mu_C^\circ + RT \ln c_C^{I} + z_C \mathscr{F} \psi^{I} = \mu_C^\circ + RT \ln c_C^{II} + z_C \mathscr{F} \psi^{II} \tag{5.30}$$

Solving for the electrical potential difference gives

$$\psi^{II} - \psi^{I} = \frac{RT}{z_C \mathscr{F}} \ln \frac{c_C^{I}}{c_C^{II}} \tag{5.31}$$

The electrical potential difference is, thus, proportional to the logarithm of the ratio of the two concentrations. For cations z is positive and the electrical potential is higher on the more dilute side of the membrane. Equilibrium is attained because the buildup of electrical potential on the dilute side of the membrane raises the electrochemical potential of the solution in the more diluted compartment to that of the more concentrated solution in the other compartment. If the membranes were only permeable to anions, we would have had the reverse situation. It is to be noted that in each compartment (I and II) the law of electrical neutrality is still valid, because the charge difference (or charge displacement) cannot be detected; it is only manifest as an electrical potential difference. Such a potential difference is often referred to as a *diffusion potential* (because it results from an apparent diffusion of ions of one sign through the membrane). Equation (5.31) often is called the *Nernst equation*. If an external electrical field is applied to a membrane which is permeable to ions of one sign only, and the membrane is separating two compartments each containing solutions of the ion, the concentrations of the ion at equilibrium are given by the Nernst equation.

Donnan Equilibrium. A particular case of ionic equilibrium across membranes, such as we have just described, is the *Donnan equilibrium.* In this case an electrical potential exists at equilibrium even when the membrane is permeable to (relatively small) ions of both signs. This happens when one of the two compartments separated by the membrane contains, in addition to a salt to which the membrane is permeable, a large molecule (a protein, for example) bearing a net charge, to which the membrane is not permeable. (Fig. 5.27). Suppose that compartment I contains a solution of a simple univalent electrolyte C^+A^- and that compartment II contains a solution of the same electrolyte together with a protein salt P, at a concentration c_P bearing a net charge z_P. The membrane separating the two compartments is permeable to both ions of the electrolyte. Suppose, furthermore, that an appropriate osmotic pressure exists such that the chemical potential of the solvent is equal in both compartments. At equilibrium the free energy change when a mole of the simple electrolyte is

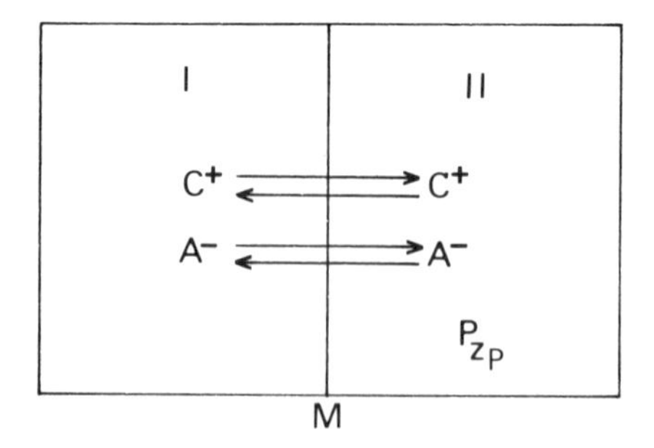

Fig. 5.27 Two compartments I and II separated by a membrane permeable to the small ions C^+ and A^- but not to a large charged molecule P_{z_P}.

transferred from one compartment to the other is zero. This means that

$$\mu^{\mathrm{I}}_{\mathrm{C^+A^-}} = \mu^{\mathrm{II}}_{\mathrm{C^+A^-}} \tag{5.32}$$

or, using the appropriate expressions for the potentials,

$$\mu^{\circ\mathrm{I}}_{\mathrm{C^+A^-}} + RT \ln c^{\mathrm{I}}_{\mathrm{C^+}} c^{\mathrm{I}}_{\mathrm{A^-}} = \mu^{\circ\mathrm{II}}_{\mathrm{C^+A^-}} + RT \ln c^{\mathrm{II}}_{\mathrm{C^+}} c^{\mathrm{II}}_{\mathrm{A^-}} \tag{5.33}$$

If we ignore the small effects of the pressure difference on the standard potentials of the salt in the two compartments then

$$\mu^{\circ\mathrm{I}}_{\mathrm{C^+A^-}} = \mu^{\circ\mathrm{II}}_{\mathrm{C^+A^-}} \tag{5.34}$$

and it follows that

$$c^{\mathrm{I}}_{\mathrm{C^+}} c^{\mathrm{I}}_{\mathrm{A^-}} = c^{\mathrm{II}}_{\mathrm{C^+}} c^{\mathrm{II}}_{\mathrm{A^-}}$$

or

$$c^{\mathrm{I}}_{\mathrm{C^+}}/c^{\mathrm{II}}_{\mathrm{C^+}} = c^{\mathrm{II}}_{\mathrm{A^-}}/c^{\mathrm{I}}_{\mathrm{A^-}} = r \tag{5.35}$$

The ratio r is called the *Donnan ratio*.

The law of electrical neutrality dictates that, in compartment I,

$$c^{\mathrm{I}}_{\mathrm{C^+}} = c^{\mathrm{I}}_{\mathrm{A^-}} \tag{5.36}$$

and in compartment II,

$$c^{\mathrm{II}}_{\mathrm{C^+}} = c^{\mathrm{II}}_{\mathrm{A^-}} - z_{\mathrm{P}} c_{\mathrm{P}} \tag{5.37}$$

From (5.35), (5.36), and (5.37) it follows that

$$r^2 = c^{\mathrm{II}}_{\mathrm{A^-}}/c^{\mathrm{II}}_{\mathrm{C^+}} = c^{\mathrm{II}}_{\mathrm{A^-}}/(c^{\mathrm{II}}_{\mathrm{A^-}} - z_{\mathrm{P}} c_{\mathrm{P}}) \tag{5.38}$$

Equation (5.38) shows us that if the net charge on the protein is negative ($z_{\mathrm{P}} < 0$), $r^2 < 1$ and, hence, $r < 1$. Consequently, according to (5.35), $c^{\mathrm{II}}_{\mathrm{A^-}} < c^{\mathrm{I}}_{\mathrm{A^-}}$ and $c^{\mathrm{I}}_{\mathrm{C^+}} < c^{\mathrm{II}}_{\mathrm{C^+}}$. The Nernst equation, Eq. (5.31), tells us then that there must be a negative electrical potential,

$$\psi^{\mathrm{II}} - \psi^{\mathrm{I}} = \frac{RT}{\mathscr{F}} \ln \frac{c^{\mathrm{II}}_{\mathrm{A^-}}}{c^{\mathrm{I}}_{\mathrm{A^-}}} = \frac{RT}{\mathscr{F}} \ln \frac{c^{\mathrm{I}}_{\mathrm{C^+}}}{c^{\mathrm{II}}_{\mathrm{C^+}}} \tag{5.39}$$

across the membrane. According to (5.35) this potential is proportional to the logarithm of the Donnan ratio:

$$\psi^{\mathrm{II}} - \psi^{\mathrm{I}} = \frac{RT}{\mathscr{F}} \ln r. \tag{5.40}$$

When the net charge on the protein is positive ($z_{\mathrm{P}} > 0$), the electrical potential evidently is positive.

The above derivation was made for a simple univalent salt. It is relatively easy to show, however, that for polyvalent electrolytes one can define a Donnan ratio for each salt k

$$r = (c_k^{\mathrm{I}}/c_k^{\mathrm{II}})^{1/z_k} \tag{5.41}$$

in which the subscript k refers to the kth ion species with a charge z_k. Equation (5.40) is, thus, applicable for the general case.

Flow across Membranes. So far only equilibrium situations in which no *net* flow of matter occurs have been discussed. The equilibrium, of course, can be a *dynamic* one, describing a steady state in which the flow in one direction equals the flow in the opposite direction. Situations in which there is a net flow of matter can be described by transport equations of the kind

$$\mathbf{J} = -L \operatorname{grad} \mu \tag{5.42}$$

or, in the one-dimensional case,

$$J = -L\, d\mu/dx \tag{5.43}$$

in which J is the *flow*, or the amount of matter passing a unit area in a unit time, L is a coefficient related to the mobility of matter through the medium (a phenomenological coefficient), and μ is a potential function. We can apply this to simple diffusion of a solute in a solution in one dimension. If v_i is the average velocity of a mole of solute i, the flow, in moles per unit area per unit area per unit time, is

$$J_i = c_i v_i \tag{5.44}$$

in which c_i is the concentration of solute i in moles per unit volume. The velocity v_i is proportional to a force F, which is equal and opposite to the driving force:

$$v_i = \omega_i F \tag{5.45}$$

in which ω_i is the mobility coefficient (the inverse of the friction coefficient). For simple diffusion the driving force is the gradient of the chemical potential μ_i. The transport equation (5.43) then becomes

$$J_i = -\omega_i c_i (d\mu_i/dx) \tag{5.46}$$

Taking the appropriate expression for the chemical potential and differentiating gives

$$\frac{d\mu_i}{dx} = \frac{d(\mu_i^\circ + RT \ln c_i)}{dx} = \frac{RT}{c_i}\frac{dc_i}{dx} \tag{5.47}$$

which, substituted in (5.46) yields Fick's law of diffusion

$$J_i = -D_i \frac{dc_i}{dx} \tag{5.48}$$

where the diffusion coefficient D_i is defined as

$$D_i = \omega_i RT \tag{5.49}$$

If the diffusing species is an electrolyte the gradient of the *electrochemical* potential, containing an electrical potential term, is the driving force. In this case, when the mobilities of the ions are different from each other, an electrical potential gradient is associated with the concentration gradient.

The Finite Thickness of Membranes. In biological systems a great many of the transport processes operate by diffusion through membranes. Up to now we have considered membranes as infinitely thin barriers between cell compartments. Membranes, however, have a finite thickness and this is one of the difficulties in applying the diffusion principles, as described above, to these structures; there is no way to experimentally obtain values of the concentration gradient within the membrane phase. The solution phases bordering the membrane on each side are accessible to experimental determinations and, therefore, approximations to membrane transport processes should refer to concentration values in these phases. Figure 5.28 indicates two solution phases I and II separated by a membrane phase M, with thickness Δx. Let the concentrations of a solute i be c_i^{I} in phase I, c_i^{II} in phase II, and c_i^{m} in phase M. According to Eq. (5.46), diffusion of the solute i through the membrane phase is given by

$$J_i = -\omega_i c_i^{\mathrm{m}} (d\mu_i^{\mathrm{m}}/dx) \tag{5.50}$$

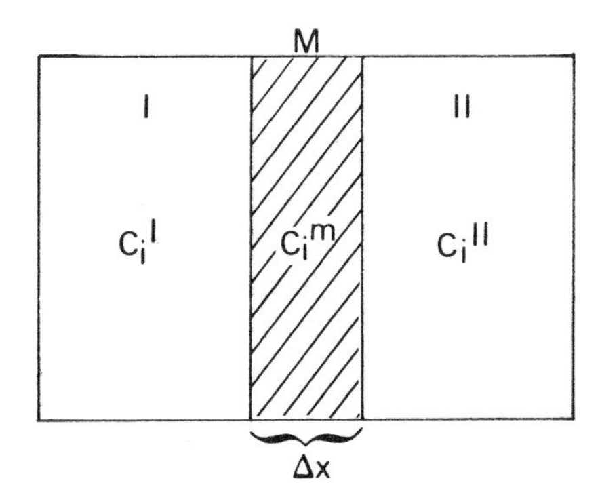

Fig. 5.28 Two compartments I and II separated by a membrane of finite thickness Δx.

Since the concentration gradient within the membrane phase cannot be determined and, therefore, neither can the potential gradient, an approximation in which the concentration in the membrane phase relates in a known way to the concentrations in the solution phases must be made. We can do this by using differences instead of differentials and approximating the potential gradient as

$$\frac{d\mu_i^{\mathrm{m}}}{dx} \approx \frac{\Delta\mu_i}{\Delta x} = \frac{\Delta(\mu_i^\circ + RT \ln c_i)}{\Delta x} \tag{5.51}$$

in which c_i relates to the concentrations in the solution phases. For small values of Δx

$$\frac{\Delta(\mu_i^\circ + RT \ln c_i)}{\Delta x} = \frac{RT}{c_i}\frac{\Delta c_i}{\Delta x} \tag{5.52}$$

With this approximation we can write

$$J_i = -\frac{\omega_i^{\mathrm{m}} c_i^{\mathrm{m}} RT}{c_i}\frac{\Delta c_i}{\Delta x} \tag{5.53}$$

for the flow. Of course, in general, c_i is not equal to c_i^{m}. We can assume, however, that at the boundaries of the membrane there exists an equilibrium concentration distribution. Thus,

$$\mu_i^{\mathrm{sol}} = \mu_i^{\mathrm{m}} \tag{5.54}$$

or

$$\mu_i^{\circ\,\mathrm{sol}} + RT \ln c_i^{\mathrm{sol}} = \mu_i^{\circ\mathrm{m}} + RT \ln c_i^{\mathrm{m}} \tag{5.55}$$

and, after rearranging,

$$\mu_i^{\circ\,\mathrm{sol}} - \mu_i^{\circ\mathrm{m}} = RT \ln (c_i^{\mathrm{m}}/c_i^{\mathrm{sol}}) \tag{5.56}$$

Since, at constant temperature, the left-hand side of Eq. (5.56) is a constant, the right-hand side must likewise be a constant. Hence,

$$c_i^{\mathrm{m}} = k_i c_i^{\mathrm{sol}} \tag{5.57}$$

in which k_i is a constant at a given temperature for a given solute. Using this in Eq. (5.53) we obtain

$$J_i = -\omega_i k_i RT(\Delta c_i/\Delta x) \tag{5.58}$$

in which we can put

$$\Delta c_i = c_i^{\mathrm{II}} - c_i^{\mathrm{I}}$$

Permeability. Equation (5.58) thus indicates that the flow of a solute through a *thin* membrane is proportional to the concentration difference across the membrane, divided by the thickness. The membrane thickness, indeed, is very small (in the order of 100 Å) and in many cases it cannot be given a precise value or meaning. Therefore, it is convenient to lump it in the proportionality coefficient and define a *permeability coefficient*

$$p_i^{\mathrm{m}} = \frac{\omega_i k_i RT}{\Delta x} \tag{5.59}$$

for a given solute, a given temperature, and a given membrane. The flow then simply becomes

$$J_i^{\mathrm{m}} = -p_i^{\mathrm{m}} \Delta c_i \tag{5.60}$$

Diffusion of ionic species through the membrane phase can be treated in the same way. The driving force in this case is the *electrochemical* potential gradient which contains an electrical component $z\mathscr{F}(d\psi/dx)$ as well. Making the same approximations as for the neutral solute, the flow of ion k through the membrane phase can be given by

$$J_k = -\omega_k c_k^{\mathrm{m}}[(RT\,\Delta c_k/c_k\,\Delta x) + z_k\mathscr{F}(\Delta\psi/\Delta x)] \tag{5.61}$$

Again assuming an equilibrium concentration distribution at the membrane boundaries,

$$J_k = -\omega_k k_k RT[(\Delta c_k/\Delta x) + (z_k\mathscr{F}/RT)c_k(\Delta\psi/\Delta x)] \tag{5.62}$$

and, using the same definition for the permeability coefficient of ion k:

$$J_k = -p_k^{\mathrm{m}}[\Delta c_k + (z_k\mathscr{F}/RT)c_k\,\Delta\psi] \tag{5.63}$$

Restrictions set by electrical neutrality prevent an electrical current from flowing across the membrane when there is no external electrical connection. If the membrane is permeable only to ion k, Eq. (5.63) can be solved for $J_k = 0$:

$$\Delta\psi = -\frac{RT}{z_k\mathscr{F}}\frac{\Delta c_k}{c_k} \tag{5.64}$$

or, after integration,

$$\psi^{\mathrm{II}} - \psi^{\mathrm{I}} = -\frac{RT}{z_k\mathscr{F}}\ln\frac{c_k^{\mathrm{II}}}{c_k^{\mathrm{I}}} \tag{5.65}$$

a result identical to that expressed in the Nernst equation (5.31) for ionic equilibrium. If the membrane were permeable to a variety of ions, with charge z_k, and each with its own permeability coefficient, p_k^{m}, electrical neutrality requires that

$$\sum_{k=1} z_k J_k = -\sum_{k=1} z_k p_k^{\mathrm{m}} \Delta c_k - \frac{\mathscr{F}}{RT} \Delta\psi \sum_{k=1} z_k^2 p_k^{\mathrm{m}} c_k = 0 \qquad (5.66)$$

Solving for $\Delta\psi$, we obtain

$$\Delta\psi = \frac{RT}{\mathscr{F}} \frac{\sum_{k=1} z_k p_k^{\mathrm{m}} \Delta c_k}{\sum_{k=1} z_k^2 p_k^{\mathrm{m}} c_k} \qquad (5.67)$$

When all ions, each with its own permeability coefficient, are taken into account this equation poses a formidable mathematical problem. Simple solutions in terms of the potential difference ($\psi^{\mathrm{II}} - \psi^{\mathrm{I}}$) are feasible, however, because in actual biological systems the permeability of a few ions largely predominate over the permeability of other ions in the solution phases. Thus, the summation needs to be carried out only over those few ions whose permeabilities are significant in the particular membrane. The membrane potential of the resting nerve, for instance, is well approximated by Eq. (5.65) applied to K^+, indicating that the permeability coefficient of this ion dominates over that of any other ion. In the conducting stage of the nerve the permeabilities of K^+ and Na^+ change dramatically, thus providing for a potential spike which is transmitted along the nerve fiber (see Section 5.7).

The permeation of a constituent, be it a neutral solute or an ion, is determined by a permeability coefficient $\mathrm{p}_i^{\mathrm{m}}$ which contains an intrinsic mobility factor ω_i^{m} of the constituent in the membrane phase and a partition coefficient $\mathrm{k}_i^{\mathrm{m}}$. The latter factor can be seen as the relative solubility of the constituent in the membrane phase with respect to its solubility in the bordering solution phases. Both factors depend upon the constituent itself as well as the membrane, and in particular upon the mechanism, or mechanisms, by which the constituent diffuses through the membrane. Our knowledge of membrane structure is still too sketchy to be able to describe precise mechanisms. It seems probable, however, that the actual process involves the simultaneous or consecutive action of a number of mechanisms.

Transport Mechanisms. If diffusion of a molecular species across the boundary between the solution phase and the membrane phase plays an important role, we would expect (in a medium as viscous as the lipid core of a membrane) a low mobility coefficient. For certain molecules this could be offset by a high value for the partition coefficient k_i^{m}. Indeed, there is a good correlation between the lipid solubility of a number of molecules and their penetration rate across cellular membranes (see Fig. 5.29). On this basis,

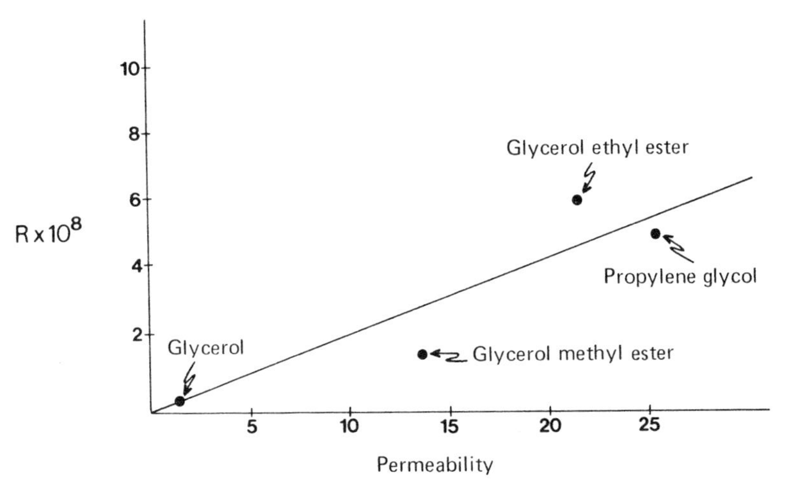

Fig. 5.29 A plot of the solubility R in oil of a number of substances and their permeability in cells of the alga *Chara*, showing the correlation between permeability and lipid solubility (data from Harris, 1960).

however, one would expect hydrophylic substances, such as ions and amino acids, to be poorly permeable, whereas in many cases they are rapidly penetrating. Moreover, the high degree of selectivity of membranes with respect to different solutes cannot be accounted for by differences in lipid solubility alone.

In another proposed mechanism the membrane is seen as more loosely structured such as to provide a number of "holes" or pores; through such channels solvent (water) may pass and one can readily visualize the diffusion of solutes which are small enough to "fit." Selectivity with respect to the sign of ions could be accounted for by assuming that the channels are structured with ionic species of a given sign; the channels, thus, could act as ion exchangers.

Several additional transport mechanisms have been suggested to account for the specificity of membranes. One such suggestion was induced by studies of the transport of disaccharide by bacterial membranes. These studies have led to the idea that the transport of disaccharide is effected by an enzyme called *permase* (cf. Section 5.4). This enzyme seems to be an *induced* enzyme; it is synthesized in response to the presence of its substrate. An interaction between a molecule or an ion to be transported and a substance within the membrane phase may be a more general phenomenon than the specific term permease would suggest; "carrier mechanisms" may play an important role in biological transport phenomena.

Chemical Association. The specificity of enzymatic reactions is due to close noncovalent interactions between the enzyme and the substrate which lead to a substrate–enzyme complex. Thus, in biological transport systems, one could postulate a substance within the membrane phase which has a high affinity for the species to be transported. The interaction leads to an association and the resulting complex can diffuse through the membrane phase. In Fig. 5.30 such a carrier mechanism is schematically diagrammed. A substance A is bound by a carrier C, forming a complex A · C. A by itself is poorly soluble in the membrane phase so that there is very little free A present; the complex A · C can readily diffuse through the membrane. We assume further, for reasons of simplicity, that the complex is electrically neutral. If the reaction between A and C is rapid, with respect to the diffusion rate of the complex, the reaction will proceed close to equilibrium. Thus, we have the relation

$$K[\mathrm{A}\cdot\mathrm{C}] = [\mathrm{A}][\mathrm{C}] = [\mathrm{A}]([\mathrm{C}]_0 - [\mathrm{A}\cdot\mathrm{C}]) \tag{5.68}$$

in which K is the equilibrium constant and $[\mathrm{C}]_0$ is the total concentration (bound + free) of the carrier. From (5.68) it follows that

$$[\mathrm{A}\cdot\mathrm{C}] = \frac{[\mathrm{A}][\mathrm{C}]_0}{[\mathrm{A}] + K} \tag{5.69}$$

To calculate the flow of the complex through the membrane phase we use Eq. (5.60) which, applied to our case, yields

$$J_{\mathrm{A\cdot C}} = -p^{\mathrm{m}}_{\mathrm{A\cdot C}}\Delta[\mathrm{A}\cdot\mathrm{C}] = -p^{\mathrm{m}}_{\mathrm{A\cdot C}}([\mathrm{A}\cdot\mathrm{C}]^{\mathrm{II}} - [\mathrm{A}\cdot\mathrm{C}]^{\mathrm{I}}) \tag{5.70}$$

If we ignore the back flow (which we could do, for instance, in an experiment in which we follow the transport of a radioactive isotope of substance A during a short time after the beginning of the experiment) we can make $[\mathrm{A}\cdot\mathrm{C}]^{\mathrm{II}} = 0$. Substituting (5.69) into (5.70) we obtain

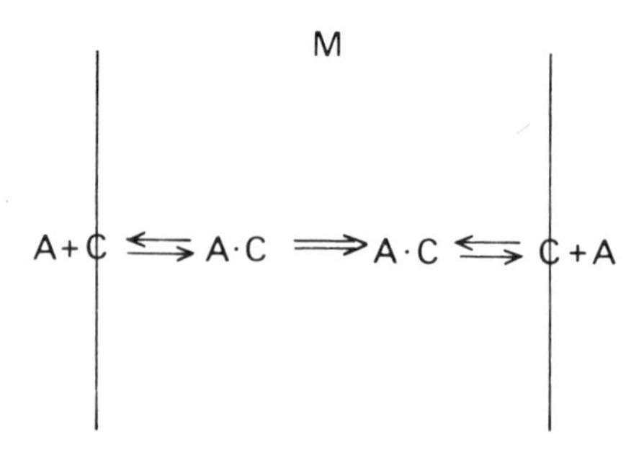

Fig. 5.30 Transport through a membrane M of a substance A by chemical association with a carrier C.

$$J_{A \cdot C} = p^{m}_{A \cdot C} \frac{[C]_0[A]^{I}}{[A]^{I} + K} \tag{5.71}$$

From (5.71) we can see that at a high concentration of A in the solution phase I the flow becomes a constant. The flow, thus, becomes saturated, just like the rate of an enzymatic reaction becomes saturated at a high concentration of the substrate. A plot of the flow as a function of the concentration of the transported substance is given in Fig. 5.31. From (5.71) it also follows that the flow is at half of its saturation value when $[A]^{I} = K$. This provides a means to determine, experimentally, the value of K.

Saturations of the rate of transport of substances through membranes have, indeed, been measured, especially in cases where there is a high degree of selectivity toward these substances. The carrier mechanism, as described above in a simplified way, offers an explanation for the selectivity which would be difficult to imagine without some form of chemical association.

Ionophores. Some antibiotics cause changes in the permeability of specific ions. One of them, the compound *valinomycin*, shows a high specificity toward K^+ and, to a lesser extent, toward Rb^+. Apparently, the antibiotic forms a complex with the ion and carries it across the membrane which is, otherwise, quite impermeable for univalent cations. Figure 5.32 shows the structure of valinomycin. Its cyclic structure enables the compound to accept the ion in its inner core, which is hydrophylic, and surround it with an outer shell, which is hydrophobic, thus permitting the ion to pass rapidly through the membrane.

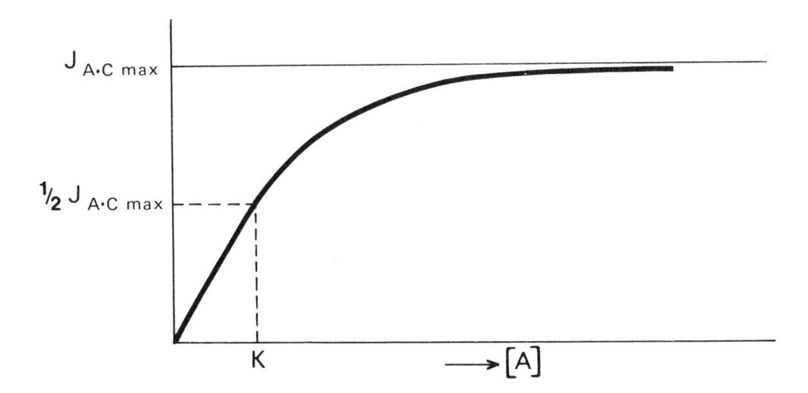

Fig. 5.31 A plot of the flow of a complex of a substance A and a carrier C through a membrane as a function of the concentration of A at one side of the membrane. At a concentration of A equal to the equilibrium constant K of the complex formation reaction, the flow is at half the saturation value.

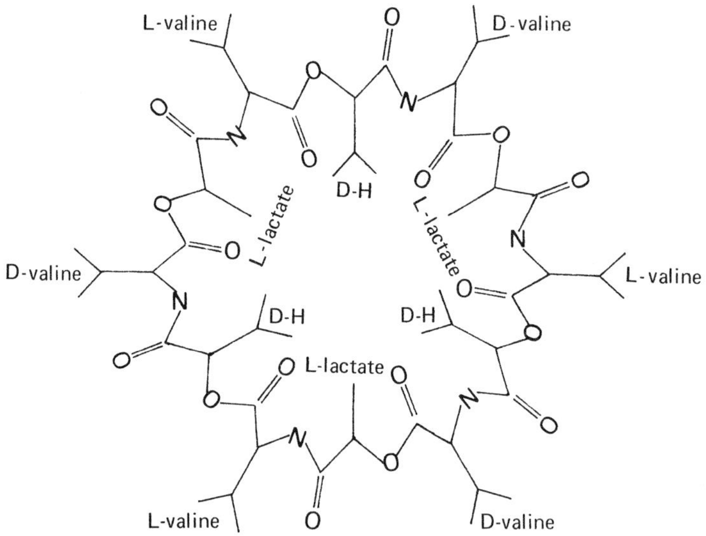

Fig. 5.32 The chemical structure of valinomycin; D-H is D-hydroxyisovalerate.

Another ion-conducting antibiotic is *nigericin*. This compound catalyzes an exchange between K^+ and H^+, so that it does not cause a change in the membrane potential. Other ion conducting compounds, known as ionophores, have been shown to affect the penetration of cations across mitochondrial, chloroplast, and cytoplasmic membranes as well as across artificial membranes. Of course, the action of such compounds does not in itself imply anything about the actual transport mechanisms in biological membranes. They do make univalent cations more soluble in the membrane phase, usually by wrapping them in a hydrophobic shell. Their specificity has to do with a more or less exact fit of the ions in their inner core. In Table 5.3 a few of these compounds are given with their different specificities for cations.

TABLE 5.3 Some Ionophoric Compounds

Compound	Ion specificity
Gramacidin A	H^+, Na^+, Li^+, K^+, Rb^+, Cs^+
Valinomycin	K^+, Rb^+
Nigericin	K^+–H^+ exchange
Dinitrophenol	H^+
Carbonyl cyanide *m*-Chlorophenylhydrazone	H^+
Fluorocarbonyl cyanide phenylhydrazone	H^+

Active Transport. Many membranes translocate molecules or ions from regions of low concentration to those of high concentration. This transport against the thermodynamic gradient can be sustained only when it is coupled with an energy-supplying process. Neither simple diffusion through the membrane phase or through pores, nor the carrier mechanism *per se* can explain this translocation by *active transport*. It is not too difficult, however, to extend the concept of transport by chemical association in order to postulate a mechanism by which it could work. In Fig. 5.33 we have a schematic representation of a carrier mechanism similar to that of Fig. 5.30. The carrier substance C in this case, however, can be converted from a configuration of high affinity for substance A to one of low affinity and vice versa. This conversion can be a chemical alteration or just a conformational change. It is essential that the conversion in one direction be coupled to an energy-yielding reaction. The complex A · C as well as the configuration of low affinity, C′, can diffuse through the membrane. The process can be visualized as follows. At surface I the configuration of high affinity, C, is formed from the configuration of low affinity, C′. Substance A then binds to C, which is continually supplied by conversion from C′. The complex A · C diffuses from I to II, where dissociation into A and C occurs. This dissociation is favored by the fact that the concentration of C is kept low because of its conversion, at surface II, to C′. C′ diffuses back to surface I because its concentration at surface II is higher as a result of the continuous conversion of C into C′ at surface II. The process, thus, has a cyclic character and it is evident that it cannot proceed unless it is driven by some energy-supplying

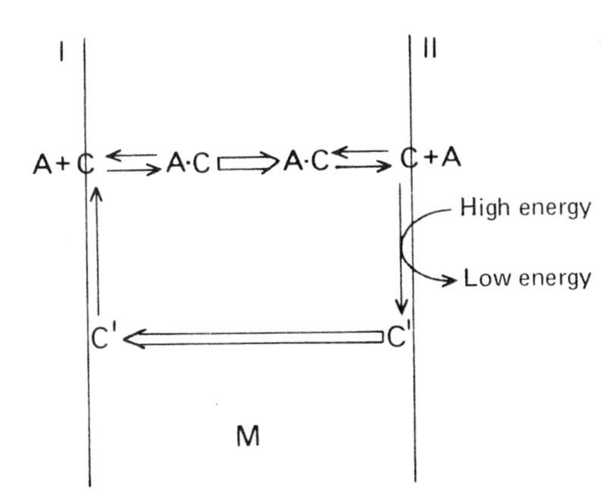

Fig. 5.33 A model for active transport by chemical association. A substance A to be transported binds to a carrier C when this carrier is in a form of high affinity for A. At each side of the membrane the carrier is converted from the high affinity form C to the low affinity form C′ and *vice versa*. Both the complex A · C and the low affinity form C′ can diffuse through the membrane, indicated by the fat arrows. The cycle is coupled to an energy-yielding reaction.

reaction. Such a reaction can drive either the conversion of C into C′ at surface II or the conversion of C′ into C at surface I.

The mechanism described above is one that could work for active transport. It should be emphasized, however, that mechanisms of this kind have not been proven to occur in actuality. At best, there is some circumstantial evidence which suggests that carrier mechanisms may be operative. Indications in this direction are observations of saturation curves for the net flow of ions across a number of membranes which do exhibit active transport (see Fig. 5.34). Moreover, as we have stated before, it seems that the total observed transport across biological membranes is not the result of a single mechanism but is, in general, the result of several mechanisms occurring simultaneously.

The dependence upon energy-yielding metabolic processes in the cell is a clear characteristic of active transport. Inhibition of such metabolic reactions, therefore, should slow down the transport. This, indeed can be observed; if red blood cells, which are actively transporting K^+ and Na^+ against their concentration gradients, are incubated at low temperatures, thus slowing down their metabolic reactions, active transport stops and the concentrations of the ions inside and outside become equal after a while.

The energy for active transport originates from the reactions in the cell that are associated with ATP production. More specifically, the high-energy intermediates of ATP synthesis are, in fact, involved and at least in some

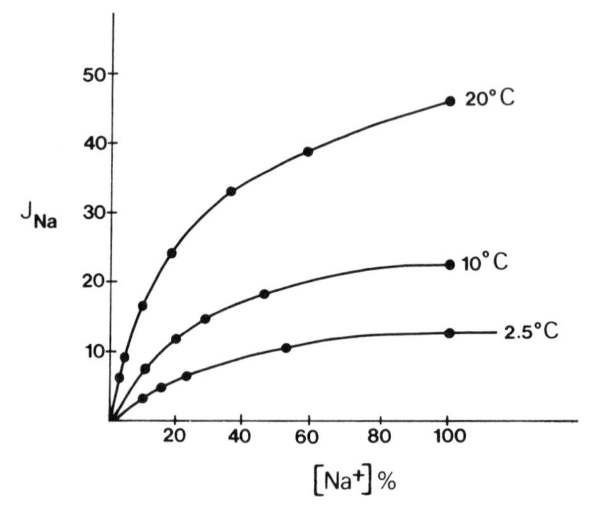

Fig. 5.34 The net flow of sodium ion J_{Na} across isolated frog skin as a function of the Na^+ concentration in the bathing medium at different temperatures. The flow diminishes at lower temperatures. From F. M. Snell *et al.*, "Biophysical Principles of Structure and Function," Addison-Wesley, Reading, Massachusetts, 1965.

instances ATP can be used to generate them. Thus, the addition of ATP to a suspension of cells or cell organelles which can actively transport solutes leads to hydrolysis of ATP in the presence of these solutes. In red blood cells, for example, this so-called *ATPase activity* can be clearly demonstrated. Red cells actively transport K^+ and Na^+ and the presence of these ions causes the hydrolysis of ATP. In fact, the uptake of K^+ and the extrusion of Na^+ are linked; it seems that the hydrolysis of one molecule of ATP yields the inward movement of two K^+ ions and the outward movement of three Na^+ ions (Fig. 5.35), and it has been shown that the enzyme responsible for this Na^+–K^+ ATPase activity is located in the cytoplasmic membrane.

5.6 Membrane-Linked Energy Transduction

The demonstration of an ion-transport-coupled ATPase system, as described in the previous section, suggests, of course, the possibility of reversibility. In other words, when active transport of ions can be induced by the *hydrolysis* of ATP by an ATPase, can transport of ions across a membrane in the direction of a thermodynamic gradient cause the *synthesis* of ATP by the same ATPase? An affirmative answer to this question has been proposed as a solution to the problem of membrane-linked synthesis of ATP, called oxidative phosphorylation for mitochondria and photophosphorylation for photosynthetic organelles.

The synthesis of ATP in the glycolytic reaction sequence (often called *substrate-linked phosphorylation*) occurs by enzymatic reactions in which common intermediates take care of the coupling of the energy-yielding

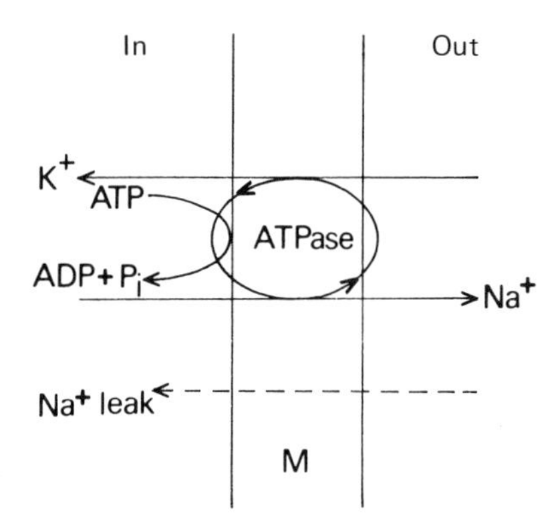

Fig. 5.35 Na^+–K^+ ATPase in a membrane M.

(oxidative) and energy-conserving (phosphorylating) reactions. These reactions take place in the cytoplasmic phase of the cell and, although some details still have to be cleared up, there are no essential difficulties in understanding the mechanism. Still no consensus exists, however, regarding the mechanism of ATP synthesis coupled to respiratory or photosynthetic electron transport. An essential aspect of this type of energy transduction is its linkage to a membrane system; no ATP synthesis occurs without an *intact* structure of the mitochondrial chloroplast or bacterial inner membrane, although electron transport may go on unimpaired in broken membrane fragments.

Chemical Hypothesis. Experimental results obtained with uncouplers (compounds which inhibit phosphorylation but leave the electron transport intact, sometimes even stimulating it) have led to the concept of the "high-energy intermediate." A hypothesis which grew up through argument by analogy from substrate-linked phosphorylation, considers this high-energy intermediate to be a chemical entity. According to this theory the coupling of the respiratory or photosynthetic oxidation–reduction reactions and the phosphorylation of ADP to ATP is accomplished as follows.

Suppose A, B, and C are consecutive components of the oxidation–reduction reaction chain. Reduction of B by AH_2 cannot proceed unless the reduced product BH_2 is complexed to an intermediate I. Thus,

$$AH_2 + B \longrightarrow A + BH_2 \tag{5.72}$$

immediately followed by

$$BH_2 + I \longrightarrow BH_2 \cdot I \tag{5.73}$$

The reduced complex, $BH_2 \cdot I$, in turn can reduce the next component in the chain in an exergonic reaction. The free energy of this reaction is then captured in a high-energy bond designated by a squiggle:

$$BH_2 \cdot I + C \longrightarrow B \sim I + CH_2 \tag{5.74}$$

thus forming the high-energy intermediate $B \sim I$. In the presence of phosphate this high-energy intermediate is phosphorylated and the high-energy phosphate can in turn phosphorylate ADP:

$$B \sim I + P_i \longrightarrow I \sim P + B \tag{5.75}$$

$$I \sim P + ADP \longrightarrow ATP + I \tag{5.76}$$

Uncouplers simply cause the hydrolysis of the high-energy intermediate $B \sim I$, thus preventing ATP from being formed while keeping the oxidation–

reduction reactions running. Inhibitors of either one or both of the phosphorylation reactions, such as the antibiotic *oligomycin*, also stop the oxidation–reduction reactions.

Although the "chemical hypothesis" outlined above seems to be compatible with any observation, it has run into difficulties. First, no one as yet has been able to demonstrate experimentally the existence of, let alone isolate, the high-energy intermediate. This by itself is no argument for rejecting the theory; the complex may be extremely unstable. More serious, however, is the fact that the theory does not account at all for the linkage to membrane structures.

Conformational Hypothesis. A modification of the chemical hypothesis is the conformational hypothesis. In this hypothesis the high-energy intermediate is a special conformation of a factor I. This theory is represented by replacement of Eqs. (5.73)–(5.76) by

$$BH_2 + I + C \longrightarrow B + I^* + CH_2 \qquad (5.77)$$

$$I^* + ADP + P_i \longrightarrow ATP + I \qquad (5.78)$$

in which I* designates the high-energy conformation of I. A proteinous fraction without which no phosphorylation coupled to electron transport appears to be possible, may play the role of I. This factor is the already-mentioned *coupling factor* (Section 5.4) which is connected to the membrane like tiny mushrooms (Fig. 5.22). Coupling factors have been detected in, and isolated from, mitochondria as well as chloroplasts and bacteria. It is possible to remove the coupling factor from the membrane, thus inhibiting the formation of ATP. Electron transport can still go on in such depleted membranes. When the purified coupling factor is returned to the depleted membrane system, phosphorylation is restored.

Evidence for an electron transport induced conformational change came from studies in which it was shown that an inhibitor of the coupled phosphorylation binds to the coupling factor only when it is "energized" by electron transport. Thus, in chloroplasts, the inhibitor *N*-ethylmaleimide inhibits photophosphorylation only when it is added to the *illuminated* particles. Furthermore, using a radioactive tracer, such as the hydrogen isotope tritium (^3H), one could demonstrate that the binding of the inhibitor to the coupling factor was markedly enhanced by light.

Although the conformational theory seems to cope with the nondetectability of a "chemical" high-energy intermediate, it still lacks sufficient detail to be satisfactory; there is no information whatsoever about the structural relation between the electron transport components and the coupling factor, and there still is no reason for the necessary integrity of a membrane system (the membrane itself may be a necessary matrix for the components of the coupled processes).

Chemiosmotic Hypothesis. A theory which rests upon a completely different kind of concept is the chemiosmotic hypothesis. This hypothesis does ascribe a functional role to an intact membrane-bounded system. In fact, the high-energy intermediate according to the hypothesis is a thermodynamic gradient formed across the inner membrane of the organelle (or bacterium) by the apparent translocation of hydrogen ions across the inner membrane. These hydrogen ions are translocated by an electric field which is generated by the transport of electrons *from one side of the membrane to the other;* the electron transport in the membrane, thus, is *vectorial.* The mechanism by which this would work is illustrated schematically in Fig. 5.36a (for the respiratory chain) and 5.36b (for the photosynthetic chain).

The oxidation–reduction components (as shown in Fig. 5.36) are arranged within the membrane in such a way that electron carriers (such as the cytochromes and nonheme iron proteins) alternate with hydrogen carriers (such as the flavoproteins and quinones) at both sides of the membrane. In such a way the electrons transported across the membrane by the electron carriers, either through reduction by substrate (in the case of respiratory electron transport) or by the light-induced primary reaction (in the case of photosynthetic electron transport), generate an electric field which causes the translocation of protons by the hydrogen carriers either from the inside to the outside (in the case of the mitochondrial inner membrane) or from the outside to the inside (in the case of the chloroplast thylakoid). The membrane in both cases, although permeable for water, is impermeable for hydrogen ions. The translocated protons thus form a concentration gradient which, in fact, is the high-energy intermediate. An ATPase system connected to the membrane (such as the coupling factor mentioned above) can then act in reverse, and by retranslocating the protons ATP is synthesized (Fig. 5.36c). A simple way in which this can be accomplished is shown in Fig. 5.37. The presence of the H^+ gradient in some unknown fashion, pulls the hydroxyl ion portion of the water, which is produced when ATP is formed from ADP and inorganic phosphate, to one side of the membrane and pushes the protons to the other side. In this way the net movement of the H^+ ions would produce ATP.

Uncouplers, according to this theory, will provide for an additional pathway for the dissipation of the proton gradient, thus preventing the ATPase from using the gradient for ATP synthesis. Indeed, all known uncouplers

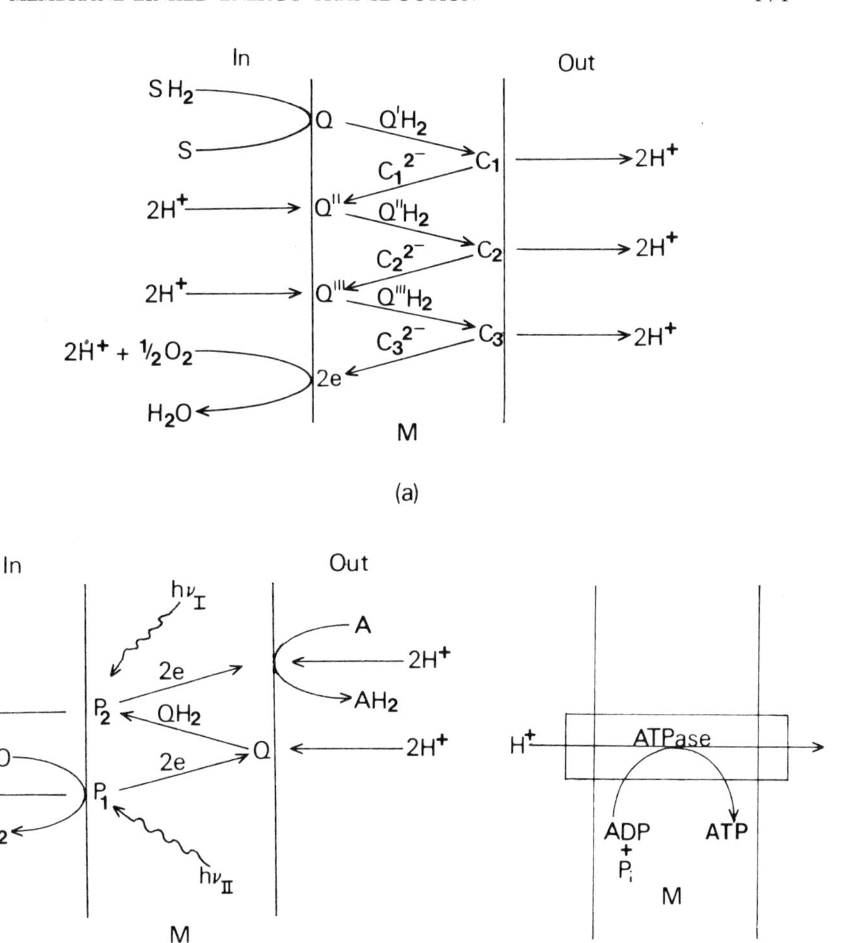

Fig. 5.36 A schematic representation of the chemiosmotic hypothesis of energy transduction. (a) In a mitochondrion; the oxidation of substrate SH_2 occurs by reduction of a hydrogen carrier Q which, in turn, reduces an electron carrier C at the other side of the membrane, releasing protons in the outer medium. The alternation of hydrogen and electron carriers at both sides of the membrane causes a net translocation of protons across the membrane and is essential for the hypothesis. (b) In a chloroplast; light causes electrons from water to flow across the membrane releasing protons in the inner medium. The reduced hydrogen carrier at the other side of the membrane picks up protons from the outer medium and donates electrons to the other light reaction. Again, the alternation of electron and hydrogen carriers at both sides of the membrane causes a net flow of protons across the membrane. (c) The reversed ATPase system in the membrane, synthesizing ATP at the expense of the collapse of the proton gradient.

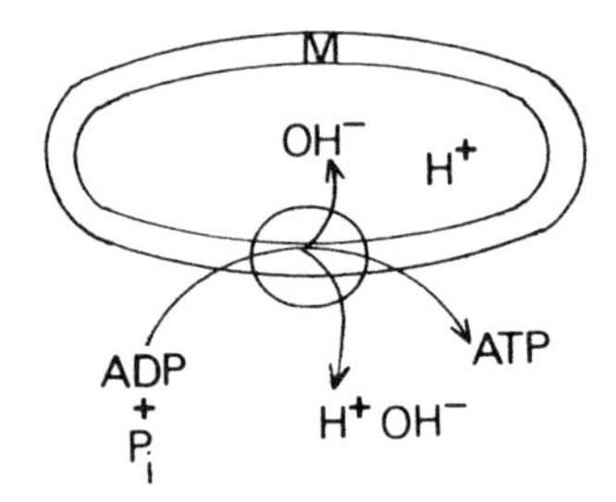

Fig. 5.37 A possible mode of operation of the reversed ATPase in the chemiosmotic hypothesis. In the reaction the components of water liberated in ATP synthesis are moved vectorially to opposite sides of the membrane by the collapsing proton gradient. This model, which assumes one proton moved per ATP synthesized, can be modified to accommodate the observed stoichiometry of two protons moved per ATP synthesized.

do have the property to "solvate" hydrogen ions in the membrane; many of them, such as dinitrophenol (DNP), carbonyl cyanide *m*-chlorophenylhydrazone (CCCP), or carbonyl cyanide *p*-trifluoromethoxyphenylhydrazone (FCCP), and many fatty acids are both weak acids and lipid soluble, and are able to carry protons across a membrane (Fig. 5.38). Thus, the protonated form of the acid can enter on one side, move across to the other side, discharge its proton, and then recross the membrane as an anion. Many of the ion-carrying antibiotics, or combinations of them, are uncouplers by virtue of their ionophoric properties.

The readiness with which the mechanism of uncoupling can be explained by the chemiosmotic theory is but one of the appealing characteristics of

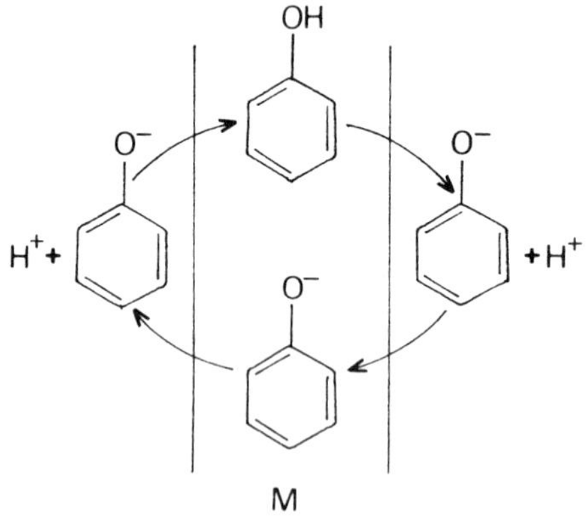

Fig. 5.38 The protonophoric action of a phenol. The phenol will carry protons in either direction as dictated by the concentration gradient. Proton transport will stop when the proton concentration is equal on both sides of the membrane.

the theory. The many predictions offered by this theory that are amenable for experimental verification have inspired a large amount of research during the 1960s and 1970s and a surprisingly high proportion of these predictions are, in fact, confirmed. Due to this success we feel that, although there are still some inconsistencies (such as the predicted stoichiometry of one proton translocated for each electron transported, which, so far, can not be experimentally verified), the probability that it provides a correct explanation for at least part of the coupling mechanism is high enough to justify a more thorough discussion.

Proton Translocation. In chloroplast membranes a reversible uptake of protons during light-induced electron transport has been detected and is now a common observation. Such a light-induced proton uptake can also be demonstrated in membrane-bounded vesicles derived from photosynthetic bacteria, although substantially lesser in extent. Mitochondria exhibit an extrusion of H^+ when supplied with respiratory chain substrates (NADH· or succinate). The transverse localization of the oxidation–reduction components in the membrane, i.e., the alternation of hydrogen and electron carriers, is a crucial requirement of the chemiosmotic hypothesis; evidence obtained, for instance, with studies using specific antibodies, confirm such an arrangement. The *sidedness* of the membranes is convincingly demonstrated by the fact that in *submitochondrial particles*, which are formed when intact mitochondria are subjected to ultrasonic vibration, the proton translocation upon the onset of respiration is toward the *inside* and not toward the outside as in the intact particles. This occurs because the disruption of the inner membrane is predominantly at the "neck" portions of the sharply folded cristae. Resealing of the fragments then results in particles which are "inside out" as compared to the intact mitochondria (Fig. 5.39). This phenomenon of reversed proton translocation after fragmentation can also be seen in photosynthetic bacteria. *Chromatophores* (membrane bounded vesicles from photosynthetic bacteria) are "inside out" as compared to whole bacteria.

Proton Motive Force. According to the chemiosmotic theory, the form in which the energy from the exergonic electron transport is primarily stabilized is as a concentration gradient of H^+. The electron transport causes the protons to be "pumped" across the membrane. This proton gradient establishes an electrochemical potential difference across the membrane which is given by

$$\Delta\tilde{\mu} = \tilde{\mu}_2 - \tilde{\mu}_1 = RT \ln([H^+]_2/[H^+]_1) + \mathscr{F}(\psi_2 - \psi_1) \qquad (5.79)$$

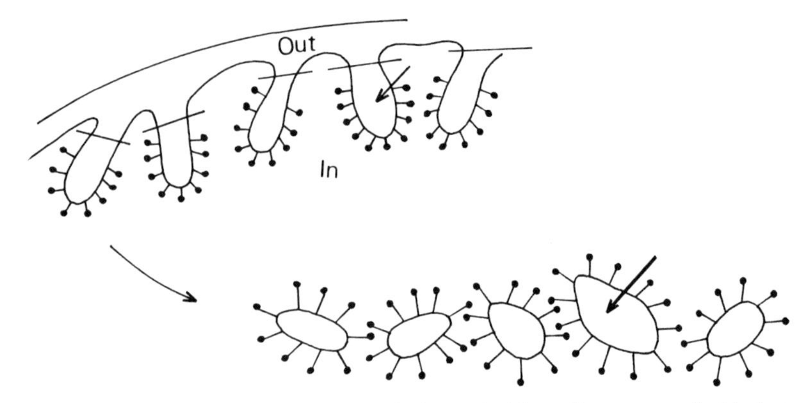

Fig. 5.39 Submitochondrial particles which are "inside out" as compared with the intact mitochondrion. Upon disruption of the organelle the inner membrane shears preferentially in the "necks" of the inward folded cristae. The coupling factor particles, which in the intact mitochondrion pointed inward, are sticking to the outside in the submitochondrial particles.

Since pH $= -\log[H^+]$ Eq. (5.79) can be written as

$$\Delta\tilde{\mu} = 2.3RT\,\Delta pH + \mathscr{F}\,\Delta\psi \tag{5.80}$$

in which $\Delta pH = pH_{(1)} - pH_{(2)}$ and $\Delta\psi = \psi_2 - \psi_1$. When the hydrogen ions move from 1 to 2 both ΔpH and $\Delta\psi$ are negative; this is the direction in which the ions are "pumped in." Thus, for the chloroplast thylakoid (1) is the outside and (2) is the inside and for the mitochondrion the reverse is true. The electrochemical potential created by the "hydrogen ion pump" or "proton pump," divided by the Faraday constant $\mathscr{F}$, is called the *proton motive force* (pmf) and is expressed in volts.

Counterion Flow. From Eq. (5.80) it can be seen that the pmf consists of a concentration term (the pH gradient) and an electrical term (a membrane potential). The latter would become excessively high with a very small number of entering protons if there were no means for compensating for the buildup electrical charge. Of course, in a phosphorylating system, with the presence of ADP and phosphate, the membrane potential is discharged by the proton movement caused by ATP synthesis which is in the opposite direction to that caused by the "proton pump." However, in the absence of phosphorylating substrates, ADP and phosphate, compensating counterion flow must occur, at least to some extent to prevent excessive rise of the membrane potential. One would, therefore, expect some cation movement in a direction opposite to that of the proton pump and/or some anion flow in the direction of the proton pump. Such counterion movements are indeed demonstrated in chloroplast preparations in the absence of ATP synthesis. In particular, a *symport* movement (movement in the same direction) of chloride ions simultaneous with proton uptake is conspicuous. *Antiport* movement (movement in the opposite direction) of K^+ and Mg^{2+} simultaneous with proton uptake has also been detected but the latter does not seem to be the important compensating counterion flow in these organelles. Antiport movement of K^+ and Na^+ seems to be the membrane potential compensating process in mitochondria.

The situation in photosynthetic bacteria is somewhat different; the bacterial membrane apparently shows a greater resistance to penetration of monovalent ions. K^+ and Na^+ have only low permeability values; the same holds for Cl^-. This is why proton uptake in membrane bounded vesicles from photosynthetic bacteria (chromatophores) is smaller than that in chloroplast thylakoids.

pH Gradient and Membrane Potential. The question arises, whether ATP synthesis by reversed action of an ATPase results from the pH gradient term of the pmf only or whether the membrane potential term can also do the job. This is far from an academic question since it is directly linked to the mechanism of the ATPase action. The effect of the ionophoric uncouplers on the membranes of different organisms can tell us something about this problem. We have seen that these uncouplers do move ions across membranes. Nigericin, for example, exchanges H^+ ions for K^+ ions. Addition of nigericin to a suspension of electron transporting membrane bounded vesicles, thus abolishes the pH gradient but leaves the membrane potential, if there is any, intact. The result of such an experiment is that, in mitochondrial preparations as well as in chloroplast thylakoids, ATP synthesis is uncoupled but in bacterial preparations it is not. Valinomycin, another ionophore, moves K^+ ions across. The effect of this antibiotic, therefore, is to destroy the membrane potential, again if there is any, leaving the pH gradient intact. Valinomycin does not uncouple ATP synthesis in any of the three types of membrane bounded vesicles unless a pH gradient destroying agent, such as nigericin, is also present. In bacterial preparations nigericin also works as an uncoupler when a *permeant anion* (such as the anion of thiocyanate, CNS^-, which easily permeates through a bacterial membrane) is present.

These results would suggest that the bacterial reversed ATPase works with a membrane potential as well. For the chloroplast or the mitochondrion, however, one cannot make such a conclusion yet; the results with the ionophoric antibiotics suggest either that there might not be a membrane potential at all or that the reversed ATPase might work only with a pH gradient.

Electrochromic Shifts. A direct measurement of the membrane potential, for instance with electrodes, is impossible for the membrane bounded vesicles we are dealing with; the largest diameter is in the order of 0.5 μm and we still cannot make electrodes small enough to get in and not destroy the membrane structure. However, there is an indirect way to do it, at least for the photosynthetic membranes. These membranes contain pigments, chlorophylls and carotenoids, which undergo spectral changes when there is an electric field across the membrane. Carotenoids especially are sensitive to these changes. In chromatophores from photosynthetic bacteria, for example, light induces a red shift of the three carotenoid absorption bands (Fig. 5.40a). This shift is due to the electrical potential created by the electron transport as well as by the membrane potential component of the proton gradient. The electrical origin of the shift can be proved by an experiment in which a solution of KCl is injected, in the dark, in a suspension of chromatophores containing valinomycin. This "pulsed" addition causes a transient change of the absorption in the carotenoid absorption spectral region (see Fig. 5.40b). After a while, diffusion of counterions will neutralize the potential and as a result the concentrations of K^+ inside and outside will become equal. Another "pulsed" addition of KCl will again cause a transient absorption change. The spectrum of the transients, as shown in Fig. 5.40c, is identical to the difference spectrum of the light-induced carotenoid absorption band shifts.

The magnitude of the transient carotenoid absorption change induced by the addition of KCl is a linear function of the logarithm of the outside concentration of K^+; this is shown in Fig. 5.41. Assuming that after equilibration the outside K^+ concentration is equal to the inside K^+ concentration, one can use this technique to calibrate the absorption band shift in terms of a membrane potential in volts by applying Eq. (5.31). Doing this, light-induced spike potentials (due to electron transport across the membrane) of 420 mV and light-induced steady-state potentials (due to the translocated H^+) as high as 240 mV have been determined in chromatophores from certain species of photosynthetic bacteria. This demonstrates that the electrical component of the pmf is substantial, at least in some species of photosynthetic bacteria.

Table 5.4 shows some results of an experiment in which the electrical component and the pH component of the pmf are measured and compared with the energy necessary to synthesize the amount of ATP actually measured in the same experiment. We can see that without additions the pmf is ample for the synthesis of ATP, with the membrane potential contributing about 56% to the pmf. With nigericin present the membrane potential by *itself* is just sufficient for the ATP synthesis; valinomycin causes a 70% inhibition of the membrane potential but the ΔpH rises to a level just high

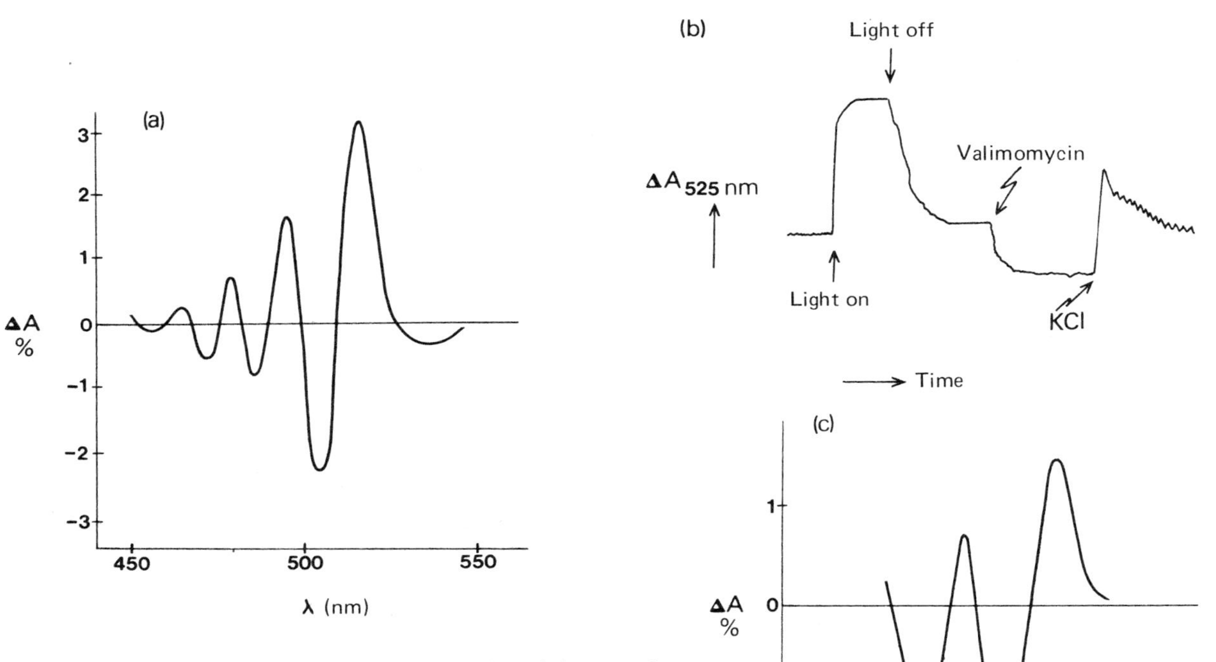

Fig. 5.40 (a) A light-minus-dark difference spectrum of a suspension of chromatophores from the purple bacterium *Rhodopseudomonas sphaeroides*. The spectrum shows the light-induced red shift of the carotenoid absorption bands. (b) A trace of the changes of the absorption of the chromatophore suspension at 525 nm. Light induces a reversible increase of the absorption; addition of valinomycin causes a small decrease, and subsequent addition of KCl causes a transient increase of the absorption. (c) The spectrum of the transient change in absorption upon addition of KCl in the presence of valinomycin to the chromatophore suspension. The spectrum shows clear similarities with the light-induced difference spectrum presented under (a) (from J. B. Jackson and A. R. Crofts, 1969).

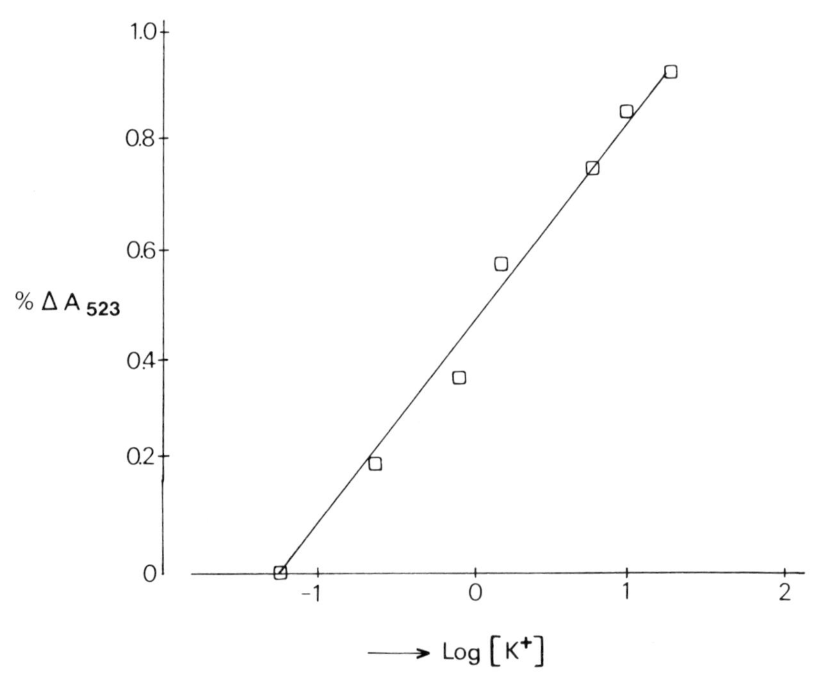

Fig. 5.41 A plot of the transient increase of the absorption of a suspension of chromatophores from *R. sphaeroides* at 523 nm as a function of the logarithm of the final outside concentration of K^+ in the presence of valinomycin. The linearity of the curve suggests the electrochromic origin of the absorption change.

TABLE 5.4 The Chemical and Electrical Components of Light-Induced Proton Motive Force (Electrochemical Potential Gradient) as the Energy Source for the Synthesis of ATP in a Photosynthetic Bacterium[a]

$\Delta\tilde{\mu}_H$ necessary to synthesize measured amount of ATP (mV)	$\Delta\psi$, measured (mV)	$(RT/\mathscr{F})\,\Delta pH$, measured (mV)	pmf, measured (mV)	Additions
325	236	183	419	None
302	272	0–36	272–308	Nigericin
285	75	208	283	Valinomycin

[a] Data from R. Casadio *et al.* (1974).

enough to make ATP synthesis possible. It is clear that, at least in these bacterial preparations, both pH and the membrane potential can drive the reverse ATPase action.

The same conclusions still cannot be made for chloroplasts and mitochondria. In chloroplasts, an absorption band shift, with a peak in the 515–530 nm spectral range, is due apparently to electric potentials across the membrane. Using the same calibration technique described as above, it can be shown that, although a substantial transient change (upon illumination with a short flash of light) can be detected, the "steady-state" membrane potential measured in continuous light is at best some 10 mV. This, undoubtedly, is due to a rapid counterion flow, probably mostly symport movement of Cl^-. Thus, in chloroplasts the electrical component of the pmf is virtually nonexistent. Hence, we cannot use the same type of experiment as that used for the bacterial chromatophores to prove that ATP synthesis in chloroplasts can be induced by a membrane potential per se. However, circumstantial evidence in this direction is suggestive. This also is true for mitochondria. As we have mentioned before, this problem has a direct bearing on the mechanism of ATP synthesis in the reverse ATPase system. The simple theory of ATP synthesis by the back movement of hydroxyl ions and protons from the water produced by the ATP synthesis from ADP and phosphate (as we have suggested before) has to be modified substantially to cope with membrane potential induced ATP synthesis. The conformational change of the coupling factor, upon energizing the membrane, may be an essential feature of the mechanism.

ATPase Activity. Some progress has been made with the discovery of the reversibility of the ATPase which, as we suggested above, is located in the mushroomlike coupling factors. The coupling factor is not only responsible for the *synthesis* of ATP, but it can also catalyze the *hydrolysis* of ATP; addition of ATP to membrane bounded vesicles containing the coupling factor energizes the membrane under certain circumstances, thereby hydrolyzing the ATP. This energization is apparent, for instance, in the induction of a proton gradient and a conformational change in the coupling factor. It is easy to trigger this forward ATPase activity in bacterial preparations without disturbing manipulations. In chloroplasts and mitochondria a subunit of the coupling factor which suppresses the forward ATPase activity, first has to be removed or inactivated.

The chemistry of the ATP synthesis itself and especially the way in which chemistry is driven by the electrochemical potential of the protons are challenging problems which are still far from being solved. At the time of this writing these problems are central in all bioenergetical research endeavors.

5.7 Action Potentials: Nerve Conduction

Excitability. A characteristic property of living organisms is *excitability*, that is the ability to react upon a stimulus from without. Changes in the immediate environment of the organism (a stimulus) evoke specific changes in the organism itself or in some parts of it (a response). A clear example is *phototaxis*, the light-triggered movement of photosynthetic microorganisms. Another example is the contraction of a muscle fiber upon an electric stimulus. In all these cases the response is rapid (in the order of milliseconds). Although an organism can also have relatively slow responses to changes in its environment (for instance the induced synthesis of enzyme in response to substrate concentration, which takes minutes or even hours), the term excitability is reserved for the fast responses. All such phenomena seem to be closely associated with the distribution of charge across membranes, particularly the cytoplasmic membrane.

Membrane Potential. When metabolism is maintained in a cell a characteristic potential difference exists across the cytoplasmic membrane. This potential difference can vary between 50 and 100 mV and is positive on the outside. By careful insertion of a microelectrode the potential difference of many types of cells can be measured and compared to the concentration distribution of permeant ions. The latter, as we have discussed in Section 5.5, must satisfy the Nernst equation (5.31). Most cytoplasmic membranes have relatively large permeabilities for K^+ and Cl^- and a much smaller permeability for Na^+. The membrane is virtually impermeable to other ions.

The origin of this membrane potential is the very distribution of the K^+ and Na^+ cations. This distribution is maintained by active transport of the cations similar in form to the ATPase activity described earlier for red blood cells. K^+ and Na^+ are actively transported in antiport, K^+ to the inside Na^+ to the outside, by one ATPase. In the absence of K^+ from the external medium no Na^+ extrusion can be detected. Since the antiport transport causes the movement of equal charges in both directions, the active transport by itself cannot result in a membrane potential. However, the passive leak of K^+ to the outside is much more rapid than the leak of Na^+ to the inside, owing to the differences in permeability of the two ions. Thus, there is a net movement of positive charge to the outside which builds up a positive potential at the outside sufficiently great to oppose further leakage of K^+. At this point there is a steady state maintained by the activity of the Na^+–K^+ ATPase and the membrane potential is given by the Nernst equation applied to K^+.

Action Potential. The membrane potential described above is called the *resting potential.* A disturbance of the membrane, in other words a stimulus (either mechanical, chemical, or electrical), upsets this balance and causes a transient change in the membrane potential, called an *action potential.* The action potential does not occur simultaneously over the entire membrane surface but is localized at the spot where the disturbance (the stimulus) took place. However, the potential, after its initiation, moves as a wave over the membrane surface away from the spot. In the case of an elongated fiber, such as the axon of a nerve cell (Fig. 5.42) the action potential passes along the fiber at a constant rate, thus constituting the nerve impulse.

Although action potentials are an essential feature of the function of nerve and muscle cells, they are by no means confined to only this type of cell. Action potentials can be evoked in virtually every type of cytoplasmic membrane; they can be considered as a means of rapid communication between different regions of the cell in order for the cell to respond as a whole to a local stimulus. However, the phenomenon is best studied in nerve and muscle cells and we shall, therefore, discuss it as it occurs in these cells.

In nerve and muscle cells action potentials are used to transmit messages. The coding of these messages is still an unsolved problem which will be discussed later. The nerve cell is called a *neuron* (Fig. 5.42) and consists of a cell body from which protrude small processes called *dendrites* and one large process called an *axon.* Many of the larger axons are "wrapped" in a fatty sheath, called the myelin sheath (see Chapter 2). The myelin sheath is periodically interrupted, forming the nodes of Ranvier. Some axons can be

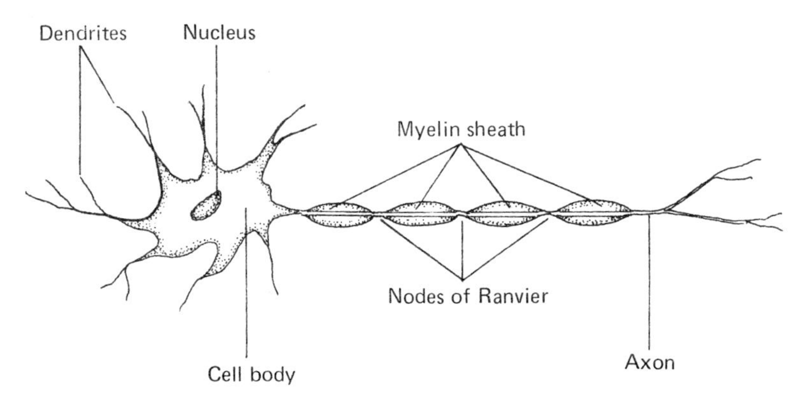

Fig. 5.42 A drawing of a neuron, showing the cell body, the myelin sheath covered axon, and the dendrites.

very long, for instance, the neurons which control the muscles in the human finger have their cell bodies in the spinal cord! Bundles of axons are called *nerves* and groups of nerve cell bodies are called *ganglia* (singular *ganglion*).

Connections between neurons are called *synapses*. They occur between branched ends of axons, dendrites, and collateral branches of axons of different neurons. Axons can also end in a synapse at *receptors*, such as those of the sense organs, or, like in the motor nerves, at the end plates of a muscle fiber. The *endplate* of a muscle fiber is a special structure at which an action potential is generated by synaptic stimulation from an axon of a motor nerve.

A single axon can be removed from an organism and, for a certain period, be examined in the laboratory. Moreover, one can uncover a nerve cell, or part of the nervous system, in a living animal and test it under *in vivo* conditions. In all such experiments it is relatively easy to stimulate the nerve with an electrical stimulus and observe the effect by electrodes placed in or on the cell membrane. A number of properties of the action potential can be demonstrated by this kind of experimentation.

1. The action potential is a potential spike with an amplitude that does not depend upon the amplitude of the stimulus. It has a typical value of about 130 mV and occurs in a regenerative way when the stimulus reaches a certain threshold value (Fig. 5.43). The spike is an *all-or-none* effect, suitable for conveying messages in a binary code (on-or-off, 1 or 0); its duration is about a millisecond.

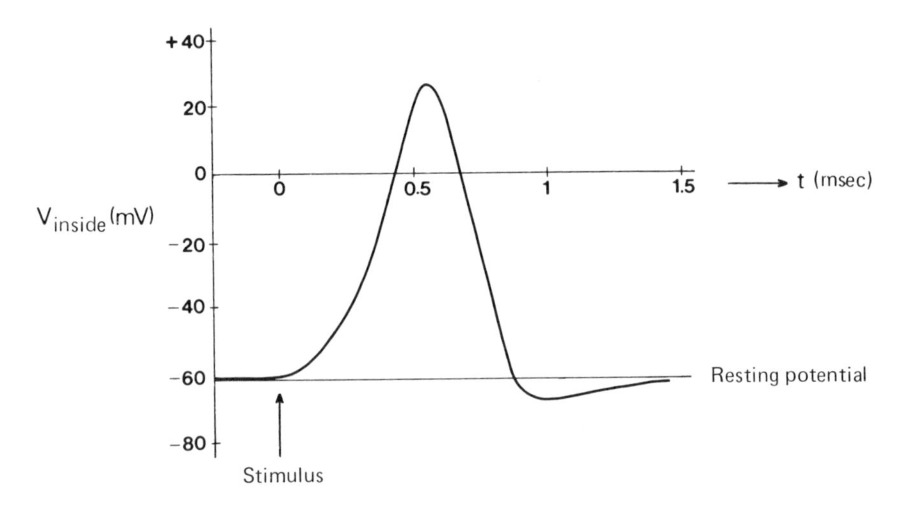

Fig. 5.43 A typical action potential.

2. The spike has a polarity which is opposite to that of the resting potential. Thus, it is negative at the outside, positive at the inside.

3. There is a refractory period immediately after the passage of the spike of about 1 msec during which a stimulus does not evoke an action potential, no matter how strong it is. Because of this feature no mixing of spikes can occur and the frequency of occurrence is limited to less than 1000/sec.

4. The potential spike evokes new spikes in adjacent regions of the membrane. Because of the refractory period, this effect is as if the spike were propagated *in both directions* from the point of stimulation, like a traveling wave. Since the occurrence of each action potential is regenerative, there is no attenuation of the "wave" as it travels along the membrane. The velocity at which the spike moves varies between 50 and 150 m/sec in vertebrates.

Na^+ and K^+ Permeabilities. It seems obvious that the action potential, which is a transient departure from the resting potential, is a transient change of the steady-state balance of the ion concentration distribution across the cytoplasmic membrane. This, indeed, appears to be true. In fact, what happens is (see Fig. 5.44) that when a stimulus exceeds a threshold value, the permeability for Na^+, which is low in the resting state, increases abruptly and Na^+ floods into the cell. The membrane potential becomes

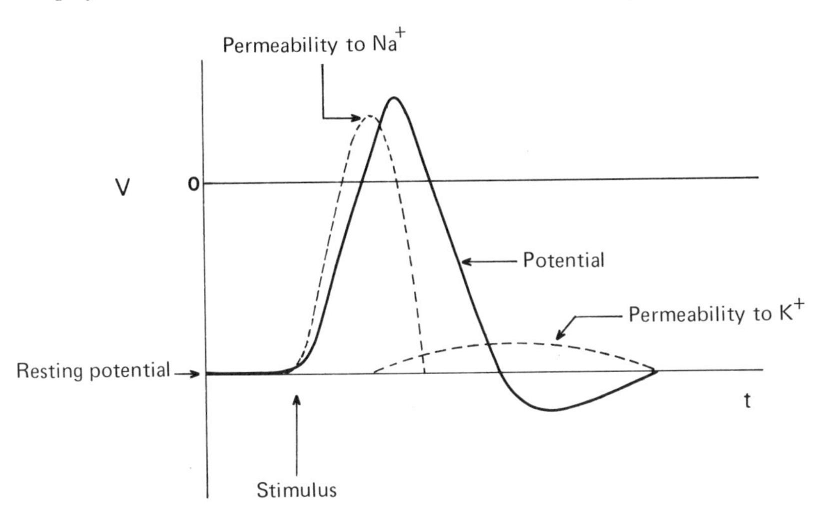

Fig. 5.44 The formation of an action potential. Upon stimulation a rapid transient increase of the permeability to Na^+ occurs which is followed by a slower increase of the K^+ permeability. As a result Na^+ will flow in and K^+ will flow out, thus creating the regenerative potential spike.

more positive inside the cell which again increases the Na^+ permeability, making the event explosive. The potential even becomes depolarized (positive at the inside) until it reaches a maximum, after which the permeability for Na^+ decreases again. Meanwhile K^+ leaks out and this tends to restore the original membrane potential. The final phase of the action potential is the operation of the K^+–Na^+ ATPase to exchange the cations to their original concentration levels. Thus, due to the increase of the Na^+ permeability, a Na^+ current starts to flow in, followed by a K^+ current to the outside. This causes the transient depolarization of the membrane potential. When the Na^+ permeability returns to the resting value, the K^+ current restores the potential to a level slightly above the resting potential. The resting state is then restored by the K^+–Na^+ ATPase activity. During this process the membrane is refractory to new stimuli.

The transient inward Na^+ current produces charge dislocations in adjacent regions, causing an increase in Na^+ permeability. The sequence of events is thus repeated in adjacent parts of the membrane and the action potential is propagated away from the point of stimulation. Because of the refractory period the action potential never will move back.

This picture of spike generation in nerve and muscle cells has emerged from meticulously carried out experiments in which action potentials are evoked, measured, and controlled by microelectrodes. Ionic fluxes across membranes of nerves have also been measured, using radioactive isotopes of Na and K. The results of such experiments are fully consistent with the mechanism of action potential generation described above.

Voltage Clamp. An important type of experimentation is that called the voltage clamp. In these experiments the membrane potential is clamped at a given value by a feedback system which introduces a counter current opposing changes in the membrane potential from the preset value. When the membrane in such an experiment is clamped at a voltage exceeding the threshold value for initiating the action potential, a transient current flows inward for a few milliseconds and then the current reverses and flows outward. The transient inward current was found to be a "sodium current" while the outward current replacing it after a short time was a "potassium current." Of course, no action potential can develop in such an experiment because the voltage is clamped at the preset value. This behavior has been empirically analyzed and a set of differential equations has been derived which are generally known as the *Hodgkin–Huxley equations*. These equations can be used to characterize the behavior of the nerve fiber (the axon) in terms of changes in the permeability of the membrane to Na^+ and K^+. Numerous checks have been carried out and the equations, as well as the molecular model of action potential generation based upon transient changes

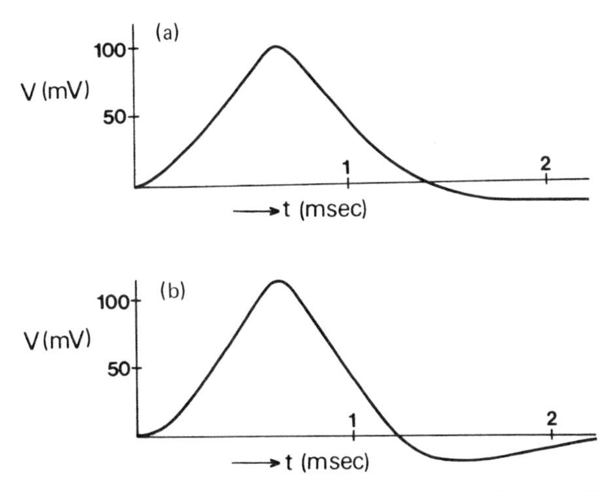

Fig. 5.45 A comparison between an action potential calculated from the Hodgkin–Huxley equations (a) and a measured action potential (b).

in permeability to Na^+ and K^+ have been found to be essentially correct (Fig. 5.45). One could also fit the transmission of action potentials between different cells in the synapses by means of chemical transmitters into these equations.

Synapse. We have stated above that the propagation of an action potential along a nerve axon occurs in both directions from the point of the stimulus. This would not make much sense in a nervous system which is meant to carry messages from one point to another unless there were some kind of rectifying system. Such rectifying systems are present at the *synapses*, junctions of one (nerve) cell and another (illustrated in Fig. 5.46).

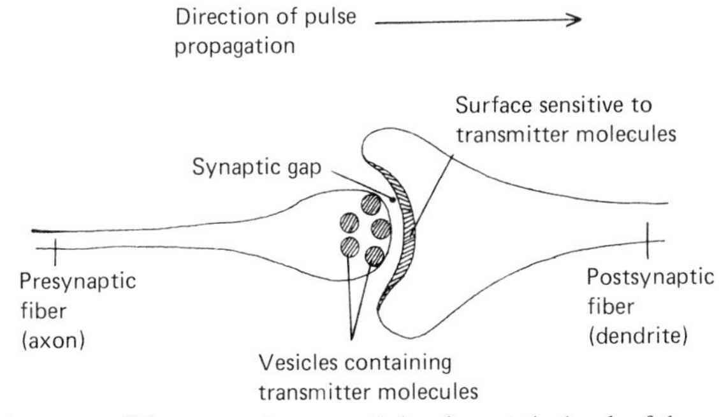

Fig. 5.46 A synapse. When an action potential arrives at the knob of the presynaptic fiber, synaptic transmitter is released from the small pockets in the knob. The transmitter diffuses rapidly across synaptic gap and stimulates an action potential at the sensitive surface of the postsynaptic fiber (if the concentration exceeds the threshold value).

This system consists of a "specialized knob" at the terminal end of the *presynaptic fiber* (usually the axon), separated by a small (about 1 μm) gap from a receptor surface of the *postsynaptic fiber* (usually a dendrite or an endplate of a muscle fiber). When an action potential arrives at the terminal of the presynaptic fiber a chemical substance, the *synapse transmitter*, is released from small pockets in the "knob" at the end of the fiber. The transmitter rapidly diffuses (within 1 msec) across the synaptic gap and alters the permeability to Na^+ and other monovalent ions in the membrane of the receptor surface at the terminal end of the postsynaptic fiber, thus generating an action potential. The transmitter is enzymatically destroyed immediately thereafter.

It has been definitely confirmed that *acetylcholine* (an ester from acetic acid and the lipid choline) is a synaptic transmitter; another transmitter, noradrenaline, has been identified in postganglionic fibers of the sympathetic nervous system (Fig. 5.47). Only very small amounts (about 10^{-18} M or some 1000 molecules) are necessary to evoke an action potential in the postsynaptic fiber; this amount is called a *threshold quantity*. Evidently the transmitter must be inactivated immediately after the occurrence of the action potential in the postsynaptic fiber; otherwise the synapse would not be ready to transmit the next pulse. The destruction of acetylcholine is accomplished by the enzyme *acetylcholinesterase*, a protein located in the receptor surfaces of postsynaptic fibers and the motor endplate of muscle fibers.

In a very few cases (for instance conduction across the giant synapse of the cray fish) the synaptic conduction only involves a charge transfer without chemical intermediates. At such synapses the impulses travel by pure electrical transfer from one axon to another with negligible time delay. The

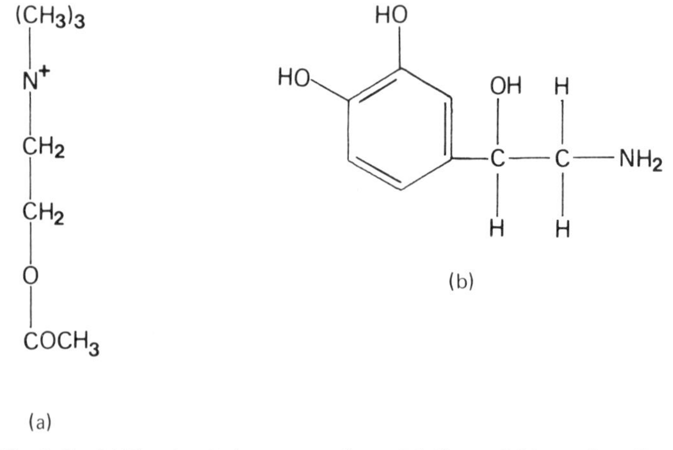

Fig. 5.47 (a) The chemical structure of acetylcholine and (b) noradrenaline.

nature of such electrical synapses is not understood at all. The molecular mechanism by which chemical transmitters can open up "Na^+ gates" is also far from being understood; this problem remains one of the most compelling questions in the study of cell and membrane function.

Information Processing in Neuronal Systems. Action potentials, thus are propagated and transmitted from neuron terminal to neuron terminal or to the endplates of muscle fibers. The transmission of information from neuron to neuron is accomplished by a binary code (the presence or absence of an impulse) which is not yet broken at all. The mechanism of the transduction of receptor signals to this binary code of impulses is also largely unknown. (These problems will be discussed briefly in later sections.) The transmission of the impulses occurs through complicated networks of neurons which are formed by literally millions of synaptic connections between axons and dendrites. One action potential on a presynaptic fiber may excite one action potential on the postsynaptic fiber. In some cases, however, the release of a transmitter by the end of a presynaptic fiber may be below the threshold. In that case more subthreshold releases, hence the firings of more than one presynaptic fiber, are necessary to evoke a synaptic response. The possibility thus exist for an adding operation. Likewise, two, three, or more firings in a short time at one synapse may be necessary to produce a transmitted spike; then the synapse acts as a divider. Multiplication can be achieved when several terminals from one neuron have synaptic connections, with different time delays, at the same second neuron. Subtraction occurs when the released transmitter in a synapse acts as an inhibitor rather than a stimulator. Such synapses have been found in the heart muscle; acetylcholine, which produces an action potential at motor endplates and other synapses inhibits the generation of a spike in heart muscle. At such synapses the transmitter increases the permeability to K^+ and larger cations but does not change the Na^+ permeability at all. The net result is a change in the membrane potential and an increase in the firings at other synapses necessary to transmit the spike. This effectively results in a subtraction of impulses from two different incoming neurons.

The nervous system thus operates with a circuitry not unlike that of a modern digital computer, only far more complex. One thing that has to be kept in mind is the fact that the system is nonlinear. In addition, the synaptic conduction is influenced by slow potential fluctuations. Not only the K^+ and Na^+ concentrations but also the concentrations of Ca^{2+} and Mg^{2+}, and in particular their ratio, can alter synaptic conduction. It has been shown that at the neuromuscular synapse the probability of a given packet of about 1000 molecules of acetylcholine (the threshold quantity) entering the intercellular fluid is a function of the Ca^{2+}/Mg^{2+} concentration ratio.

Many, in fact most, aspects of the functioning of nervous systems, nerves, and nerve conduction are still not understood. Although we have a notion now of how an action potential can arise, we still have no idea how transmitter molecules, or other membrane disturbing agencies, can change the ion permeabilities. Still larger looms the problem of understanding the way in which the information from the environment is coded and processed. More knowledge about this may eventually lead us to the understanding of things like perception, memory, and learning.

5.8 Contractility

Mechanical Work. Mechanical work is performed by all living organisms. Moving of cells or cell aggregates (organs like muscles or organisms as a whole) is the most obvious form of mechanical work; it gives organisms the ability to move away from a noxious environment or to move toward beneficient regions of space. However, movements *inside* the cells are also forms of mechanical work performed by most if not all living cells. These include processes like protoplasmic streaming, mitosis (cell division), swelling and contraction of organelles, and pinocytosis. Very little is known about the mechanisms of these motions; they may differ considerably from case to case. Some common factors seem to prevail in all these cases, however; first, the energy source is always the hydrolysis of ATP to ADP and inorganic phosphate. Furthermore, it appears likely that most mechanisms share a common underlying molecular basis. Wherever cell movement is studied on the molecular level, it turns out to involve one particular class of proteins of which *actin* and *myosin* from muscle tissue are the best known. These proteins, or combinations of them, generally exhibit ATPase activity, which clearly indicates their importance in the transduction of chemical energy into mechanical work.

Cilia and Flagella. Cells that are able to move through a liquid medium or to move the medium across their surface do so by means of whiplike structures called *cilia* or *flagella*. Cilia are short fibers occurring in bundles. Flagella are longer and usually occur in much smaller numbers together or as single fibers. Eukaryotic cilia and flagella are all built according to the same regular patterns and are of about the same diameter (about 0.5 μm). Figure 5.48 shows the flagellum of a sperm cell. The flagellum originates in a *basal granule*, or midpiece, which is found in the cytoplasm of the cell. The basal granule is covered with long mitochondria which are helically wound around the granule. The flagellum itself consists of nine pairs of outer fibrils around two inner tubules. Extractions of flagella show the

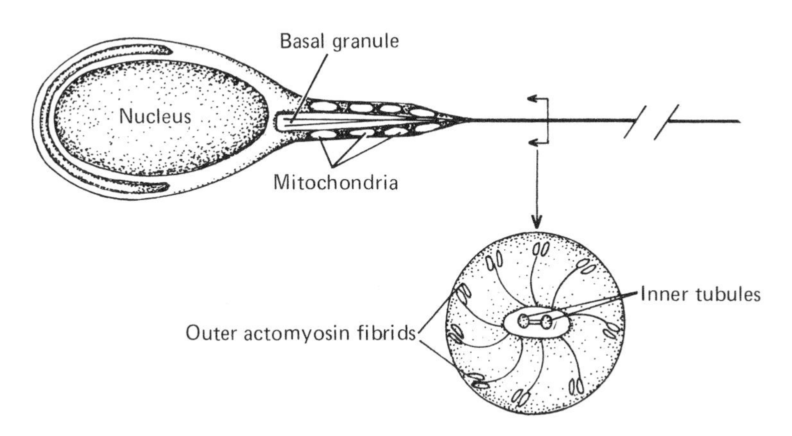

Fig. 5.48 A sperm cell with flagellum. The flagellum originates from the basal granule and shows the nine double actomyosin fibers and the set of two inner tubules in cross section.

presence of a complex protein very similar to the complex actomyosin found in muscle. It seems probable that the outer fibrils are composed of actomyosin; the inner tubules could serve as channels for the diffusion of ATP to be hydrolyzed at the actomyosin sites.

The cells move by sweeping beats of the flagella or cilia, sometimes with a slightly circular component like the cilia of protozoa, or in a single plane, like those of the epithelium cells of the human respiratory tract. Flagella of sperm cells beat in a wavelike fashion. The regular arrangements of the fibers in the cilia and flagella suggest that ciliar or flagellar beat could be based on the contraction of the fibers on one side of the array, drawing the tip of the cilium in that direction. Thus, the ciliar or flagellar movement to be accounted for is rather complicated. No theory, so far, has been put forward which would give a satisfactory explanation for the way these movements are produced.

Bacterial flagella do not show the tubular array of the eukaryotic flagella or cilia. They are much smaller, about 0.15 μm in diameter, and in electron micrographs show a helical structure with a pitch that seems typical of each particular species. The mechanism of bacterial flagellar motion is not known, although their movement is less complicated than that of eukaryotic cells. The photosynthetic bacteria *Rhodospirillum rubrum*, for example, has flagella at each polar end. When the cell moves the flagella rotate around the cell describing a cone, so that the bacterium rotates in the opposite direction. Since the bacterium is a spiral itself, its rotation drives it through the liquid medium. The cell can reverse direction very rapidly by simply flipping its flagella at each polar end in the appropriate direction and apparent coordination with each other. This reversal of direction occurs during

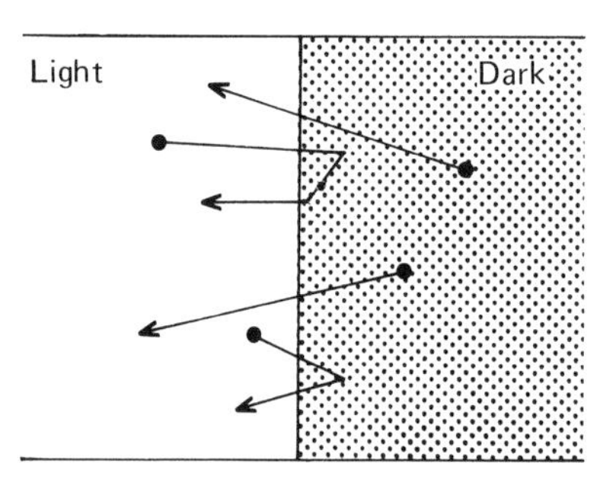

Fig. 5.49 Phototaxis of the photosynthetic bacterium *Rhodospirillum rubrum*. The cells reverse direction when crossing the boundary from light to dark. No change in direction occurs when the cells cross the boundary from dark to light.

phototaxis, the light-induced movement of many photosynthetic organisms (see Fig. 5.49). Crossing boundaries from light to darkness causes *Rhodospirillum rubrum* cells to abruptly change their swimming direction. If a cell crosses the boundary from darkness to light no reaction occurs. The cells, thus, have the tendency to accumulate in the light. The light-induced synthesis of ATP in photophosphorylation is connected with this motile reaction.

Muscles. In higher organisms of the animal kingdom, the capacity to perform mechanical work is exhibited by highly specialized structures called *muscles*, groups of cells which move as a unit by means of contracting intracellular filaments. Histologically, we can distinguish three types of muscles: the *smooth muscles*, tissue which forms the walls of internal organs like intestines, blood vessels (particularly the arteries), esophagus, and bladder; *cardiac muscle*, the muscle tissue comprising the *myocardium* or heart muscle; and finally, *striated* or *skeletal* muscle. Smooth muscle consists of spindle-shaped cells about 0.5-mm long and about 0.02-mm diameter (Fig. 5.50). Contraction of smooth muscle is not under voluntary control and proceeds in a slow and generalized manner. The cells of cardiac muscle are fibrous and slightly striated (Fig. 5.51); its periodic contraction and relaxation goes on

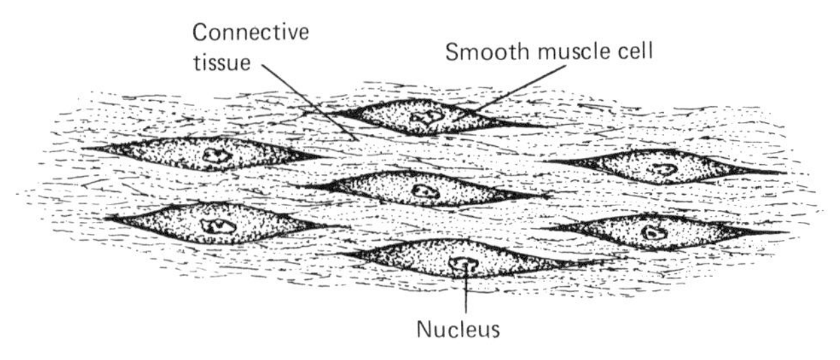

Fig. 5.50 Smooth muscle cells.

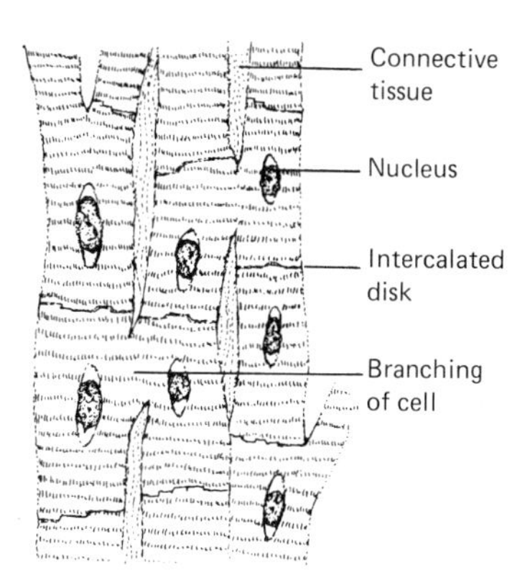

Fig. 5.51 Cardiac muscle.

automatically, although the sympathetic nervous system can influence the rate of the heart beat. Most of what is known about cellular motion comes from studies with striated muscle and it is the only contractile organ about which a notion of the mechanism of contraction exists. Striated muscle is highly organized and has many elongated cells, the muscle fibers, which contain the contractile organelles, the *myofibrils* (see Fig. 5.52). Myofibrils show characteristic repeating structures, about 2.5 μm in length, which give

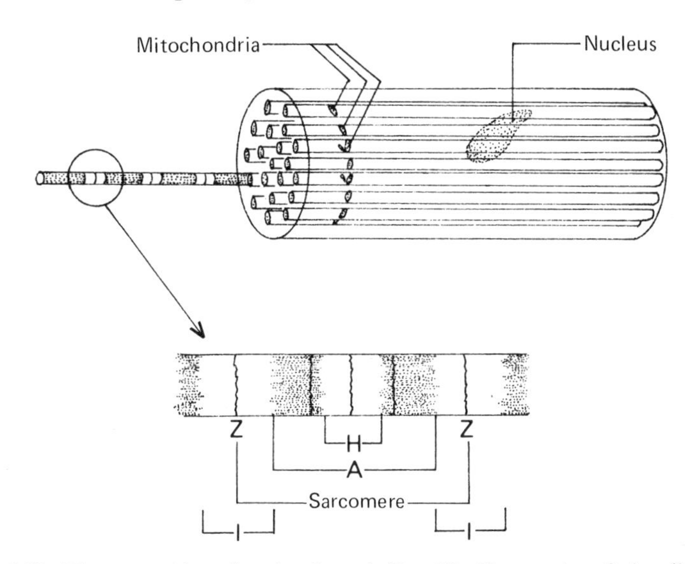

Fig. 5.52 The composition of a striated muscle fiber. The fiber consists of a bundle of many myofibrils which have the striped appearance. The myofibril is a line-up of many sarcomeres; in the sarcomere one can distinguish the isotropic I bands, the anisotropic A bands, the Z lines, and the central H disk.

them the striated appearance. The repeating unit is called a *sarcomere* and is clearly visible in the electron micrograph of part of the myocardium of a dog, shown in Fig. 5.53.

The Sarcomere. A sarcomere has a banded structure formed by regions of sharply different optical properties. The boundaries of the sarcomere, the *Z lines*, run in the middle of light bands which are called *I bands* because they are optically isotropic regions. At each side of an I band is a darker region which shows strong birefringence. This is the *A band* (A for anisotropic). The middle of the A band is somewhat lighter, forming an *H disk* within which a sharp darker line, the *M line* can be seen.

This optical appearance is caused by an array of filaments which make up the different bands. Figure 5.54 shows a schematic drawing of a blown-up sarcomere. There are two kinds of filaments; the *thick filaments* making up the A band, and the *thin filaments* responsible for the I band which are

Fig. 5.53 Electron micrograph of a section through the myocardium of a dog. The sarcomere structure is clearly visible. The I bands, optically isotropic regions, are at each side of a Z line. The darker regions at each side of an I band are the optically anisotropic A bands. In the middle of an A band is the H disk with a sharper M line. (Courtesy of Dr. W. Jacob and Dr. A. Van Laer, Electron Microscopy Laboratory, Universitaire Instellingen Antwerpen, Antwerp, Belgium.)

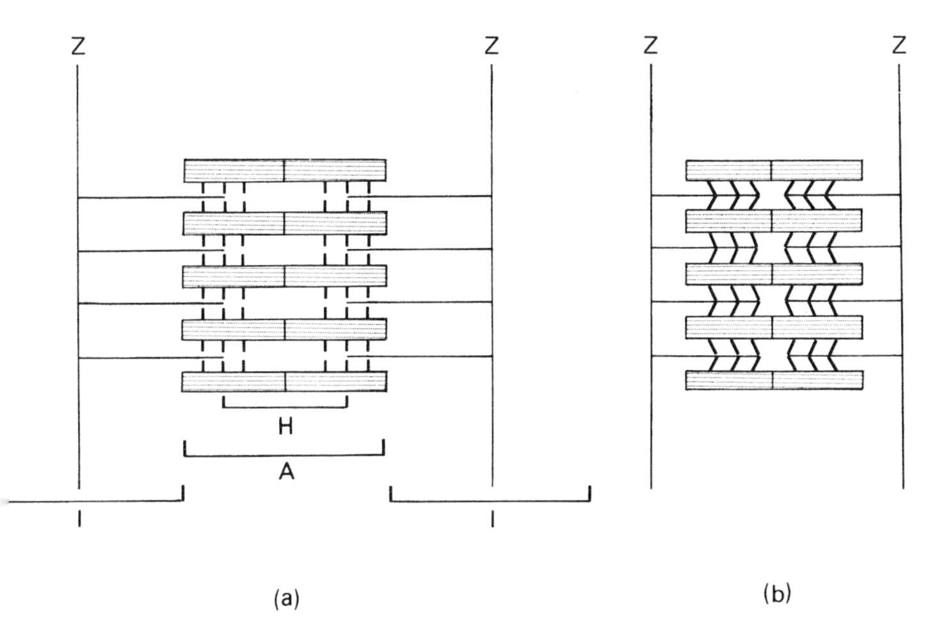

Fig. 5.54 A schematic representation of a sarcomere; (a) muscle relaxed and (b) muscle contracted. The bars represent the thick filaments, the lines the thin filaments connected to the thick filaments by cross bridges. The H disk is the area with only thick filaments, the A band includes the H disk and the overlap area between the thick and the thin filaments. The I band is the area in which there are only thin filaments. The sliding-filament model considers contraction as resulting from a sliding of the thick and thin filaments along each other, thus shortening the distance between two Z lines and causing the I bands to disappear.

attached at one terminal end to a plate called the Z line. Where the thick and thin filaments overlap is found the darker part of the A band. The H disk is the region where only thick filaments are found. The array is extremely regular in space. Each thick filament is surrounded, in the region of overlap, by six thin filaments (see Fig. 5.55). The filaments are connected to each other by tiny, so-called *cross bridges*.

Carefully designed experiments, in which observations were made with an electron microscope, and optical measurements have shown that when the muscle contracts the I zones shrink and finally disappear in the fully contracted muscle; the H disks also disappear. These observations have led to the *sliding-filament* model for the contraction mechanism. This model, which does have substantial molecular justification considers contraction as resulting from a sliding of the filaments along each other, thus making the region of overlap of the two kinds of filaments larger. The sliding of the filaments could be visualized as being caused by a repeated detachment and reattachment of the cross bridges between the two kinds of filaments. The

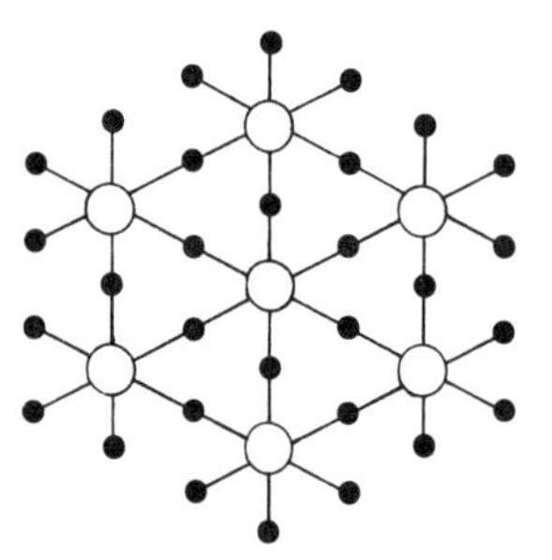

Fig. 5.55 The extremely regular arrangement of the thick (open circles) and the thin (dots) filaments in a cross section of a sarcomere.

cross bridges change angle during each cycle of attachment and detachment, pulling a thick filament along a thin one. The sliding-filament model is consistent with all information available thus far about muscular contraction and muscular morphology. For instance, when a muscle is stretched it loses the ability to exert maximum force. This result is a direct consequence of the geometry described above. The force of the muscle, according to the sliding-filament model, depends on the interaction of the two kinds of filaments and this interaction depends on the degree of overlap. In stretched muscle, the overlap is diminished and thus the force is lessened.

Contractile Proteins. The two major protein components in myofibrils were found to be *myosin*, which makes up about 50%–55% of the total protein, and *actin*, which accounts for 20%–25%. Actin is the major protein of the thin filaments. When it is extracted from the muscle it comes out as a globular protein, the so-called G-actin, with a molecular weight of about 60,000. When G-actin is incubated with ATP in the presence of Mg^{2+} the protein polymerizes to a larger aggregate, the fibrous F-actin. In the muscle fibril the actin globules are arranged in double helical strands like two intertwined strands of beads (Fig. 5.56). Two other proteins are closely associated with actin in the thin filaments, *tropomyocin* and *troponin*. Tropomyosin is a long threadlike molecule. The tropomyosin molecules attach end to end to each other in long filaments on the surface of an actin strand. Each actin strand carries its own filament which lies near the groove between the two

Fig. 5.56 A thin filament consisting of two intertwisted strings of globular actin molecules. The troponin molecules (black ovals) and tropomyosin molecules (lines near the grooves between the two actin strands) are seen around the actin filament. Each tropomyosin molecule is connected to one troponin molecule, covering seven globular actin molecules.

pairs of strands. Each tropomyosin molecule covers seven G-actin monomers. Troponin, on the other hand, has a more globular shape. It has recently been shown that troponin consists of three subunits. It is attached to the tropomyosin filament, one on each molecule of tropomyosin, at about a third of the way from the end. The tropomyosin–troponin system plays an important role in the regulation of contraction by Ca^{2+} ions.

The thick filaments contain all the myosin of the muscle. Myosin is a large molecule with a molecular weight of about 500,000. This protein appears to be an elongated filament, about 60% helix, having two globular heads (Fig. 5.57). A thick filament of a normal muscle contains several hundred myosin molecules. The molecule can be cleaved at a specific place by a protolytic enzyme (trypsin), yielding a light fraction called *light meromyosin*, which is the filamentous end of the molecule, and a heavy fraction called *heavy meromyosin*, which contains the globular heads. The heads can be isolated from the tail by the treatment of myosin with the enzyme papain. Myosin has ATPase activity and the cleavage experiments have shown that this ability to hydrolyze ATP is located at the globular heads. The heads also appear to be the site of greatest affinity for actin when the two proteins mix and form a complex. This strongly indicates that the globular heads are the part of the myosin molecule that form the cross bridges with actin in the myofibril.

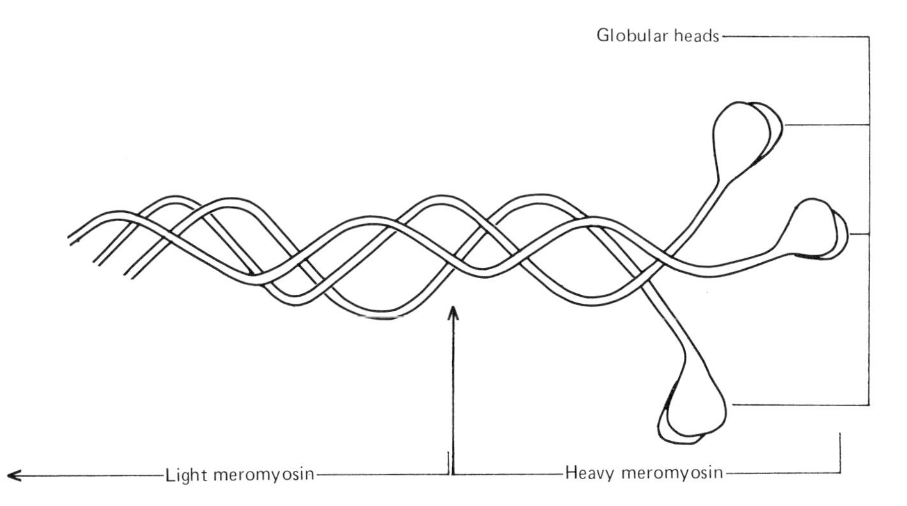

Fig. 5.57 A schematic representation of three myosin molecules of the several hundred found in a thick filament. The molecule is filamentous, having a double globular head. The arrow designates cleaving by trypsin, yielding light meromyosin (the filamentous end) and heavy meromyosin (the globular end).

Actomyosin. *In vitro*, the two proteins form a complex when they are mixed together: this complex is called *actomyosin*. Complexes are also formed when heavy meromyosin or isolated globular heads are used instead of myosin. The complexes can be used as model systems in order to find out more about the properties of the system. The complex, for instance, can be precipitated into an insoluble sheet that contracts upon addition of ATP. The ATP is actually hydrolyzed in the experiment, clearly showing the coupling of ATPase activity to the performance of mechanical work.

Contraction Mechanism. Most suggestive for the sliding-filament model is that the binding of ATP to actomyosin leads to a sharp decrease in the viscosity of a suspension of the complex. The effect, however, is reversible (as shown in Fig. 5.58). Addition of ATP to a suspension of actomyosin which is quite viscous and shows birefringence, causes a sharp decrease of the viscosity. When the ATP is being hydrolyzed (in the presence of Mg^{2+}) a slow recovery to the original viscosity occurs, after which the experiment can be repeated. It has clearly been shown that the viscosity change in the

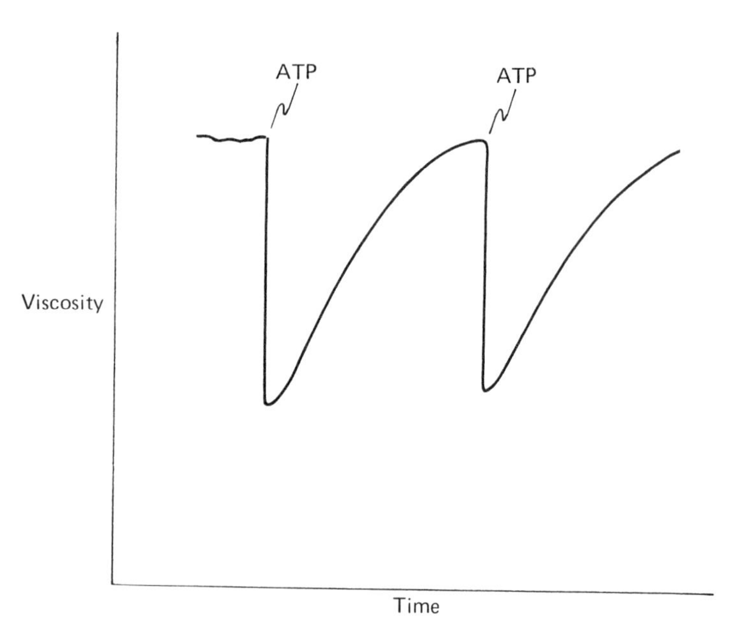

Fig. 5.58 An experiment showing the drastic change of the viscosity of a suspension of actomyosin upon addition of ATP. The binding of ATP to the actomyosin complex causes dissociation into actin and myosin · ATP complex. When ATP is hydrolyzed actin and myosin recombine into actomyosin and the viscosity increases.

presence of bound ATP is the result of a dissociation of the actomyosin into free actin and myosin · ATP complex. It thus follows that not the *formation* but the *breakage* of the cross-bridges between the myosin heads and the actin in the thin filaments requires the binding of ATP. The interaction between the two kinds of filaments during contraction is a cyclic sequence of steps which can be summarized as follows (Fig. 5.59).

1. Binding of ATP to the myosin heads.
2. Activation of the myosin · ATP complex.
3. Binding of the activated myosin · ATP complex to actin in the thin filaments; this reaction occurs *only* in the presence of Ca^{2+} ions.
4. Hydrolysis of ATP, which changes the angle of the myosin head in respect to the filament.
5. Binding of an ATP to the myosin heads, dissociating the myosin · ATP complex from the actin filament, after which the cycle is repeated from step 2.

According to the model just described, the actomyosin complex can exist in two forms, one activated and the other inactivated. The activated form is generated when Ca^{2+} triggers the binding of activated myosin · ATP to the thin filaments. The inactivated form is that part left after the hydrolysis of the ATP, which, in the absence of ATP, is very stable. This is easily produced in the laboratory by mixing myosin with actin in the absence of ATP. Binding of ATP, however, immediately causes the detachment of the myosin · ATP complex from the thin filaments, as shown by the viscosity experiment described above.

The stability of the inactivated actomyosin complex accounts for the well-known phenomenon of *rigor mortis*, the extreme rigidity that develops in the muscles after death. The gradual disappearance of ATP following death more and more prevents the binding of ATP to the myosin heads, thus leading to an increasing amount of inactivated complex. The inactivated complex, therefore, is sometimes called the *rigor complex*.

Role of Ca^{2+}. Recent evidence strongly suggests that the onset of contraction is mediated by Ca^{2+} ions and that this Ca^{2+} trigger acts through the tropomyosin–troponin system. Contraction and hydrolysis of ATP in suspensions of actomyosin prepared from muscle is dependent on the presence of Ca^{2+}. If, instead of the thin filaments, a purified preparation of F-actin is used, contraction becomes insensitive to Ca^{2+} and ATP hydrolysis goes on randomly until all of the ATP is used. From these and other results it appears that the two proteins tropomyosin and troponin are required for the control of the contraction mechanism and that Ca^{2+} is the agent by which such control is exerted.

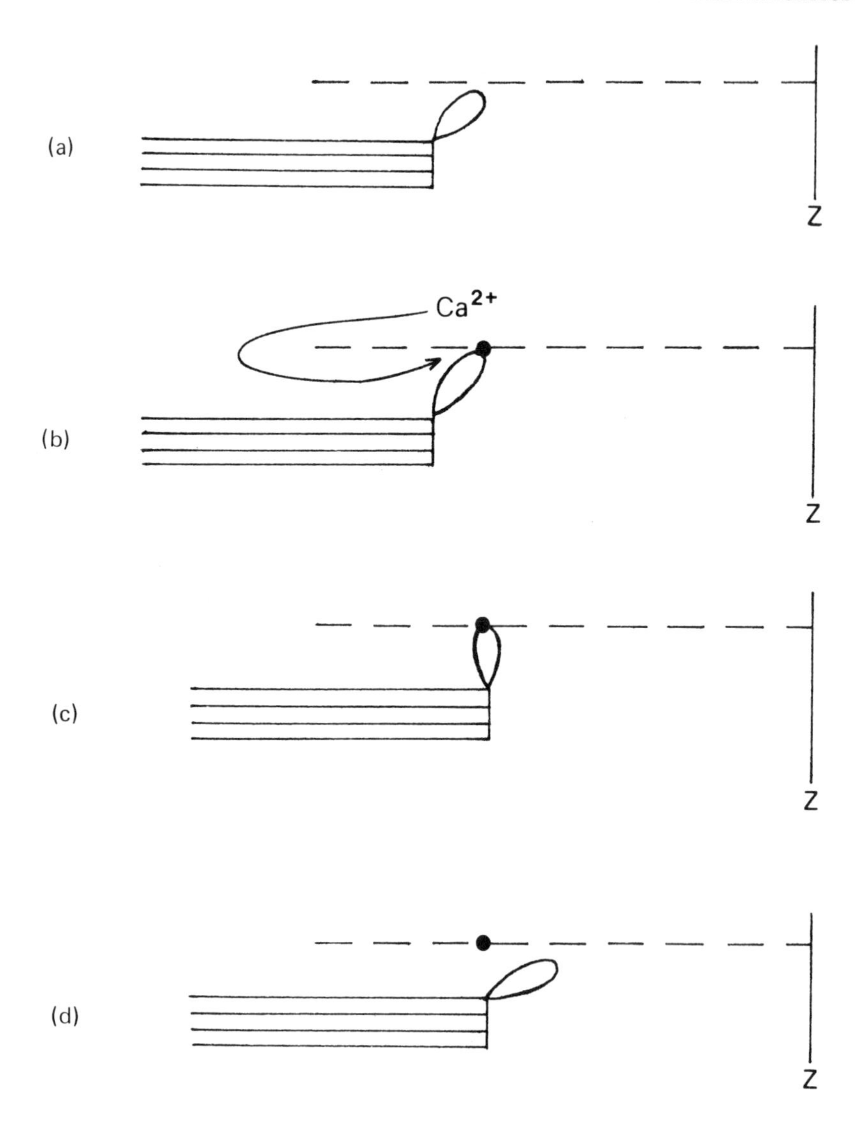

Fig. 5.59 The mechanism of contraction. (a) ATP binds to the myosin heads and the myosin · ATP complex is activated. (b) Ca^{2+} then triggers the binding of the myosin · ATP complex to actin; (c) ATP hydrolyzes and this results in a change of angle of the myosin heads in respect to the thick filaments. This pulls the thick filament along the thin filament. (d) Binding of another ATP molecule to the myosin heads then causes the dissociation of the actomyosin complex and the cycle is repeated.

Of all the components of the contractile system in striated muscle only troponin can bind calcium. All available evidence shows that the step that is sensitive to Ca^{2+} is the binding of the activated myosin · ATP complex to actin (step 3). It looks then as if the two proteins tropomyosin and troponin are responsible for blocking a site on the actin for the binding of the activated myosin · ATP and that the binding of Ca^{2+} to troponin causes the site to be exposed so that binding of the activated myosin heads can occur. A model in which these assumptions are incorporated is shown in Fig. 5.60. In the absence of calcium the troponin molecules hold the tropomyosin strands on the actin filament in such a way that a site for the binding of the activated myosin heads is blocked. The binding of Ca^{2+} to the troponin causes conformational changes in the molecule in such a fashion that the tropomyosin thread is pushed deeper into the grooves of the actin strands, away from the blocking site (thus allowing the activated myosin heads to "catch in"). The binding of the Ca^{2+} ions can be pictured as leading to a "tightening" of the bonds among the subunits of the troponin and a weakening of the interaction of the molecule with actin; lowering the calcium level "loosens" the troponin complex and makes it bind more strongly to actin. The tropomyosin strand amplifies this action over seven actin molecules. Recent X-ray diffraction studies of troponin and tropomyosin under different binding conditions have supported this model.

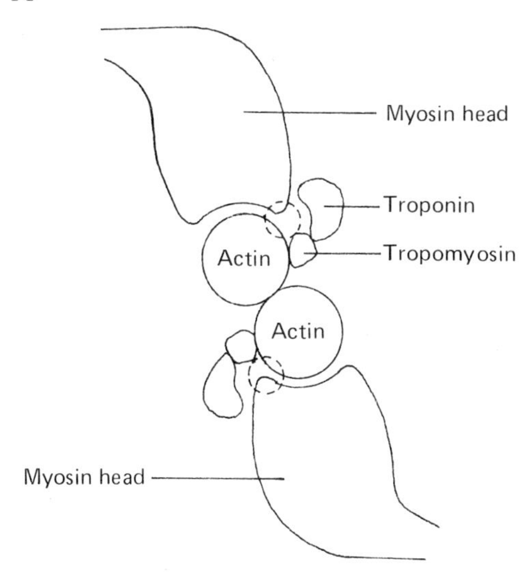

Fig. 5.60 A possible model for the Ca^{2+} trigger of muscle contraction. The binding of Ca^{2+} to troponin causes a conformational change in the molecule which results in the tropomyosin being pushed deeper into the grooves between the two actin strands, thereby exposing the binding site for the binding of the myosin head to the actin.

The Energy Source of Contraction. It is presently beyond all doubt that ATP is the direct source of energy for the contractile process. This is also the case *in vivo*, although it does not appear so at first sight. Muscles can also contract under complete anaerobic conditions or after poisoning with cyanide blocking respiratory electron transport. ATP then can be supplied by glycolysis and, indeed, the glucose concentration decreases with a proportional increase of lactic acid in these cells. But even when glycolysis is inhibited, for instance by inhibiting the enzyme phosphoglyceraldehyde dehydrogenase by iodoacetate, contraction of the muscle still goes on. Subsequent chemical analysis of the muscle reveals that the amount of ATP in the cell has remained pretty much constant but that the concentration of another "energy rich" phosphate, phosphocreatine, has declined considerably with a proportional increase of the concentration of creatine. Phosphocreatine by itself, however, cannot make muscles contract if there is no ATP or ADP present. Thus, it looks as if phosphocreatine is providing a supply of ATP and, indeed, the only known enzymatic reaction in which this compound participates is

$$\text{phosphocreatine} + \text{ADP} \rightleftharpoons \text{creatine} + \text{ATP} \tag{5.81}$$

a reversible reaction catalyzed by the enzyme *creatine phosphokinase*. We have already seen in Section 5.2 (Table 5.1) that the standard free energy change of hydrolysis of phosphocreatine is substantially more negative than that of ATP. Equilibrium of reaction (5.81) is, therefore, shifted to the right. Moreover, the use of ATP in muscle contraction tends to diminish the concentration of ATP, thus letting reaction (5.81) proceed to the right and keeping a constant level of ATP. Poisoning the creatine phosphokinase by dinitrofluorobenzene results in a rapid decline of the ATP concentration and a constant level of phosphocreatine.

Phosphocreatine is regenerated from creatine and ATP; the latter is formed by oxidative phosphorylation during the recovery period. The phosphocreatine pool, thus provides for a reservoir of "high-energy" phosphate groups; the ATP concentration in the muscle is always maintained at high and constant steady-state level by this system. The system allows the muscle to operate even in a short time when not enough oxygen is present to produce sufficient ATP.

All these arguments provide plenty of evidence for the direct involvement of ATP in muscle contraction, *in vivo* as well as *in vitro*. This is, of course, entirely consistent with the universal role of ATP in biological energy transformations and, for the muscle in particular, with the striking juxtaposition of numerous mitochondria to the myofibrils.

Striated muscle contracts upon stimulation by a motor nerve. The communication between nerve and muscle cell, as we have seen in the previous section, is a synapse between the nerve end and an endplate or *myoneural junction*. The endplate is a postsynaptic surface, sensitive to the synaptic transmitter acetylcholine. Smooth muscle as well as heart muscle are contracting in an autonomous way; they contract rythmically even when all nerve connections are removed.

It seems clear that in striated muscle the initiation of the contraction is mediated by Ca^{2+} ions. How Ca^{2+} becomes available for the control of contraction is in fact unknown, although some suggestive hypotheses do exist. The stimulation of the motor nerve causes the release of acetylcholine in the synapse between nerve and endplate. This produces a local potential change at the endplate surface which, when the threshold value is exceeded, causes a rapidly spreading action potential in the cell membrane of the muscle cell, and the cell contracts. It seems that changes in the ion distributions across the cell membrane, in particular that of Ca^{2+}, caused by the action potential make the connection between the action potential, which is a property of the cytoplasmic membrane, and the contractile process, which is a property of the myofibrils. It is likely that the action potential at the cytoplasmic membrane is linked to the release of Ca^{2+} in the immediate region of the contractile proteins, by means of a system of tubules which is continuous with the plasma membrane as well as the endoplasmic reticulum. This tubule system is called the *sarcoplasmic reticulum* and the close proximity of this system of channels to the contractile apparatus leads to the possibility that the release of Ca^{2+} is "felt" simultaneously by the entire bundle of myofibrils, so that they can contract synchronously.

Chapter

6 Biophysics of the Sensory Systems

6.1 The Transmission of Information

The Senses. Multicellular organisms, especially the higher animals, are stimulated by the environment through sensory systems; specialized organ systems which transmit information from the environment to the brain. We traditionally speak of them as the five senses: vision, hearing, olfaction, taste, and touch. But they, in fact, include more. Sensations of pain and temperature, and even sensations of hunger and thirst can be thought of as being evoked by sensory detectors. We can, in general, classify sensory detectors as follows. First, there are the special senses of vision and hearing, involving the highly specialized organ systems of the eye and ear. These are called *teleceptors*, receiving information from distant objects. Second, we have the *chemical receptors*, those receptors that are excited by chemical stimulation; olfaction and taste belong to this group. Third, *somatic receptors* are those responsible for the sensations of the body, namely touch, pressure, pain, and temperature. Finally, there are the *visceral receptors* that keep the brain informed about conditions inside the body and are responsible for the sensations of hunger, thirst, and the urge to discharge.

By means of the senses, organisms are able to respond appropriately to outside stimuli; there is a coordination between sensory information and muscle action (we jump out of the way if we see an oncoming automobile; we turn away if something smells bad; we cover our ears if we hear loud noises). Dramatic examples of more refined coordination are someone doing a fine piece of woodcarving or someone driving an automobile. Apparently, the senses are part of complicated *feedback control systems*.

Neuronal Coding of Receptor Signals. Although the senses are quite different from each other, as are the results of the different types of sensory perception (we know very well the difference between what we see and what we hear or smell), there are some common features which justify a generalized discussion. An important aspect of the senses is that they are all *transducers:* they transform one form of energy into another, or rather, translate information from one code into another. Information contained in electromagnetic radiation (vision), mechanical vibration (hearing), mechanical pressure and temperature (touch), and chemical structures (olfaction and taste) is translated into a code that can be handled by the nervous system. This information is collected in *receptors* or *receptor organs*, where the translation is then carried out. The receptor organs for vision and hearing, the eye and the ear, are highly specialized and quite elaborate. The interaction of electromagnetic radiation and the receptor cells in the eye is reasonably well-known (see Section 6.2). The state of our knowledge of the interaction of sound waves and the auditory parts of the ear (see Section 6.3) is less certain. In both cases, however, the transduction process proper is still an unsolved problem. The receptors for olfaction (the olfactory lobe in the upper part of the nose), taste (the taste buds in the tongue), and the receptors for touch, pain, and temperature are less elaborate. The transduction process in these cases is also unknown.

When stimulated, each of these receptors will initiate nerve discharges in the form of action potentials. Each of them, of course, is particularly sensitive to specific stimuli, such as light for the eye and sound for the ear. Nerve discharge, however, is also initiated in each receptor when it is subjected to excessive pressure, damage, or high temperature. The resulting sensation, however, is always that characteristic for the particular receptor, even when the stimulus is abnormal. The common experience of "seeing stars" when the eyes are struck is an example of this. Thus, it seems that the recognition of a particular kind of stimulus by the brain is linked to the origin of the stimulus, rather than to a particular kind of code transmitted through the nervous system. The answer to the question of how the brain does distinguish stimuli from one type of receptor from that from another type of receptor is not easily discerned. The anatomical relation between certain loci in the brain and the sensory organs is only part of the answer. The high degree of processing that is going on in the organs themselves as well as in the brain contributes to the tremendous complexity of the problem.

The code transmitted through the nervous system is a binary code; as stated in Chapter 5, action potentials are "on-or-off" events. This fact would make concepts developed in disciplines like information theory and cybernetics applicable to this subject. After all, digital computors, the rapid development of which would have been unthinkable without information theory and cybernetics, operate with a binary code, make use of electrical pulses, and are systems of networks in which feedback control is essential. Indeed, concepts of information theory and cybernetics have been used, but so far with only limited success. It has to be borne in mind that both of these disciplines do not treat *things* but *ways of behaving;* they do not ask "what *is* this thing" but "what *does it do.*" They are sciences in their own right, not depending on or being derived from any other branch of science such as physics or biology. Therefore, they never can provide answers to questions about mechanisms. One could still hold, however, that applications of cybernetics, and perhaps even more so with information theory, help express ideas and hypotheses and, more important, help to realize similarities and analogies between widely diverse fields. In such a way it can lead to a *direction* in which certain explanations must go. In itself, however, it never can offer such explanations or lead to new hypotheses about mechanisms.

Sensory Signal Processing. As we have stated before, the way in which sensory information, such as the image on a retina or a tone of a certain pitch at the entrance of the cochlea, is translated into the binary code of action potentials is unknown. For higher, more developed organisms it is certain that, with perhaps the exception of the sense of touch, it is not by a simple correspondence between a receptor cell and a neuron. In the case of hearing, for example, the old Helmholz theory in which tuned resonators in the cochlea were thought to excite their own neuron cannot hold because the frequency range of hearing (spanning 10 octaves from 20 Hz to 20 kHz) is orders of magnitude larger than the maximum firing frequency of a neuron (less than 1000/sec). It seems clear that the cooperation of many neurons is necessary in order to cope with the frequency capacity of the auditory sense.

This is even more obvious with vision, in which the cooperation of many nerve cells has actually been demonstrated. Figure 6.1 shows a schematic drawing of the arrangement of cells in the retina. Receptor cells (rods and cones) form synaptic connections with intermediary cells which in turn have synaptic connections with ganglion cells, the axons of which together form the optic nerve (for a complete picture of the eye, see Fig. 6.4). The intermediary cells can be distinguished as *horizontal cells*, *bipolar cells*, and *amacrine cells.* These cells form a network of communication, processing

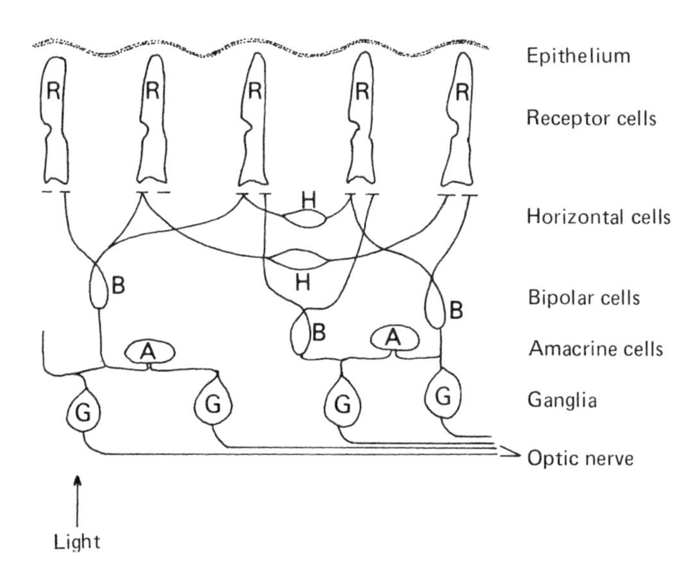

Fig. 6.1 A schematic drawing of the different layers of cells and their interconnections in a vertebrate retina.

the visual information and delivering it to the optic nerve. The signals from the receptor cells are transmitted to the bipolar cells which in turn stimulate the ganglion cells, either directly or through the amacrine cells. The horizontal cells make cross connections among receptor and bipolar cells, and the amacrine cells provide contacts between bipolar cells and ganglion cells. The synapses can be stimulatory or inhibitory. In this way, one ganglion cell is "served" by a number of receptor cells. The area covered by these receptor cells is called the *receptive field* of the ganglion.

One can map the receptive field of a ganglion by measuring the action potentials evoked in a single fiber of the optic nerve by moving a tiny spot of light across the surface of the eye. A number of experiments of this type have been carried out with vertebrates as well as invertebrates. The results have shown that, in general, three patterns of recordings of action potentials can be found in different fibers of a single optic nerve (Fig. 6.2). These *on*, *on–off*, and *off* responses are the basis of a complicated information processing system and the basis of many different patterns of response. *Movement* and *contrast* detection involve different ganglions. Many ganglions, movement detectors as well as contrast detectors, show a pattern of antagonism between central and peripheral regions of the receptive field. When illumination of the center of the receptive fields evokes an *on* response in such a ganglion, additional illumination of the peripheral region suppresses it. The reverse can also be true. It seems that the inhibitory action against

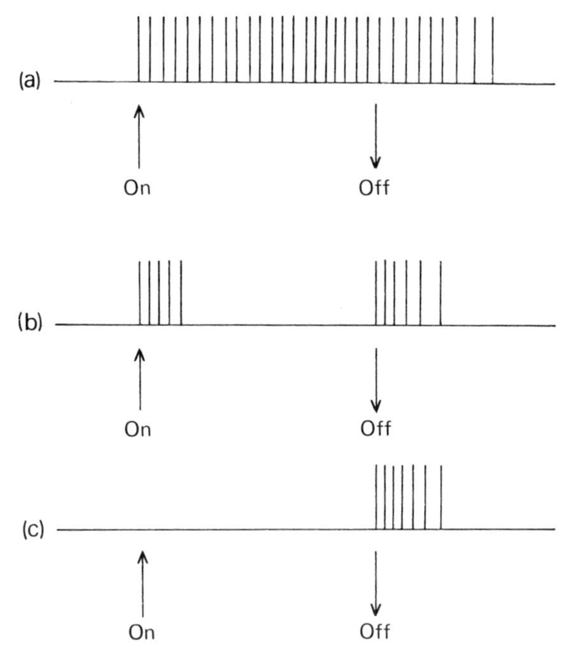

Fig. 6.2 Three patterns of responses of different fibers in a single optic nerve. (a) The "on-response" is a volley of action potentials triggered by turning on the light without any effect by turning the light off. (b) The "on–off-response" is a short burst of action potentials when the light is turned on and a similar burst when the light is turned off. (c) The "off-response" is a volley of action potentials only when the light is turned off. These responses are a result of a combination of excitatory and inhibitory synaptic connections.

either an *on* response or an *off* response involves messages reaching the connecting bipolar cell from a receptor cell through intervening horizontal cells.

Most of the details of the signal manipulation of this network of nerve cells remain obscure. It is clear, however, that transmission and processing of sensory information, at least as far as vision and hearing are concerned, occur at the level of the receptor organs as well as in the brain itself.

Generator Potentials. The action potentials generated in the sensory nerves (the optic nerve, the auditory nerve, the olfactory nerve, the dendrites inside the taste buds, and perhaps also the fiber endings connected to the receptors for pressure, temperature, and pain) are usually evoked by synaptic stimulation. A presynaptic cell may itself carry action potentials or it may only generate a change in its membrane potential, lacking the regenerative and traveling aspects of an action potential. These slower "generator potentials" are often observed in cells comprising sensory systems. They have magnitudes that vary with the intensity of the stimulation of the receptors, although this variation may not be linear. They can evoke action potentials in postsynaptic fibers at a frequency that is related to their size. Thus, a generator potential *modulates* the frequency of a sequence of action potentials in a way that reflects the intensity of the stimulus. In the squid eye these events all take place in a single receptor cell; light causes an intensity-dependent change in the polarization of the membrane and this generator potential modulates the frequency of a continuous volley of action potentials that are carried through the optic nerve fibers (the fibers being processions of the receptor cells). Such generator potentials have also been detected in other sensory systems. They seem to involve ion translocations across receptor membranes, similar to those found in the cell membrane of the rods and the cones of the visual receptor and perhaps also the basilar membrane of the cochlea of the auditory receptor, or the cell membrane of the taste bud receptor cells.

The basic question in sensory transduction is, how does a stimulus such as light, sound, or a chemical substance initiate a change in the electrical polarization of a membrane involving the movement of ions? Some clues to the answer exist for the visual process and are given in Section 6.2. This transduction process in the other sensory receptor organs still is poorly understood.

6.2 The Visual Receptor

Light-Sensitive Receptors. Light-sensitive receptors of various forms, ranging from very primitive systems to very elaborate ones, are found in most living organisms. The most primitive forms include receptors for phototaxis and photokinesis (light-induced motion) in plants and some algae and also the light-sensitive cells or clusters of cells found on the surface of worms and molluscs. More highly developed animals possess the light-sensitive organs known as eyes. In eyes an image of the environment is formed on a "screen" made up of light-sensitive cells. The image formation can be simple like that in the pinhole eyes of some simple marine animals (Fig. 6.3a). Because of the small aperture, relatively little light is admitted and the sensitivity is not very high. The *compound eye* of insects and spiders is more elaborate (Fig. 6.3b). Compound eyes consist of many separate channels, called *rhabdomeres*, which give the eye its faceted appearance. Light coming from a certain direction can enter only those rhabdomeres that point in that direction. Finally, there is the vertebrate eye which is very much like a conventional camera; an image of the environment is formed by a system of lenses on a screen called the *retina*.

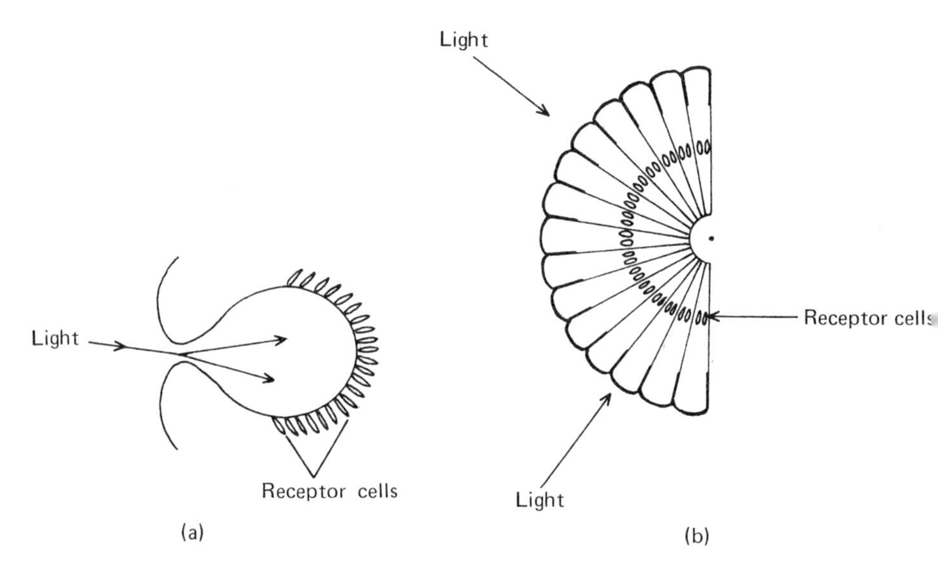

Fig. 6.3 (a) A "pinhole" eye, as found in some simple marine animals. (b) A compound eye, found in insects and spiders.

Figure 6.4 is a schematic representation of a horizontal cross section of a vertebrate eye. The lens system has three components: first, there is the cornea with the aqueous humor behind it; second, the lens itself which has a convexity that can be changed by the ciliary muscle, thus providing an adjustable focus; and finally, the vitreous humor in the eyeball. This system projects an image of the environment onto a layer containing the receptor cells in the back of the eye; this layer is the retina. The amount of light entering the eye is automatically controlled by a circular muscle called the *iris*.

The Vertebrate Retina. The receptor cells in the retina are embedded in a pigmented epithelium layer and are covered by a layer of intermediary cells, nerve fibers, and blood vessels (see also Fig. 6.1). Thus, light must pass through a layer of tissue before it reaches the receptors. The evolution of the vertebrate eye has tolerated this imperfection apparently because, so far, it does not have much selective disadvantage. The epithelium bed contains a black pigment which serves to reduce the haze of scattered light. The central part of the retina is called the *fovea;* it is here that acuity of vision is sharpest. The spot at which the optic nerve fibers pass out of the eye is a "blind spot" because it cannot contain receptor cells.

There are two kinds of receptor cells in the vertebrate eye (see Fig. 6.5). The *rods* are elongated cylindrical cells with a narrow "waist" separating an *outer* and an *inner* segment. In the outer segment are found several

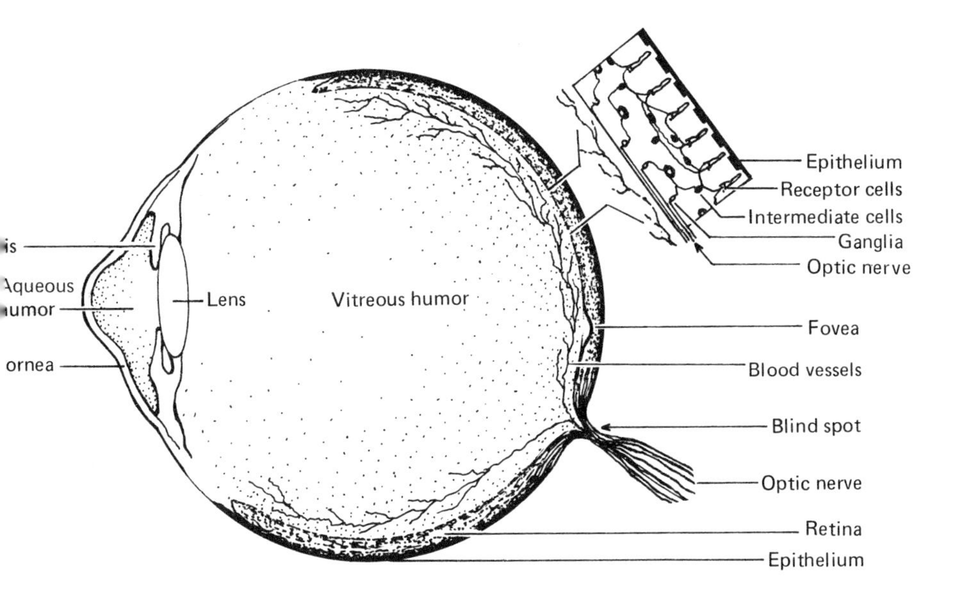

Fig. 6.4 A schematic diagram of the human eye.

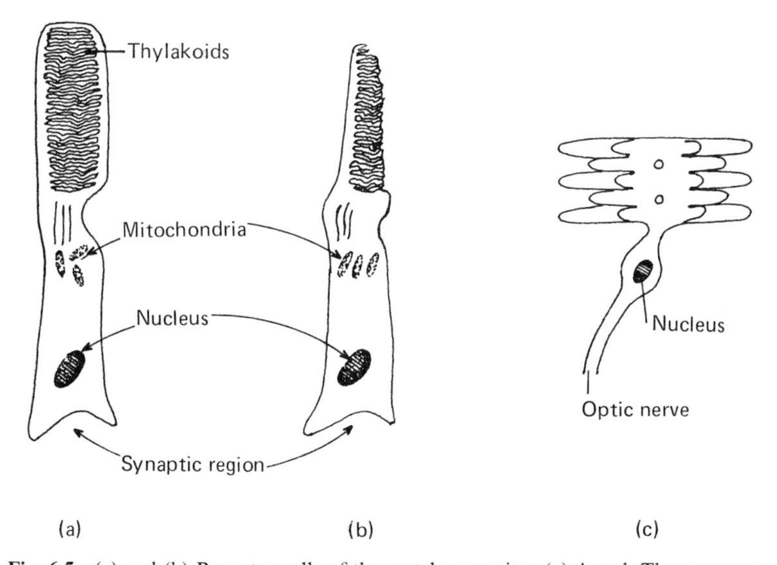

Fig. 6.5 (a) and (b) Receptor cells of the vertebrate retina. (a) A rod. The outer segment contains several hundred thylakoids, each about 200-Å thick, that contain the visual pigment. The length of the outer segment is about 10 μm in man. (b) A cone. The cytoplasmic membrane, folded-in repeatedly, contains the visual pigment of the cone. (c) A photoreceptor cell of a squid. Thousands of tiny outgrowths (only a few are shown) are sticking out of a cylindrical cell. The cell is a specialized neuron; the axons of many photoreceptor cells form the optic nerve.

hundred thylakoids, the flat membrane-bound sacs which we have also seen in the grana of the green leaves of plants. These thylakoids contain the visual pigment. The inner segment contains mitochondria and the nucleus; the synaptic regions are at the far end of the cell membrane. The *cones* have a conical outer segment. The visual pigment in the cones is also confined to membranes, but not in separate thylakoids as in the rod. It is contained in the cell membrane itself which is folded repeatedly at one side, forming a stacked structure somewhat similar to the thylakoid stacks in the rod.

Cones are more concentrated in the central part of the retina and rods are more abundant in the periphery. Cones are specialized for color vision and rods for vision in weak light. The acuity in the fovea is a result of the fact that it contains only cones. Cones are less sensitive than rods but since for the most part each cone is connected, through a bipolar cell, to a separate ganglion cell, they make for high acuity; stimulation of two adjacent cones can lead to the firing of separate fibers in the optic nerve.

In more primitive eyes the receptor apparatus is far less organized. In the squid eye, for example, the receptor cells have thousands of little processions called *microvilli*, which contain the visual pigment (Fig. 6.5c). These cells are actually neurons with axons forming the optic nerve.

Visual Pigments. The principal visual pigment is *rhodopsin*, a compound consisting of a protein called *opsin* and a chromophore which is responsible for the absorption of light in the visible spectral region. The chromophore is *retinal*, a chain of conjugated double bonds (thus an extensive π system) attached to an ionine ring at one end and an aldehyde group at the other. Retinal is a derivate of *vitamin A* obtained by oxidation of the alcohol residue to aldehyde (see Fig. 6.6). *In vivo* the oxidation is coupled

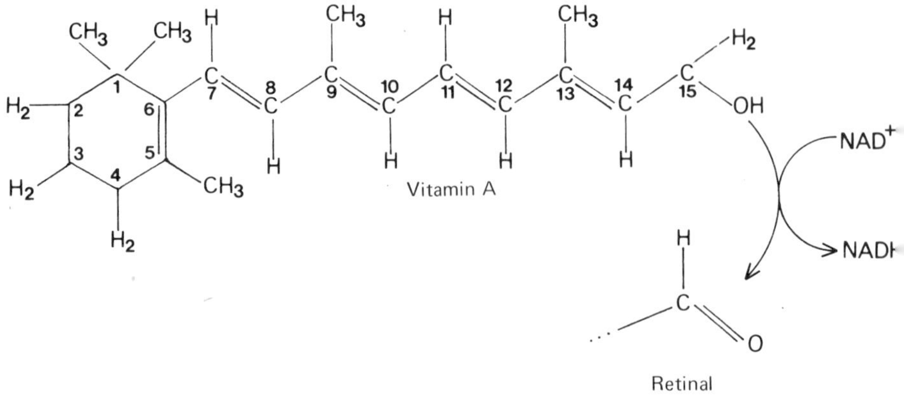

Fig. 6.6 The chemical structure of vitamin A, which is oxidized to retinal by NAD^+.

to the reduction of NAD^+ to NADH. Although it is the most abundant, rhodopsin is not the only visual pigment found in nature. Other pigments are iodopsin, the pigment found in cones of most vertebrates having cones, porphyropsin which is found in the rods of many fish, and cyanopsin, a pigment found in cones of tadpoles. All these pigments are complexes of a specific opsin and retinal. Retinal is found in two forms, corresponding to two forms of vitamin A. The most common form is retinal$_1$ (R_1) which is part of rhodopsin as well as iodopsin. The other form, R_2, differs from R_1 in that its π system is extended to carbons 3 and 4 in the ionine ring and is found in porphyropsin and cyanopsin. Table 6.1 summarizes some properties of these four pigments.

The absorption spectrum of rhodopsin is given in Fig. 6.7. The main absorption band in the visible spectral region peaks at about 500 nm which is also the most effective wavelength for human vision in a dark-adapted eye exposed to weak light. In strong light the peak wavelength of such a visibility curve is shifted to about 550 nm (which is close to the absorption maximum of iodopsin). The agreement of the visibility curves with the absorption spectra provides evidence that in weak light, when rod vision is predominant, rhodopsin is responsible for vision while in strong light, when rod vision changes to cone vision, iodopsin is the pigment involved.

Light causes the bleaching of rhodopsin in both the intact eye and *in vitro*, in excised retinas as well as in solution. Recovery of the absorption occurs if the retina is intact and metabolically active. The bleaching occurs through a detachment of the retinal from the opsin. The recovery involves the reduction of retinal to vitamin A, the diffusion of the vitamin A out of the cell and into the pigmented epithelium, the reconversion of the vitamin A, and the reattachment of the chromophore to the protein in the outer segment of the receptor cell. The detachment and reattachment of the retinal is the essential part of the photochemistry of vision and is accomplished as follows.

TABLE 6.1 Visual Pigments

Pigment	Composition	Absorption maximum (nm)	Occurrence
Rhodopsin	Retinal$_1$ + rod opsin	500	Rods of most vertebrates
Iodopsin	Retinal$_1$ + cone opsin	560	Cones of most vertebrates
Porphyropsin	Retinal$_2$ + rod opsin	520	Most fish
Cyanopsin	Retinal$_2$ + cone opsin	620	Tadpole, but not frog

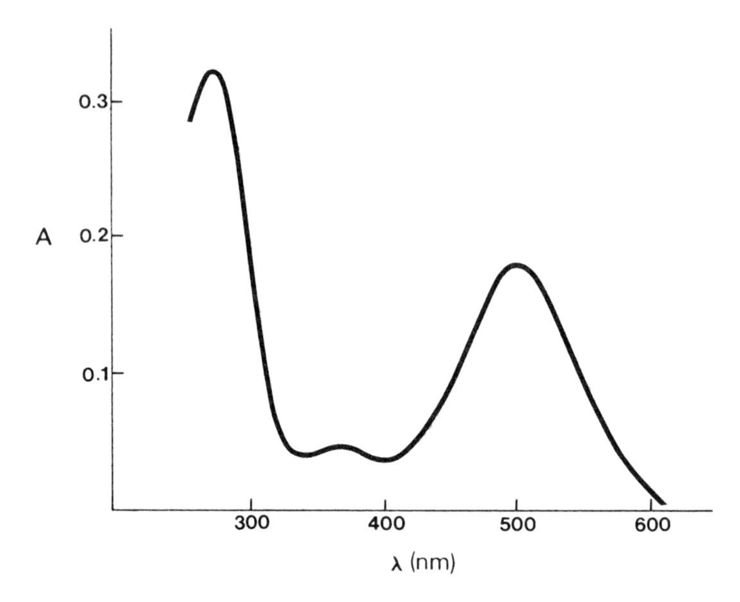

Fig. 6.7 The absorption spectrum of rhodopsin.

Retinal, and also vitamin A, can exist in various stereoisomeric forms. The *11-cis isomer* and the *all-trans isomer* are important for vision (Fig. 6.8). These stereoisomers are interconvertible by a photochemical reaction. The 11-cis isomer spontaneously combines with opsin; the all-trans isomer does not. Thus, in rhodopsin the chromophore is bound to the protein in its 11-cis configuration. This bond is by a so-called Schiff-base linkage; it is formed when the aldehyde group of the retinal reacts with the amino group of an amino acid (lycine) of the opsin, expelling a water molecule:

$$\text{Ret-}\overset{\text{H}}{\overset{|}{\text{C}}}{=}\text{O} + \text{H}_2\text{N-lys-Opsin} \longrightarrow \text{Ret-}\overset{\text{H}}{\overset{|}{\text{C}}}{=}\text{N-lys-Opsin} + \text{H}_2\text{O} \tag{6.1}$$

Photochemistry of Vision. When rhodopsin is excited, the 11-*cis*-retinal converts to all-*trans*-retinal. The shape of the chromophore is now incompatible with the opsin and the protein probably goes through a number of configuration changes before the all-*trans*-retinal comes off. It is then reduced to all-*trans*-vitamin A and, after diffusion into the pigmented epithelium, is reconverted either by light or enzymatically (involving the enzyme isomerase) into the 11-cis form. The 11-*cis*-vitamin A is then oxidized to 11-*cis*-retinal and linked by a Schiff-base linkage to the opsin back in the outer segment of the receptor cell. The protein then assumes its original conformation characteristic of rhodopsin.

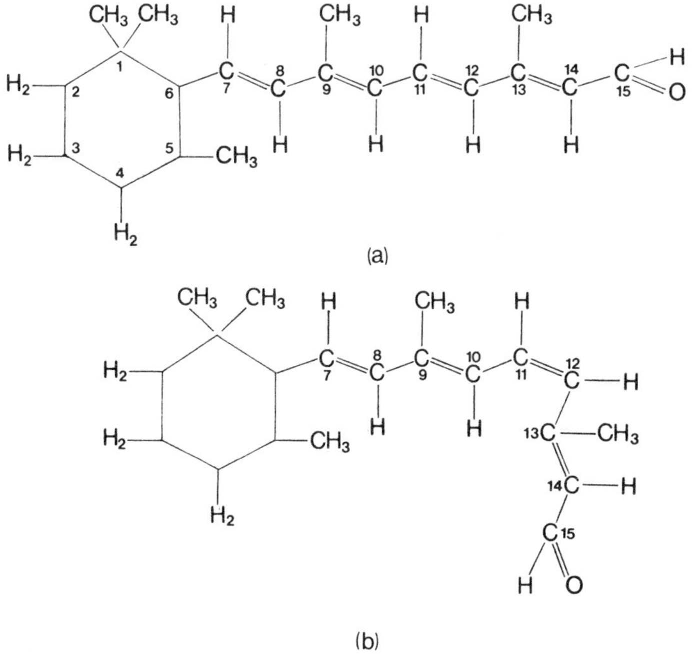

(a)

(b)

Fig. 6.8 The two isomeric forms of retinal. (a) The all-trans form; (b) the 11-cis form.

The intermediate stages which we have assumed to be different conformations of the opsin can be followed by spectrophotometric experiments at different temperatures; except for the first step, the photoisomerization proper, all successive steps are temperature dependent. If a suspension of rhodopsin is brought to liquid nitrogen temperature (77 K = −196°C) and excited by a flash of light its absorption spectrum changes. Some of the rhodopsin with an absorption maximum at 500 nm will disappear and a substance with a maximum absorption at 543 nm will be formed. This first product of the photochemical reaction sequence is all-*trans*-retinal bound to opsin, called *prelumirhodopsin.* Warming up the suspension gradually allows us to follow the subsequent steps in the series. The next substance formed, appearing at temperatures above −140°C, is *lumirhodopsin.* We then find *metarhodopsin I*, above −40°C, and metarhodopsin II, above −15°C. Metarhodopsin II decays either to *pararhodopsin* or to free all-*trans*-retinal and opsin at temperatures above 0°C. Pararhodopsin itself also decays into free all-*trans*-retinal and opsin. Because of the photochemical interconvertibility of the isomers of retinal, light can reconvert every intermediate in the sequence back to rhodopsin. This sequence of events is shown schematically in Fig. 6.9.

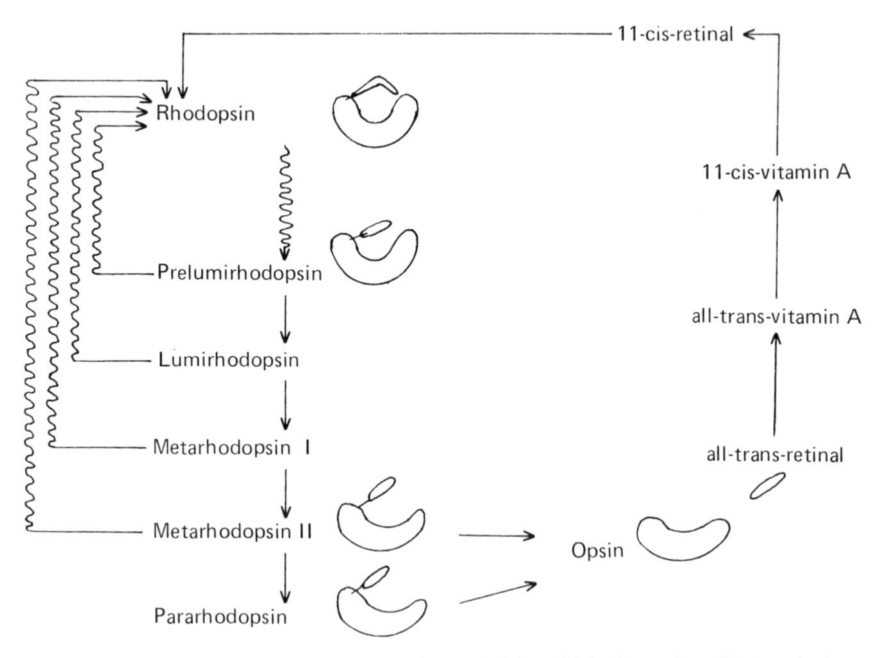

Fig. 6.9 A diagram of the photochemistry of vision. Light isomerizes 11-*cis*-retinal to all-*trans*-retinal in the first intermediate product prelumirhodopsin. The protein then goes through a number of intermediate configurations until the all-*trans*-retinal comes off. After oxidation, isomerization into the 11-cis form and reduction rhodopsin is again formed. Since the photoisomerization is reversible, as long as the retinal stays on the opsin, rhodopsin can be reproduced by light (designated by the wavy arrows).

The photointerconvertibility of the chromophores causes the phenomenon of the *photostationary* state. Suppose a solution of rhodopsin at liquid nitrogen temperature is exposed to a brief flash of light, so that many molecules of rhodopsin are converted to prelumirhodopsin. A second flash of light will excite more rhodopsin and also some prelumirhodopsin formed during the first flash. The excited molecules of both types, can thus decay partly to one isomeric form and partly to the other; that is, rhodopsin to prelumirhodopsin and prelumirhodopsin to rhodopsin. Consequently, under continuous illumination a steady state can be established in which the rates of conversion of one pigment to the other are equal. This steady-state condition is called the photostationary state; the relative proportions of the two pigments in this state are determined by the relative rates at which the pigments absorb light (their absorption coefficients and their concentrations) and the relative probability that each kind will decay to its photochemical product (rhodopsin for prelumirhodopsin and prelumirhodopsin for rhodopsin). If the absorption spectra of both pigments are different (which is true in the case of rhodopsin and prelumirhodopsin), one can let the concentration of one predominate over the other by a suitable choice of the wavelength of the exciting light. Light of 600 nm, which is absorbed far more by prelumirhodopsin than by rhodopsin, will cause the concentration of rhodopsin to predominate over that of prelumirhodopsin in the photostationary state. Conversely, the use of 450 nm light, absorbed more strongly by rhodopsin, causes a higher concentration of prelumirhodopsin in the photostationary state. Photostationary states are set *in vivo* and are also important, for instance, in the setting of light-influenced daily circadian and other periodic rhythms.

ERG and ERP. The sequence of steps in the photochemical cycle of vision can also be followed by measuring electrical potential changes caused by light entering an intact eye. If an electrode is placed on the cornea and another behind the retina (or in the mouth) a voltage change will occur between them when light is allowed to enter the eye. A recording of such changes is known as the *electroretinogram* (ERG). The changes start a few milliseconds after the onset of illumination and show typical waves labeled a, b, c, and d in Fig. 6.10. These potential waves apparently reflect changes in the electrical activity (mobility of ions) of the membranes of the receptor cells and interneurons. When strong but short flashes are used instead of continuous illumination, a smaller electrical effect can be detected that begins with no detectable lag period and lasts a few milliseconds, thus filling the initial gap of the ERG. This more rapid response is called the *early receptor potential* (ERP) or *fast photovoltage* (FPV) and most probably reflects the formation of electric dipoles attending the photochemical steps. It shows a positive peak, labeled R_1, immediately followed by a negative peak R_2, after which the ERG starts to develop. At temperatures below -15°C the R_2 peak disappears. It has been established that the rise of R_1 is associated with the formation of metarhodopsin I and that of R_2 is associated with the conversion of metarhodopsin I to metarhodopsin II. By using closely spaced pairs of flashes the conversions of the intermediates of the cycle back to rhodopsin can be followed, and the amounts of the intermediates formed together with the kinetics of their formation can be revealed. Experimentation of this kind has contributed substantially to our knowledge about the visual photochemical cycle.

Color Vision. Color vision seems to be possible when two different wavelengths of light cause different relative stimulation in two or more types of receptor cells. Although two kinds of receptors with different absorption spectra are enough to provide some sense of color, the existence of three different types is suggested by J. C. Maxwell's demonstration that nearly

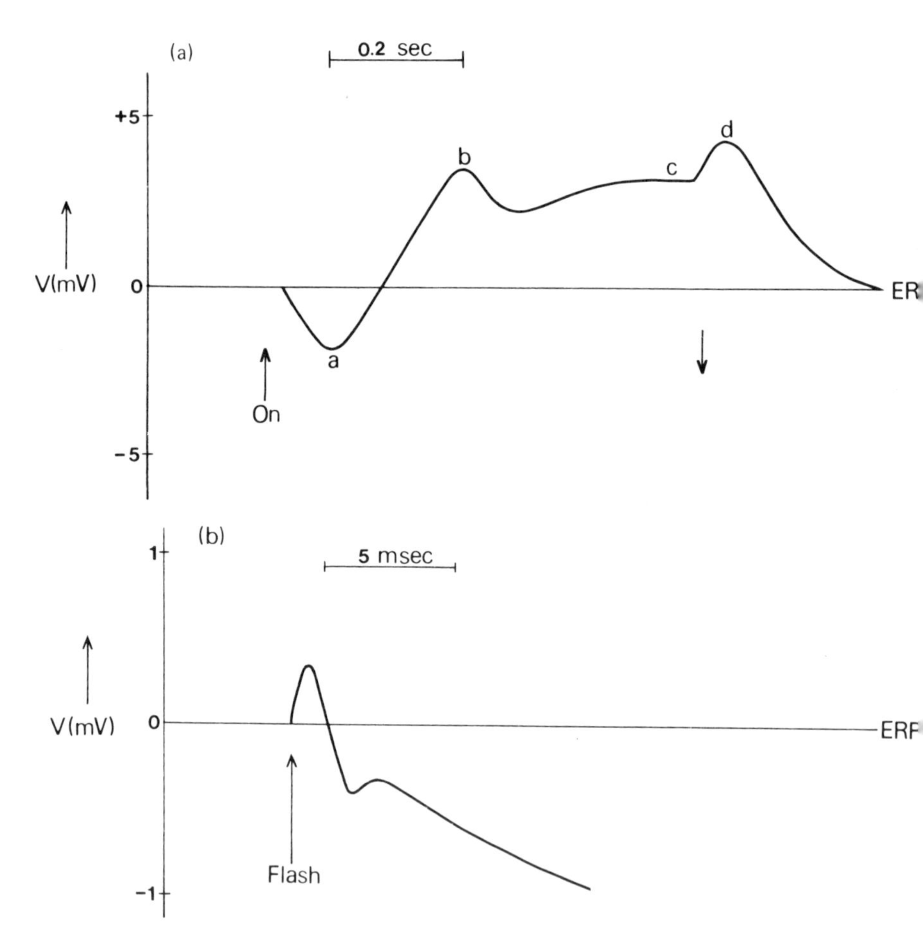

Fig. 6.10 The electrical responses of the retina. (a) The electroretinogram (ERG) occurring a few milliseconds after the onset of illumination, showing the a, b, c, and d waves. (b) The early receptor potential (ERP) occurring immediately after flash illumination, showing a positive peak followed by a negative one.

all colors can be reproduced by mixing of three "primary" colors chosen in such a way that none of the three can be made by mixing the other two. There is evidence for two different pigments in the cones in the human retina. These pigments have been labeled *chlorolabe* and *erythrolabe* and absorb maximally at 540 and 580 nm, respectively. Direct evidence for a far less abundant blue pigment (*cyanolabe*, absorbing at 440 nm) is not yet convincing. Experiments in which attempts have been made to identify distinct color sensors in the eye by measuring electrical responses in single receptor cells (cones) have shown that, in the carp, there are three kinds of cones each responding to different colors of light. It seems, however, that the perception of color requires a good deal of processing in the interacting neurons as well as receptors with different pigments.

The Visual Generator Potential. Far less is known about the way in which the visual photochemical cycle causes generator potentials which then give rise to the firing of action potentials in optic nerve fibers. Information about this problem has been derived from experiments in which microelectrodes are implanted in selected places so as to study electrical activities of single cells or small groups of cells during stimulation by light. In this way it has been established that vertebrate rods possess a current loop maintained by Na^+ ions which flow *inward* in the outer segment, where the pigment is found, and *outward* in the inner segment, where communication with the intermediate cells is located. When light strikes the outer segment this current decreases, in other words the membrane of the inner segment becomes more polarized (more positive at the outside). This hyperpolarization is communicated to the horizontal cells, but in the bipolar cells it may be either a hyperpolarization or a depolarization. This seems to be related to the patterns of antagonism between central and peripheral regions of a receptive field (see Section 6.1). The amacrine cells are all depolarized and occasionally show an action potential. The ganglion cells respond with numerous action potentials and only a slight depolarization.

The way in which excitation of a rhodopsin molecule anywhere in the outer segment can cause the development of generator potentials is still an intriguing riddle. One could think of the following as a plausible sequence of events for the transduction. The isomerization of the 11-*cis*-retinal to all-*trans*-retinal exposes a membrane area with a diameter of about 10 Å. Through this area a leak of Na^+ can occur, either by opening a pore, closing a preexisting pore, or else by exposing a site for active transport. This pore or site causes the inward sodium current that is electrically compensated for by the outward current in the inner segment. Such a model is feasible when the pigment is embedded in the external cell membrane, such as the case of invertebrate receptor cells and vertebrate cones. In the vertebrate rods, however, we still have to deal with the fact that the pigment is in a thylakoid membrane which, although very close to the external membrane is still a membrane bounding the interior of a separate thylakoid and different from the cell membrane across which the generator potential develops.

6.3 The Auditory Receptor

Hearing. Hearing, like vision, is a major sensory system. It enables many animals, including humans, to perceive their environment through mechanical vibrations within given frequency ranges. Although the study of hearing is one of the oldest fields in biophysics, little is known about its transduction mechanism. The reason is most probably the poor accessibility of the system to experimental approach. To obtain relevant information the system must be pretty much intact, and under such conditions it is extremely difficult to reach the inner ear, where the transduction organ is located. The system is enclosed in deep cavities of the skull; to reach the inner ear, parts of the temporal bone have to be cut away, which is difficult to do without extensively damaging the system. Anatomical and histological studies have revealed much about the structure of the ear, but not many clues to the mechanism of transduction. The early biophysical studies were predominantly limited to measurements of impedance and impedance matching devices of the sonic energy transmission system. Although data collected from these studies may eventually help us in throwing some light on the transduction problem, they have not yet revealed anything in that direction.

Hearing is the perception of pressure waves that are propagated by a vibrating medium. The waves of vibration enter the hearing apparatus which in all vertebrates is very homologous to the human ear, although the frequency characteristics may vary appreciably from species to species. For the human ear the frequency range is between about 20 Hz and about 20 kHz; the exact limits depend on the individual and are influenced by age and environment.

The Ear. Figure 6.11 is a sketch of the human ear. It consists of three parts. First, there is the *outer ear* with the *pinna* or *auricle* and the outer *auditory canal.* At the end of the auditory canal is the *tympanic membrane*, or ear drum, which separates the outer ear from the *middle ear* cavity. In this air-filled space are three tiny bones called the *middle ear ossicles;* these are the *malleus* (hammer), the *incus* (anvil), and the *stapes* (stirrup). The outermost of the three is the malleus which is pressed against the tympanic membrane. The innermost of the bones, the stapes, pushes against a membrane called the *oval window* which separates the air-filled middle ear cavity from the liquid-filled channels of the inner ear. These channels are part of a coiled tube-like organ called the *cochlea.* The cochlea contains the auditory transduction organ, called the *organ of Corti.* A narrow channel called the *Eustachian tube* connects the air-filled space of the middle ear with the pharynx. This connection is necessary in order to maintain equal average pressure on both sides of the tympanic membrane.

When mechanical vibrations of air enter the auditory canal the tympanic membrane is set in motion. This motion is transmitted to the middle ear ossicles which together act as an amplifier system. The bones, in fact, form a lever system that decreases the amplitude of the vibrations of the tympanic membrane but increases the force exerted on the oval window, thus perform-

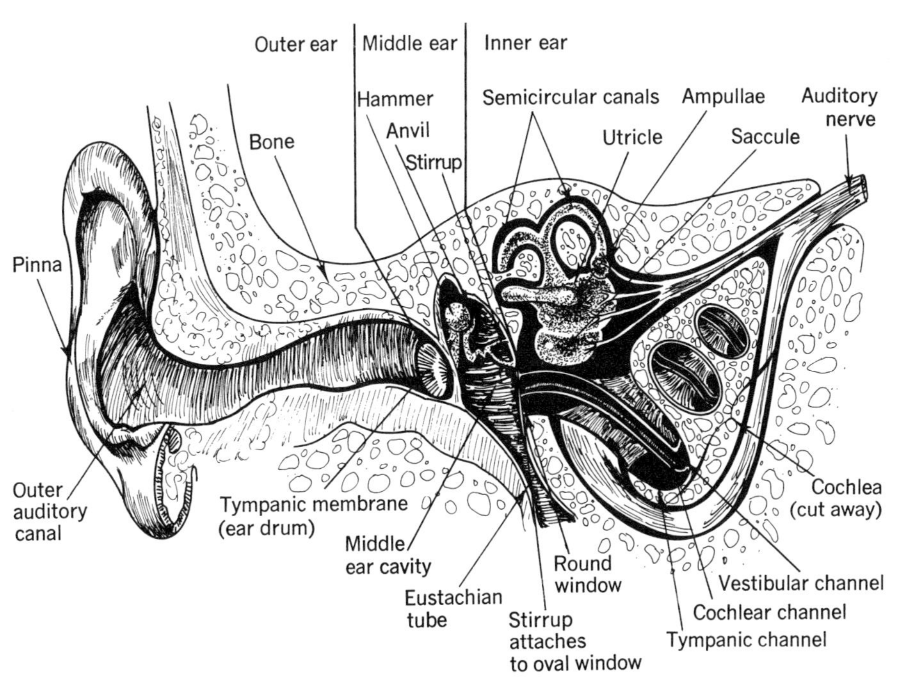

Fig. 6.11 A drawing of the human ear showing the outer, middle, and inner ears and their related parts (from Nason, 1965).

ing an impedance matching between the tympanic membrane and the fluid in the cochlear channels. The amplification amounts to about a 25-db gain in acoustic pressure. Both the malleus and the stapes have muscles which, upon contraction, can change the elasticity of the system, thus providing a volume control.

The Cochlea. The inner ear is a multichambered cavity. The *saccule*, *utricle*, and the *semicircular canals* have nothing to do with the sense of hearing, but they are involved in sensing balance. The only auditory part in the inner ear is the *cochlea*, which is a tube that is coiled into a helix of two and a half turns, decreasing in size as it turns away from the middle ear. Two longitudinal membranes divide the "tube" into three ducts (Fig. 6.12) all filled with fluid. Below the *basilar membrane* is the *tympanic channel* which is filled with a fluid called *perilymph*. Above the basilar membrane are two ducts separated by *Reisner's membrane*. These ducts are the *vestibular channel* and the *cochlear channel*. The vestibular channel connects with the tympanic channel through an opening called the *helicotrema* and, thus, also contains perilymph. The fluid in the cochlear channel is somewhat different and is termed *endolymph*. One could visualize the three ducts as being

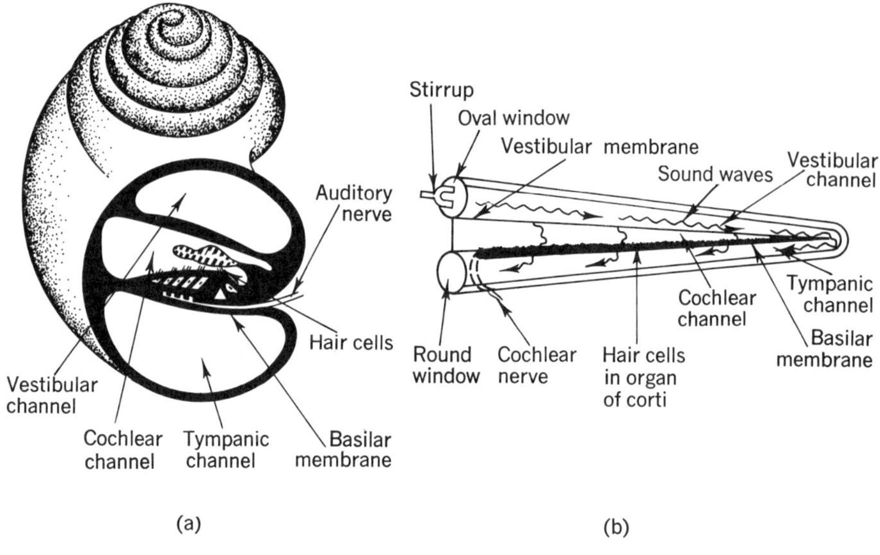

Fig. 6.12 A cross-sectional view of the cochlea (a) and a schematic representation of the organization of the three liquid filled channels in the cochlea (from Nason, 1965).

formed by a long tube which is folded, thus forming two interconnected ducts with a third one pressed in between (see Fig. 6.12b). The oval window is found at one end of the vestibular channel. The footplate of the stapes fits against this window; vibrations picked up by the middle ear ossicles are transmitted to the perilymph across the oval window. The end of the tympanic channel is called the *round window*.

The Auditory Receptor Cells. Seated on the basilar membrane and sticking out into the endolymph is the transducing organ, called the *organ of Corti*. This organ is a complicated system of cells, including the *hair cells* which seem to be the site at which the transduction proper takes place. It is believed that bending of the hair cells somehow excites the nerve endings also located in the organ of Corti.

As we have indicated above, little is known about the transduction mechanism in hearing. Some pertinent information has been obtained by observing the responses of the basilar membrane to sound of different frequencies and by using microelectrodes placed in the cochlear spaces and in fibers of the auditory nerve. The basilar membrane shows displacements in response to sound. This movement is actually a traveling wave having an amplitude maximum at a distance from the oval window which is a function of the frequency of the exciting sound wave. Some typical frequency responses are shown in Fig. 6.13.

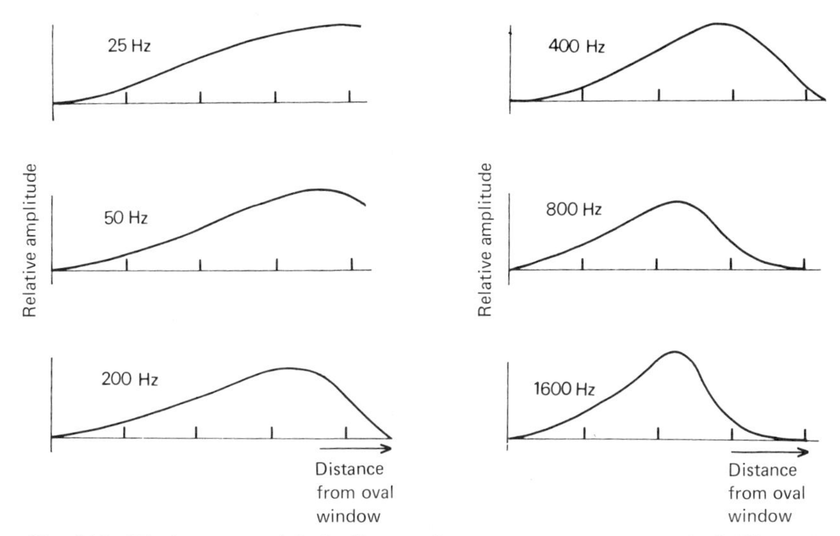

Fig. 6.13 Displacements of the basilar membrane as a response to sound of different frequencies (from von Békèsy, 1953).

Microphonic Potential. Measurements of electrical activity using microelectrodes inserted in the cochlear spaces have shown that sound causes a potential difference across the basilar membrane. This potential difference is correlated with both the frequency and amplitude of the incident sound wave. This is sometimes termed the *microphonic potential* and a comparison with generator potentials similar to those manifest in vision immediately suggests itself. Thus far, however, there is no evidence for any ion translocation associated with this potential change. Action potentials in auditory nerve fibers do occur concommittantly with the microphonic potential and there is some evidence for an association between the volleys of action potentials and the microphonic potential. A unique relation could not, however, be established, although it seems certain that groups of neurons are involved with the pitch discrimination. Processing of information undoubtedly takes place, as with vision, but it is not known whether part of this processing occurs in the ear (like that part of visual information processing that occurs in the retina) or whether all of it occurs in the central nervous system.

6.4 The Chemical, Somatic, and Visceral Receptors

Chemical Stimulation. The chemical receptors are those of olfaction and taste. Although they both respond to chemical stimulation, they are

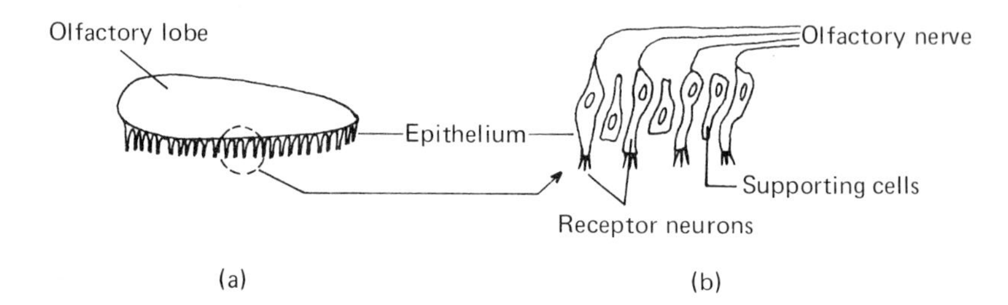

Fig. 6.14 (a) A drawing of the human olfactory receptor organ and (b) a part of the olfactory epithelium showing the receptor neurons.

quite different otherwise, involving different sensory pathways to different areas of the brain. Olfaction requires that the stimulating chemical substance be volatile and soluble in the mucous secretions of the upper membranes of the nose. Taste requires that the chemical be soluble in water.

The olfactory receptor is located in the mucous membrane of the upper part of the nasal passage. The receptor cells are part of the so-called *olfactory epithelium* and are, in fact, specialized nerve cells; they have fibers which together form the olfactory nerve (Fig. 6.14) at one end and have little hairy outgrowths at the other end. There is no information about the reactions which evoke action potentials in the olfactory nerve fibers.

The receptor for the sense of taste is the *taste bud* (Fig. 6.15). There seem to be four fundamental sensations of taste each of which has a separate kind of taste bud located on different parts of the tongue. Taste buds for the sensation of sweetness and saltiness are mostly located at the tip of the tongue. Sourness is detected by taste buds distributed along the sides of the tongue and the back of the tongue contains the buds for the taste of bitterness. All taste buds have the ovoid structure shown in Fig. 6.15 but they can differ substantially in sensitivity and specificity of taste. Furthermore, the receptor cells in taste buds are specialized neurons; their outgrowths at one end are axons together forming the gustatory nerve.

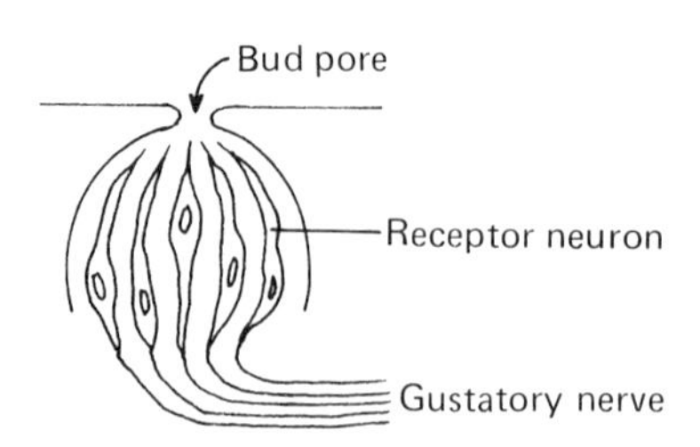

Fig. 6.15 A drawing of the human taste bud showing the receptor neurons.

Somatic Stimulation. Somatic receptors can be distinguished as those responsible for touch and pressure, those for heat and cold, and those for pain. The sensation of touch seems to be provided by three types of receptors. The bending of a hair can cause a single discharge from a myelinated fiber within an area of 1.5–5 cm^2. Continuous discharge can be maintained by another type of receptor under pressure. The "field" for such mechanoreceptors is very small, something like 1–2 mm in diameter. The rate of the discharges can be very high (330 discharges/sec under high pressure). Finally, there are receptors which are sensitive to both pressure and cooling. The rate of discharge from these receptors is usually much smaller than that of the pure pressure-sensitive type.

Two different sets of receptors are responsible for the sensation of temperature, one initiating firing more rapidly when the temperature of the skin drops and the other more rapidly when the temperature is raised. The skin seems to be divided into spots that are sensitive only to cooling and spots that are sensitive only to heating. There is little information about the anatomical, let alone the physiological, character of these receptors.

Pain receptors do not have the high specificity that other receptors have; they are spread throughout the body. Electrical and mechanical disturbances, extremes in temperature, and chemical reactions can all initiate pain responses. Although the sensations of pain, touch and pressure, and temperature are different and thus must involve different types of receptors, they quite often go together. Moreover, the firing of the nerve fibers is also initiated directly by touch, cold, and heat or by combinations of these stimulants. This is the reason why it is so extremely difficult to identify the individual receptors.

Visceral Sensory Signals. Very little can be said about the mechanisms of visceral sensations. For most of them it is apparent that they originate in the visceral (internal) organs (hunger, for example, may be initiated by rhythmical contraction of the muscles in the wall of the stomach; thirst is related to a reduction of the body fluid content). The mechanisms by which all these sensations are evoked are totally unknown.

From a biophysical point of view the study of the "lower" sensory systems has not been very rewarding. This is largely due to the fact that little is known about the anatomy and histology of these systems and hardly anything about their physiology. The approach to these systems is extremely difficult and it may very well be that more understanding about their function can only be expected from an increased knowledge of the functions of the central nervous system.

Chapter

7 Theoretical Biology

7.1 Physical Concepts and Biology

Theoretical biology is a subdiscipline of biology that may have a significant impact on the further development of biology. Its meaning for the biological sciences is comparable to the meaning of theoretical physics for the physical sciences. Theoretical physics, and especially mathematical physics, is concerned with the description of physical phenomena in mathematical terms. Beyond this, however, it makes abstractions from physical reality and puts them into a conceptual framework in such a way that from abstract models predictions can be made about the phenomenology of real systems. Theoretical biology is trying to do just that; and the fact that in doing so many concepts of the physical sciences are "taken over" and used to describe biological phenomena in an abstract manner provides an argument for considering theoretical biology as a legitimate domain in biophysics. Its aim, of course, is the unification of assemblies of facts and their simplification through the logic of a (abstract) concept, thus arriving at a deeper understanding and greater predictability.

In the preceding chapters of this book we have seen physical concepts being used to explain biological phenomena as well as put them into a framework wherein closer investigation becomes more systematic. Examples of this are the uses of quantum mechanics and thermodynamics in our discussions.

This is not really theoretical biology, but in some of the subjects discussed (i.e., membrane transport) the application of physical concepts can lead to a much more theoretical treatment. In this chapter we will briefly discuss the way in which some of these physical concepts are used in theoretical biology and to what kind of results this application may lead. The purpose of this discussion is to *introduce* the reader to these problems, rather than to treat them *in extenso*. Moreover, the choice of topics is somewhat arbitrary and far from complete. Readers who desire to pursue this domain more thoroughly are referred to the literature.

The concepts which have been used in theoretical biology are those of quantum mechanics, statistical mechanics, thermodynamics, and, more recently, cybernetics and information theory. Some applications of quantum mechanics have already been discussed in Chapter 4. Other applications involve attempts to deal with the stability of a (living) system operating far from equilibrium and with the collective behavior of a many-unit system (such as an organ consisting of a large number of cells). However, the usefulness of quantum mechanics for the description of such systems is much debated (see, for instance Fröhlich, 1969). The application of (classical) statistical mechanics and statistical thermodynamics sometimes seems easier to justify, but results have so far given only limited information. An example is a study by Oosawa and Higashi (1967) on the polymerization and polymorphism of proteins (such as actin, for example), based on principles of statistical thermodynamics. Kornacker (1972) tried to devise a generalization of statistical mechanics to provide a basis for understanding how the microscopic behavior of nonliving atoms can generate the macroscopic appearance of a "living aggregate."

7.2 Nonequilibrium Thermodynamics

A recent addition to the methods of theoretical biology is the approach based on *nonequilibrium thermodynamics* or the *thermodynamics of irreversible processes*. This newly emerging branch of thermodynamics (see for instance Prigogine, 1955, or Katchalsky and Curran, 1967) is a generalization which makes possible a quantitative description of biological phenomena such as the transport of matter, nerve conduction, and muscle contraction. The common factor of all such phenomena is that they are essentially nonequilibrium open systems in which irreversible processes are occuring. Without the formalism of nonequilibrium thermodynamics, a quantitative description of these phenomena could only be based on kinetic equations derived from specific models or mechanisms. However, the detailed information required for an adequate kinetic description is not often available or is difficult to obtain. The thermodynamic treatment is independent of such kinetic or statistical models. This method thus provides an enrichment which is at the same time a limitation: while the method offers additional insights into the factors influencing the phenomena in question, it never can lead directly to the mechanisms of the phenomena. By using nonequilibrium thermodynamics one could determine, for instance, whether or not a model is feasible from a physicochemical standpoint, but one could never arrive at such a model just by using the method.

Thermodynamics deals with macroscopic quantities, such as pressure, temperature, etc. From a macroscopic point of view we can distinguish between two types of structure: *equilibrium structures* and *dissipative structures*. Equilibrium structures are maintained without an exchange of energy or matter. They form the domain of equilibrium or classical thermodynamics. Dissipative structures are maintained *only* through the exchange of energy and in some cases also that of matter. A living system is typically a dissipative structure. Nonequilibrium thermodynamics is an extension (or rather a generalization) of equilibrium thermodynamics which include dissipative structures. Its basis is the splitting of the entropy term (see Appendix II) dS into a part describing the flow of entropy due to interactions with the environment, d_eS and a part describing the production of entropy inside the system, d_iS. Thus,

$$dS = d_eS + d_iS \tag{7.1}$$

or

$$\frac{dS}{dt} = \frac{d_eS}{dt} + \frac{d_iS}{dt}$$

The second law implies that for all physical processes

$$d_iS \geq 0 \tag{7.2}$$

Equilibrium thermodynamics deals with equilibrium situations in which the entropy production term d_iS vanishes. In nonequilibrium thermodynamics we study macroscopic states on the basis of their entropy production.

We can define the entropy production per unit time and volume σ by

$$\frac{d_iS}{dt} = \int \sigma \, dV \tag{7.3}$$

in which σ is related to flows and forces of the irreversible processes according to

$$\sigma = \sum_i J_i X_i \tag{7.4}$$

In Eq. (7.4) J_i is a flow, or a reaction rate, of an irreversible process i and X_i is the conjugated generalized force driving process i. If such a process is a chemical reaction, J_i is the reaction rate v and X_i is the chemical affinity A divided by the absolute temperature, $X = A/T$. The chemical affinity is defined as

$$A = \sum_j v_j \mu_j \tag{7.5}$$

in which v_j is the stoichiometric coefficient and μ_j is the chemical potential of reaction component j.

Elaboration of Eq. (7.4) can lead to a quantitative description of processes like the transport of matter through biological membranes and energy conversion reactions. Interesting implications follow when the theory is applied to nonlinear systems. Prigogine (1969) has shown that in such cases new structures which are the result of instabilities can bring the system to a new steady state which has a higher order of organization than the original state. In physical chemistry and in hydrodynamics there exist a wealth of systems which show this type of behavior. This, evidently would have tremendous implications for the relation between the physical sciences and biology, if it can be shown that biological dissipative structures can arise from such events.

7.3 Modeling

Physical Modeling. The application of physical concepts to biology has often led to the design of models. Many different kinds of models of living entities can be created. Although in most cases quantitative information is difficult or impossible to obtain from models, they are very useful from a conceptual point of view, giving insight into the cperational strategy of organisms. In this respect they have become an important tool of theoretical biology. We can distinguish physical analog models from mathematical models. A physical analog is a construction that simulates specific functions and/or properties of a real counterpart. A physical analog need not be, but could be constructed in actuality. It could be a model on paper which simplifies the real counterpart in the sense that only the function or property to be studied is represented in the model. A mathematical analysis of the model could then yield information about the real system. An example is a hydrodynamic model of the bloodstream. From such a model useful information can be obtained, for instance, about such matters as pulsate blood flow in elastic tubes, reflections of pulse waves at complex vascular bifurcations, and the effects of viscosity on pulsative flow. Other examples are electrical analogs in which organic functions are simulated by electrical circuits.

An organic function can often be described by the same set of differential equations that describe specific electrical circuits. Such electrical analogs have been made for the cardiovascular system, the myocardium itself, the lung gas flow, and also for things like natural biological (circadian) rhythms. Physical modeling has also become increasingly important in the areas of agronomy and ecology. It should be emphasized, however, that, although such analogs are of considerable interest to physiologists, the models are usually at best very complex but *qualitative* simulations of the real system. In only very few instances have physical models allowed the prediction of unknown quantitative data.

Mathematical Modeling. In physical modeling there is usually some geometric or functional similarity between the model and the real system. In mathematical modeling this similarity is abandoned; in other words, a mathematical model is a much further extended abstraction from the real system. A simple type of mathematical modeling is *numerical kinetics simulation*. This is often used to obtain data about metabolic processes, to establish time-dependent distributions of substances (drugs, for example) in organisms, and to evaluate data obtained from experiments with radioactive or heavy isotopes. The method involves the design of a numerical, so-called *compartment model*. The compartments in such a model represent pools of the substance under study. Transfer of the substance between the pools is described by rate coefficients. This type of model leads to a set of simultaneous differential equations which can usually be solved after application of a few approximations. A generalized compartment model comprising three compartments is illustrated in Fig. 7.1. If it is assumed that the rate

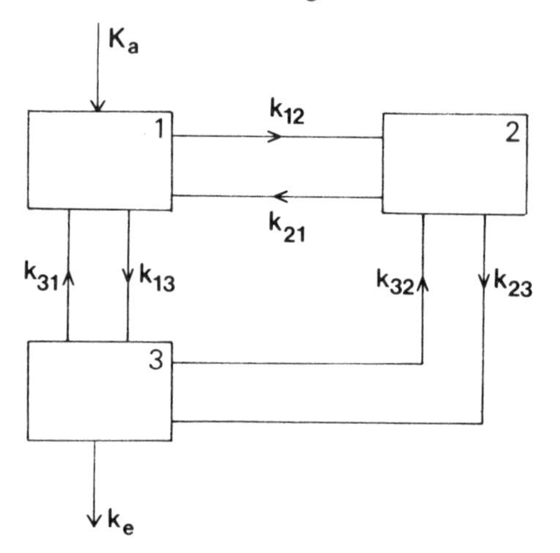

Fig. 7.1 A three-compartment model as described in the text.

coefficients are constants, the set of linear differential equations describing the time-dependent distribution of a substance N among the three compartments is as follows:

$$
\begin{aligned}
dN_1/dt &= -(k_{12} + k_{13})N_1 + k_{21}N_2 + k_{31}N_3 + K_a \\
dN_2/dt &= -(k_{21} + k_{23})N_2 + k_{12}N_1 + k_{32}N_3 \qquad (7.6) \\
dN_3/dt &= -(k_{31} + k_{32} + k_e)N_3 + k_{13}N_1 + k_{23}N_2
\end{aligned}
$$

in which the k_{ij} are the transfer rate constants for substance N, K_a is the rate of intake of N (from an indefinite large reservoir), k_e is the excretion rate constant (from compartment 3), and N_i is the concentration of substance N in compartment i. These compartments can be defined in anyway convenient for the problem at hand. They need not to be actual reservoirs; they can also be states of the substance. For example, to examine iodine metabolism one can define one compartment for the thyroid, another for free or inorganic bound iodine, and still another for protein-bound iodine. By using iterative curve-fitting techniques one can compare calculations using the compartment model with measurements of the disappearance rate of radioactive iodine from the thyroid, the rate of excretion of radioactive iodine, and the time-dependent concentration of radioactive iodine bound to proteins in the bloodstream.

7.4 Cybernetics

Cybernetics and Information Theory. A large class of mathematical models can be described by the branch of science called *cybernetics*. Cybernetics is defined by Wiener (1961) as "the science of control and communication in the animal and in the machine." The term originates from the Greek word κυβερνήτης, which means *steersmanship*. Cybernetics, indeed is a theory of machines, but not of the kind which would contain information about cogs and levers or resistors and transistors. It treats behavior, rather than the things that behave. Its terminology would contain statements such as "this variable is undergoing an harmonic oscillation" without being concerned with whether the variable is actually a point of a turning wheel or a potential in an electric circuit. Its power, therefore, is the fact that it establishes a common language, expressing the common aspects of such diverse subjects as a room thermostate and the central nervous system. For biology in particular, the development of cybernetics may have great significance because it provides a method for the scientific treatment of systems of which complexity is an essential part. Living organisms are complex and their complexity is essential to their function; we know now that living systems are so dynamic and interconnected that alteration of one factor immediately causes alterations in a great many others. With cybernetics it may be possible to study this complexity as a subject in its own right.

Cybernetics and Brain Function. Cybernetics deals with automatons, sometimes called "switching systems" or just "robots." A computer is an automaton. Concepts developed out of the abstract theory underlying computing have been widely used to investigate the interaction of animals and machines with their environment. In biology it is especially the study of the nervous system that has drawn heavily on such concepts. This is not too amazing since one can hardly escape noticing the analogy between pulse conducting neuronal networks and the circuitry of a computer. Already in the early 1940s the idea of the "formal neuron" emerged. Formal neurons, an abstraction of what was known at that time about neuronal operation, operated upon and emitted at specified times binary (all-or-none) signals. Their junctions were either excitatory or completely inhibitory. They functioned by summing up their binary inputs algebraically and subsequently giving an output signal if, and only if, this sum exceeded a certain specified threshold. Networks made up of formal neurons could represent complicated logical formulas and, thus, symbolize the process by which the brain responds to and gives off stimuli. Moreover, a formal neuronal network provided with "receptors" and "effectors" is equivalent to an abstract representation of a computing system. Analogies like these, computing systems and formal neuronal networks on one hand and brain function on the other, have been developed by Wiener (1961).

Genetic Information and Brain Structure. The analogy between formal neuronal networks and brain function implies that real neuronal networks in a brain are circuits assembled according to "wiring diagrams." This conclusion has some interesting consequences, among which is that the circuitry in a brain, involving some 10^{10} neurons, represents in its structure an amount of information that far exceeds the genetic information of a genome (the total of genetic information contained in the triplets of the nucleotide sequence of DNA). One could conclude then that brain structure, at least as far as its "wiring diagrams" are concerned, is only in part determined genetically. However, one has to be careful in using the term information in this context. H. Bremerman (1967) made this calculation, comparing the number of triplet combinations in a typical genome (4×10^9 nucleotide pairs) with the number of ways in which n neurons with m dentrites can make connections with each other. By putting $n = 10^{10}$ and $m = 10^2$ and introducing the constraint that each neuron can only make connections with a subset n' ($= 10^4$) of neurons (arising from the limiting length of dendrites), he arrives at slightly less than 10^{13} bits of information in brain structure, which compares to about 10^{10} bits of information in the genome (the number of bits of information in this case is equal to the logarithm to the base 2 of the number of all possible combinations). In this theory, however, the fact that the information content in bits of brain structure cannot be compared fully with the content of genome information in bits based on base sequences alone is completely ignored. In addition to this base-sequence determined *genetic* information there is also *epigenetic* information, something which is more subtle and much harder to quantitatively determine. The meaning of epigenetic information becomes clear when we look at protein, or more specifically, enzyme synthesis. We have seen in Chapter 3 that enzymes catalize and control biological reactions by their tertiary, and often quaternary structure. Enzymes loose all their activity when by mild heating, or by changing the pH, quaternary and tertiary structure are destroyed; activity is recovered when the original conditions are restored. But genetic information (the base sequences in DNA) determines only the primary structure (the sequence of amino acids) of the enzymes. Enzyme function (bound to higher order structure) is not *genetically* but *epigenetically* determined. It is important to note, however, that under the proper conditions the active higher-order structures *are fully determined* by the sequence of amino acids and, thus, ultimately by the information from DNA. Epigenetic information does not, therefore, come from sources other than DNA.

The Neuronal Code. It may be too simplistic to state, *a priori*, that the "wiring diagrams" of neuronal networks in a brain are determined the same way as the higher-order structures of enzymes are. In a sense a "wiring diagram" is comparable with the higher-order structure of an allosteric enzyme (its "structure" serves a specific function). It is, however, a far more complex system. A specific "wiring" of a neuronal network should serve to process a specific type of information. However, ablation studies on the brains of animals have shown that signal processing, such as visual integration, is not dependent on the specific details of the "wiring" in the visual cortex; even extensive destruction did not result in marked disintegration of function. A number of cybernetic models (based on formal neuronal networks) have been proposed and elaborated to cope with this problem. One of these models assumes very complicated and extended networks in which incoming signals are coded in so-called error-insensitive codes (codes with a sufficient amount of redundancy, such that the "message" can be recognized even after extensive mutilation of the code).

The difficulty in applying cybernetic concepts to real neuronal systems is that the functional units are never identified by directly observing neurons and their interconnections. The successful application of cybernetics requires a precise knowledge of the *ensemble* of possibilities upon which the real system operates; in other words, what one needs to know is the code (or codes) of the nervous system. Thus, any real application of the concepts of cybernetics can only follow from a knowledge of what the neurons do and how they do it. This information can only be obtained from measurements of the firing patterns of neurons, the changing potentials of the electroencephalogram (EEG), and the generator potentials from the sensory systems.

7.5 Generalizations in Biology

Generalizations. We conclude this chapter, and this book, by a look at the validity of the generalizations made by the biophysical approach described in this and in the preceding chapters. This approach is one among many possible approaches to biology. It differs from many others in that it is a predominantly molecular approach. We looked at the biological macromolecules and concluded that their structure has a pure physical basis. We looked at their function and found that their function is a logical consequence of their structure and is, thus, based on pure physical or physicochemical principles as well. Furthermore, we found, that to describe the way in which an enzyme catalyzes and regulates a metabolic process we neet not specify what metabolic process really was catalyzed and regulated. Neither did we need to know in what cell of what organ in what organism this process took place. We found also that the enzyme came about by processes, involving nucleic acids and a protein synthesizing apparatus, which are exactly similar in all known forms of life. Finally, we found that these processes follow from and are based on well-known physicochemical principles. On a molecular basis then, generalization seems to be possible and quite legitimate.

The Principles of Biology. What do these generalizations tell us of the fundamental principles of biology? Are these fundamentals nothing other than the physics and chemistry of these structures, which arose out of inanimate matter by a process that "captures" a rare fluctuation away from the average evolution toward disorder and randomness and which, subsequently, evolved necessarily and deterministically to the self-preserving and self-reproducing entities we know as living systems? Or is it that living systems are essentially different and that there is, perhaps, another as yet unknown physical principle, a complementarity principle, that would generalize physics to include these strange entities?

This latter opinion is held by the theoretical physicist W. Elsasser (1966). His argument for the singularity of living systems is that they are members of of a class which is radically different from those classes formed by physical systems. This is a consequence of their complexity. Because of the complexity, of structures as well as processes, a living system is a rare realization out of an immense number of possible realizations (possible states) of matter in this kind of organization. All these possible states are determined according to pure physical principles, but the fact that (because of this immense number of possibilities) living system are members of a strictly heterogeneous class, predictions based on sampling of the class is fundamentally impossible. Physical systems, on the other hand are members of homogeneous classes in which one member is indistinguishable from another. Sampling, and, therefore, statistical predictions are quite possible.

The idea that the progress of knowledge about the properties of matter is tied to a "loss of explanation" was originally proposed by N. Bohr (1933) who applied it to quantum mechanics. Elsasser asserts, with reference to the abstraction of the inhomogeneous classes, that uncertainties analogous to the quantum mechanical uncertainty are basic in biology. In this he sees an important tool for the design of a new theoretical biology.

These ideas seem to be somewhat contradictory to the universality of many processes in molecular biology. Moreover, many *physical* systems do belong to classes which satisfy Elsasser's definition of an inhomogeneous class quite well. What could be stated is, that the complexity of living systems makes an *a priori* prediction impossible but does not at all put into question an *a posteriori* explanation based on pure physical principles. But this is not a prerogative of living systems only; the shape and look of every rock picked up from a field has this property.

The Frontiers. Of course, the last word has not yet been spoken in biology. There are still frontiers of knowledge beyond which properties may be revealed that may or may not radically change our views. Two of these frontiers seem to be the beginning stage and the present end stage of evolution. At the beginning stage there is the origin of the self-reproducing self-preserving system which we call a living organism. The theory of evolution is no longer a hypothesis full of speculations. Many aspects of it are amenable to experimentation and much is revealed in that way. There still is, however, the nagging problem of the evolution of the genetic code and the protein synthesizing apparatus. The latter consists of some fifty macromolecular components which are all coded by DNA. Thus, *the code can be translated only by translation products of the code.* How and when has this closed system developed? How did the code develop and why is it universal?

Is there a direct relation between the development of the code, its universality, and the evolution of the genetic apparatus?

At the other end is the central nervous system. Although, as has been shown in previous chapters, we are beginning to understand how a single neuron fires its spikes, we still have not the slightest idea of how this firing relates to the operation of even a simple neuronal network. Even the structure of relatively simple nervous systems, let alone that of brains, is unknown and as yet inaccesable to physiological experimentation. Thus, we are still far from having a notion about the operation of processes like perception, learning, and memory.

These frontiers are challenging, especially for those who, led by previous successes, want to approach biology with the conceptuality of the physical sciences.

Appendix

I Elements of Quantum Mechanics and the Electronic Structure of Atoms

AI.1 The Principle of Quantization and the Uncertainty Principle

The introduction of quantum mechanics at the beginning of the twentieth century was a major development of physics and, for that matter, science in general. Before that, physical theories seemed to be settled and in a rather satisfactory state. Mechanics obeyed the Newtonian laws, optical phenomena were explicable by the wave theory developed by Huygens, Young, Fresnel, and Hertz, and electricity and magnetism were ruled by the laws of Maxwell. There were, however, a few unsolved problems. One of these was the fact that the theoretical derivation of the density of black-body radiation as a function of its wavelength did not agree with the experimental determination of this quantity. In order to overcome this difficulty, Max Planck proposed (in 1900) that energy is not transferred continuously but in integral multiples of a fundamental amount called a *quantum*. If ν is the frequency of the radiated energy, this quantum of energy is

$$E = h\nu$$

in which h is a universal constant known as Planck's constant.

It soon became apparent that this principle of quantization was indeed a universal one. Difficulties in the reconciliation of experimental data with theory could now be solved. An elegant example is the theory of specific heat which Debye developed by the application of Planck's ideas. Another is Bohr's explanation of the typical atomic line spectra, such as the Balmer lines in the hydrogen spectrum, by quantization of the energy states in atoms.

Einstein extended Planck's theory by proposing that electromagnetic radiation actually consists of little parcels of energy $h\nu$, called photons, which travel in a vacuum with velocity c. This assumption explained the photoelectric effect and the absorption and emission of electromagnetic radiation. A particulate description of light was proposed a long time before by Newton, but it was discarded in view of the overwhelming evidence for the electromagnetic wave theory. The new element brought in by Einstein was not so much a revival of Newton's photon theory but rather a recognition that electromagnetic radiation sometimes appears as a wave phenomenon and at other times as a stream of particles. These two aspects of light do not, as may appear at first sight, *exclude* each other as a description, but rather *complement* each other.

Particles such as electrons, protons, or neutrons also have this dualistic character. A wave is associated with the movement of particles which has a wavelength

$$\lambda = h/p \tag{AI.1}$$

in which p is the linear momentum of the particles and h, again, is Planck's constant. These "matter waves" or "de Broglie waves" (named after their original proponent) can be demonstrated in phenomena such as electron diffraction.

The theoretical justification of this duality is *quantum mechanics*, a theoretical concept set up by Erwin Schrödinger and Werner Heisenberg which links the continuous characteristics of a wave with the discrete characteristics of a particle. This is done by replacing the equation-of-motion of a system by a wave equation, the so-called *Schrödinger wave equation*. The solution of this wave equation is a function of space and time, the *wave function* (the mathematical expression of quantum mechanics). This function represents the *state* of a system (for instance, an electron in an atom or a molecule); physical quantities such as energy, angular momentum, etc., can be calculated relatively easily once the wave function is known.

The Schrödinger wave equation is a differential equation generated by replacing a physical quantity β by a *mathematical operator* $\boldsymbol{\beta}$ and having the operator act on a function ψ in equations of the type

$$\boldsymbol{\beta}\psi = \beta\psi \tag{AI.2}$$

A mathematical operator is a *prescription* for one or a combination of mathematical operations, such as multiplication, division, and differentiation. In (AI.2) it can mean, for example: "Differentiate the function ψ with respect to the space coordinates and multiply it with a number of constants." In (AI.2) the operation is put into an equation; usually, solutions for such equations are possible only for a number of discrete values of β, the so-called *eigenvalues* (from the German word *eigen* which means "own" or "belonging") of the equation. The solutions belonging to the eigenvalues are called *eigenfunctions*.

If the operator in (AI.2) is derived from the total energy of the system, it is called the *Hamiltonian* **H** of the system. The Schrödinger wave equation then becomes

$$\mathbf{H}\psi = E\psi \tag{AI.3}$$

The eigenvalues E of this equation are the values of the energy in each of the states determined by the eigenfunctions ψ, the so-called wave functions. This operation turns the wave equation (AI.3) into an equation which is analogous to the equation-of-motion of, for instance, a vibrating string.

The solution of the motion equation of a vibrating string, when solved for a stationary (time-independent) case, is the *amplitude* of a standing wave. The square of the amplitude is the *intensity* of the wave. In analogy, the solution of wave equation (AI.3), for example, for a moving particle, is a wave function ψ, the square of which (rather of its absolute value) is the "intensity" of the particle or in other words, the *probability* of finding the particle at a given point. We can illustrate this by the following consideration: in diffraction experiments light behaves as a wave. In a typical experiment, such as the one illustrated in Fig. AI.1a, light from a point or line source passes through a slit and a diffraction pattern is recorded on a photographic plate. The process by which the photographic plate is sensitized can be explained, much like the photoelectric effect, by considering light as moving particles. From this point of view, then, the experiment can be considered as the passage of a stream of photons from the source to the plate. If it were possible to perform the experiment with one single photon (Fig. AI.1b), we would never obtain a diffraction pattern; at most one grain of the plate would be sensitized and we would not be able to predict the position of that grain in the photographic plate. We could, however, conduct such a single photon experiment many times, thus obtaining the equivalent of an experiment with many photons at one time. The diffraction pattern, thus, is an expression of the *probability* that a photon emitted from the source will strike the photographic plate at a given point. The diffraction pattern gives the intensity of the diffracted wave and the intensity is the square of the amplitude of the wave. So the square of the wave function (or rather of its absolute value) is then a measure of the probability of finding a particle in a given place. We could have followed exactly the same reasoning had we considered a stream of particles (electrons for instance) passing through a

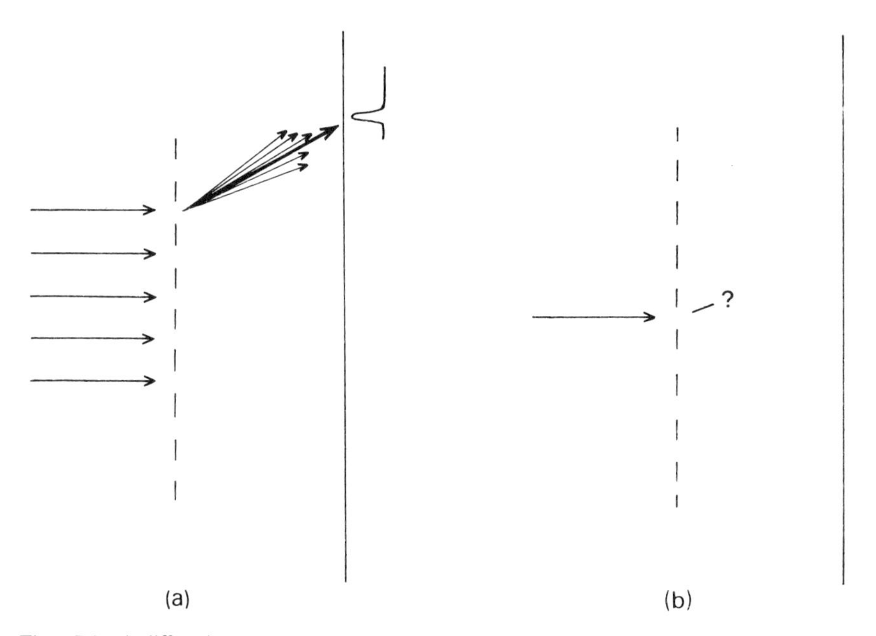

Fig. AI.1 A diffraction experiment (a) with a plane wavefront and (b) with a single photon.

narrow slit. Diffraction patterns can actually be obtained from such a stream of electrons if the "slit" is small enough, which is the case for the openings between the molecules in a crystal.

In experiments of this sort, with light as well as with particles, the waves themselves are not observed. In fact, we never *observe* light as a wave but rather as a quantized phenomenon, whether we detect the light with a photographic plate, a photocell, or our own eyes. We might, therefore, conclude (although this statement is inaccurate) that light "really" is a stream of photons and that the "waves" are the mathematical expressions of the way in which the photons move, just like the matter waves are the mathematical expressions of the way a stream of particles, such as electrons, moves. The similarity of such expressions with the expressions used for describing the motions in a vibrating string does not mean that light or a stream of electrons *are* vibrations of some mysterious medium, as "ether," or that the "wave" aspect and the "particulate" aspects are contradictory.

An important corollary of quantum mechanics is the *uncertainty principle of Heisenberg*. This principle states that there are various pairs of variables (called canonically conjugated variables) which cannot be known with unlimited accuracy at the same time. This is not a technical but a fundamental limitation. Momentum and position are canonically conjugated variables; if the minimal uncertainty in the momentum is Δp and the minimal uncertainty in the position is Δx, their product is, according to Heisenberg's principle, a *finite* value equal to Planck's constant

$$\Delta x \cdot \Delta p = h \tag{AI.4}$$

[When rather than Cartesian coordinates a polar coordinate system is used (which is the case in many quantum mechanical calculations) this value is $h/2\pi$.] In "classical" mechanics it was assumed that it was possible to simultaneously determine these variables to any degree of accuracy. The uncertainty principle now states that this specification cannot be carried out beyond a certain limit. Suppose we set out to measure the position and the momentum of an electron in order to determine its motion. We can, in principle, observe the electron by a setup similar to that illustrated in Fig. AI.2. In such a setup we "illuminate" the electron with "light" and let the "light" scatter into our detection instrument. It is obvious that the shorter the wavelength of this "light," the more accurately we can observe the electron in a certain position. However, "light" quanta, now considered as particles with a momentum, have more energy and thus more momentum with shorter wavelength. Part of this energy is taken over by the electron after the collision, thus giving it an uncertainty in its momentum which can be calculated from the laws of conservation of energy and momentum (this is an effect which can be demonstrated experimentally and is known as the Compton effect). The attempt to gain accuracy in position by using "light" of shorter wavelength is, therefore, defeated by the loss of accuracy with

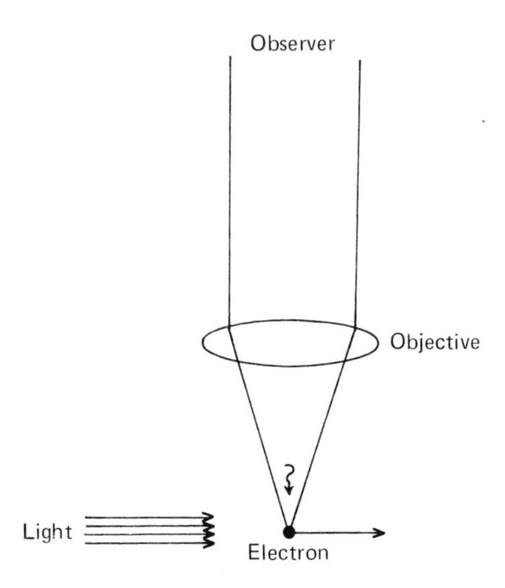

Fig. AI.2 An experimental setup to measure the position and momentum of an electron.

the determination of momentum. A mathematical elaboration of this effect leads to relation (AI.4).

The "wave character" of a stream of particles is *complimentary* to its "particle character." If we design an experiment to determine the position of a particle, its momentum escapes our observation. If we want its momentum, we can only give a probability distribution (given by the wave function) for its position. This is the reason why the Bohr–Rutherford picture of an atom in which an electron revolves around the nucleus like a planet around the sun is incorrect. In a stable atom the energy of the electron, and its momentum, are determined. The electron's position with respect to the nucleus, therefore, cannot be determined. The best we can do is to state that the wave function, which in the stationary case is time independent, gives us the *probability distribution* of the electron's position. The electrons, therefore, are often described as being "smeared out" in moving "clouds." The higher the probability in a certain area, the denser the cloud in that area.

AI.2 The Electronic Structure of Atoms

According to the old Rutherford–Bohr theory an atom consists of a heavy nucleus with a charge Ze around which Z electrons rotate. Z is the atomic number of the atom in the periodic system of the elements. In order to explain the typical line spectra of emission in atoms, Bohr postulated two basic assumptions:

1. Of the infinite number of orbits of the electrons about an atomic nucleus *only a discrete number* of orbits actually occur. These so-called *quantum states* are stationary; this means that in spite of the accelerated motion of the electrons and in contradiction to Maxwell's theory, no electromagnetic energy is emitted by the electrons while in these orbits.
2. Radiation is emitted (or absorbed) only by the transition of an electron from one quantum state n_1 to another n_2. The energy difference between the two quantum states then appears as an emitted light quantum, whose wavelength is given by

$$h\nu = \frac{hc}{\lambda} = E_{n2} - E_{n1} \qquad \text{(AI.5)}$$

By making another postulate that in the quantum states the angular momentum of the electron in its orbit is an integral multiple of Planck's constant divided by 2π,

$$mvr = n\frac{h}{2\pi}, \qquad n = 1, 2, 3, \ldots \qquad \text{(AI.6)}$$

Bohr was able to develop a theory which gave a surprisingly accurate explanation of the spectra of atoms and ions with a single electron. For atoms with more electrons, however, serious discrepancies showed up. Moreover, the quantum states themselves were hard to understand. Quantum mechanics solved these problems, at least in principle. The mathematical expression for Bohr's quantum states are the wave functions for the electrons in the atoms. These wave functions often are called *atomic orbitals.*

For a hydrogen atom the solutions of the wave equation for the electron are determined by three parameters. One is the total energy given by

$$E = \frac{2\pi^2 me^4}{n^2 h^2} \tag{AI.7}$$

in which m and e are the mass and the electrical charge of the electron, h is Planck's constant, and n is an integer which can have the value 1, 2, 3, ... This integer n is called the *principal quantum number* and specifies the total energy of the electron. Another parameter is the angular momentum $\vec{l}$ associated with the orbital motion of the electron. This is specified by another quantum number, the azimuthal (or orbital) quantum number l, which can assume values of 0, 1, 2, ..., $n - 1$. Thus, when the principal quantum number $n = 1$, the azimuthal quantum number can only have the value $l = 0$. Electrons for which $l = 0$ (and thus with zero angular momentum) are called s electrons. The wave functions for these electrons, as shown in Fig. AI.3a, have spherical symmetry and are functions only of the radial coordinate. Electrons for which $l = 1, 2, 3, 4, 5, \ldots$ are p, d, f, g, h, ... electrons. The wave functions for these electrons also depend upon the angular coordinates (Figs. AI.3b and AI.3c).

Then there is, as a third parameter, the *orientation* of the angular momentum. The movement of electrons with an angular momentum different from zero is associated with a magnetic moment in a particular direction. This, of course, only has meaning when we specify a reference, for instance by the direction of an external magnetic field. The direction of the magnetic moment with respect to such a field can only have a discrete set of values specified by a *magnetic quantum number* m_l. In another sense we can view this quantization as the number of independent orientations that an "orbit" can have in three-dimensional space, or, in other words, the number of different directions of the angular moment (see Fig. AI.4). The physical quantity expressed by $\vec{m}_l$ and determined by the magnetic quantum number m_l, is then the projection of the angular momentum vector $\vec{l}$ along the direction of the external field. The magnetic quantum number m_l can have the values $0, \pm 1, \pm 2, \ldots, \pm l$.

These three parameters suffice to describe the spectral properties of atoms with one electron. For atoms (and molecules) with more electrons,

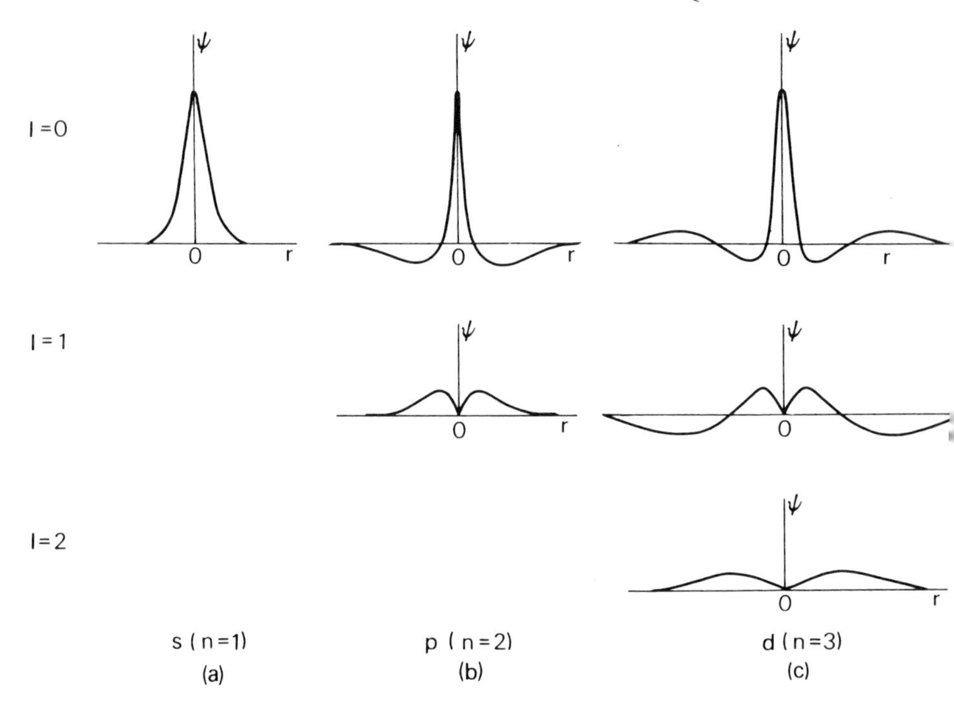

Fig. AI.3 Wave functions of s, p, and d electrons in hydrogen.

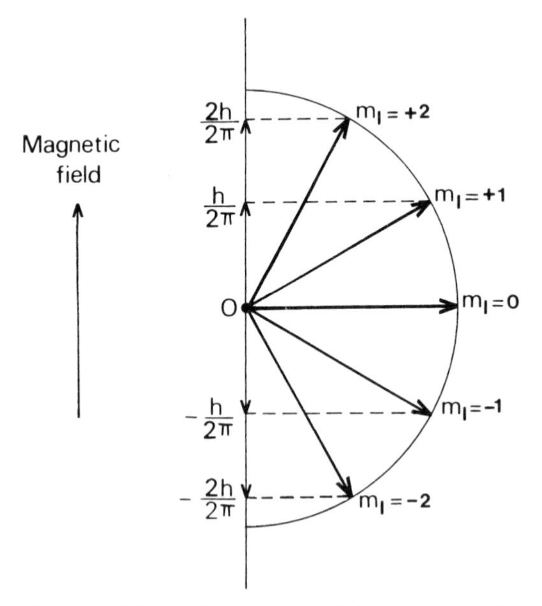

Fig. AI.4 Quantization of the axial component of the orbital angular momentum of an electron in an atom. The projection along the axis of an external magnetic field is determined by the magnetic quantum number m_l.

however, a fourth quantity is necessary. This quantity is related to an *intrinsic angular momentum* of the electron itself and could be visualized as the angular momentum associated with the spinning of the electron about its own axis. This spin angular momentum $\vec{s}$ has only one value but it can have, again in respect to an external magnetic field, two orientations which are roughly parallel or antiparallel. These orientations are specified by a spin magnetic quantum number m_s with possible values of $+\frac{1}{2}$ or $-\frac{1}{2}$.

These four quantum numbers determine the state of an electron in an atom or a molecule as described by its wave function. The *Pauli exclusion principle* dictates furthermore, that *no two electrons in a system can be in the same detailed state*. This means that no two electrons in a system can have the same set of quantum numbers. If two electrons are in the same orbital, having the same values for n, l, and m_l, they must differ in the spin quantum number m_s, or, speaking in a classical analogy, their spins must be antiparallel to each other.

In the *ground state* the electrons of an atom occupy the lowest energy levels allowed by the Pauli exclusion principle. In this way the electrons in the atomic orbitals distribute themselves over the quantum states as shown in Table AI.1. Quantum states differing from the ground state are called *excited states*.

The promotion (or degradation) of an electron from one quantum state to another is possible by the absorption (or emission) of a quantum of energy, satisfying relation (AI.5). Such an event is called an *electronic transition*, and involves the change of one or more of the quantum numbers. There is a restriction, however, which follows from the quantum mechanical theory that dictates that the probability of the transition be extremely small unless the azimuthal quantum number l changes by $+1$ or by -1. This restriction is given by the *selection rule* $\Delta l = \pm 1$. Absorption and emission spectra of atoms such as hydrogen or the alkali vapors can be precisely explained by a detailed description along the lines pointed out above.

TABLE AI.1 Distribution of Electrons in Atomic Orbitals

Atomic shell	Quantum number				Designation	Number of electrons
	n	l	m_l	m_s		
K	1	0	0	$-\frac{1}{2}, +\frac{1}{2}$	1s	2
L	2	0	0	$-\frac{1}{2}, +\frac{1}{2}$	2s	2
		1	$+1, 0, -1$	$-\frac{1}{2}, +\frac{1}{2}$	2p	6
M	3	0	0	$-\frac{1}{2}, +\frac{1}{2}$	3s	2
		1	$+1, 0, -1$	$-\frac{1}{2}, +\frac{1}{2}$	3p	6
		2	$+2, +1, 0, -1, -2$	$-\frac{1}{2}, +\frac{1}{2}$	3d	10

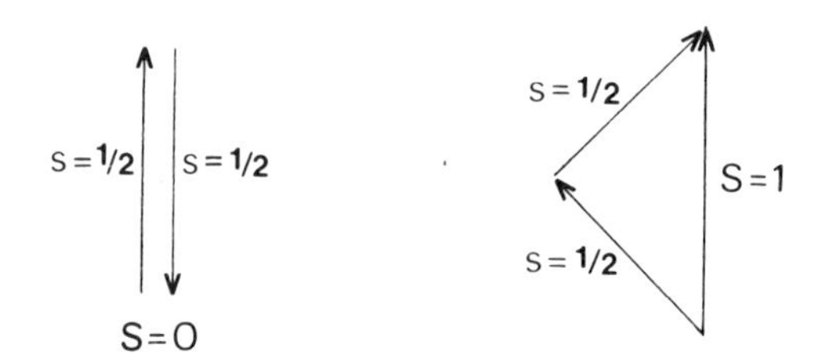

Fig. AI.5 Vector addition of antiparallel and "parallel" electron spin vectors.

In atoms with *more* than one electron, the angular momenta associated with the electron spin can be added vectorially (Fig. AI.5). The resulting total spin angular momentum is also quantized and determined by a total spin quantum number S. S is zero or an integer when we have an *even* number of electrons and is a half-integer when we have an *odd* number of electrons. The *orientation* of this total spin angular momentum, again with respect to a reference direction (for instance a magnetic field), is quantized in the same way as the orientation of the orbital angular momentum (see Fig. AI.4); it is determined by a quantum number M_s which can have the values 0, ± 1, $\pm 2, \ldots, \pm S$ when S is an integer, or the values of $\pm\frac{1}{2}, \pm\frac{3}{2}, \pm\frac{5}{2}, \ldots, \pm S$ when S is a half-integer. The total number of possible values of M_s, $2S + 1$, is called the *multiplicity* of the system. Designations of the multiplicity are given in Table AI.2. Thus, in an atom with two electrons the total spin can be 0 when the electrons are paired (have antiparallel spins) and the multiplicity is 1. This is called a *singlet state*. The total spin also can be 1 when the electrons are unpaired (have parallel spins*). Then three orientations of the total spin vector in respect to an external magnetic field are possible and the state is a triplet. In atoms with an even number of electrons we can have singlet, triplet, etc. states; the states in atoms with an odd number of electrons can be a doublet, quartet, etc.

Evidently, the energy levels associated with these differences in orientation (differences in magnetic momentum) are equal in the absence of a magnetic field because the differences themselves have no meaning in this case. In the presence of a magnetic field, however, differences in orientation of the spin lead to different interaction energies with the field and thus to different energy levels. More transitions, therefore, become possible and the result is a splitting of a spectral line, such as a threefold splitting in the case of triplet state. The energy differences between these split levels are relatively small, corresponding to frequencies in the microwave region. In the case of a single, hence unpaired, electron (as occurring in chemical radicals) transitions between the two doublet levels in a magnetic field can

* The electrons, in this case must occupy different orbitals to obey Pauli's exclusion principle.

TABLE AI.2 Multiplicity of Orientations of the Spin Vector

S	M_S	Multiplicity
0	0	Singlet
$\frac{1}{2}$	$-\frac{1}{2}, +\frac{1}{2}$	Doublet
1	$+1, 0, -1$	Triplet
$\frac{3}{2}$	$+\frac{3}{2}, +\frac{1}{2}, -\frac{1}{2}, -\frac{3}{2}$	Quartet

be observed by the resonance absorption of a microwave. This phenomenon is the basis of electron spin resonance (ESR) [also called electron paramagnetic resonance (EPR)] spectroscopy.

Just as the spin angular momenta of the electrons in an atom can be coupled to each other to result in a total spin angular momentum, the *orbital angular momentum* of the electrons can also be added. The resultant $\vec{L}$ is is quantized and determined by a total orbital quantum number L. Finally, we can add the total orbital angular momentum to the total spin angular momentum to give a *total angular momentum* $\vec{J}$ which is quantized and determined by a total angular momentum quantum number J. Since L is always an integer and S can be an integer or a half-integer, J is an integer when S is an integer and a half-integer when S is a half-integer. As with the partial angular momenta, the total angular momentum can only assume a discrete set of orientations with respect to external field. The splitting in the energy levels in a magnetic field due to these different orientations (which are spaced over a larger energy than the multiplet splitting) can be observed as a splitting of the spectral line in a magnetic field. This phenomenon is known as the *Zeeman effect*. It shows the quantization of the interaction of the magnetic moment associated with the total angular momentum of the electrons and the external magnetic field.

An *electric field* does not act on the magnetic moment associated with the total angular momentum $\vec{J}$. The result of an electric field is a *polarization* of the atom. The resulting dipole moment is proportional to the electric field

$$\vec{\mu} = \alpha \vec{E} \tag{AI.8}$$

The polarizability α depends on the orientation of the "orbit," that is on the orientation of the total angular momentum $\vec{J}$. The space quantization with respect to a magnetic field as described above also takes place in an electric field. This results in a shift of the energy which manifests itself in a shift of the spectral lines. The effect is known as the *Stark effect*. Its magnitude is proportional to both the dipole moment and the field strength. However, since the dipole moment is also proportional to the electric field strength [Eq. (AI.8)], the spectral shift is proportional to the *square* of the electric field strength. This, together with the fact that the *sign* of the space quantization (the sign of M_J) has no effect on the energy level (the dipole moment is always induced in the direction of the field), makes the Stark effect qualitatively different from the more simple Zeeman effect.

Appendix

II Elements of Equilibrium Thermodynamics

AII.1 Definitions

Thermodynamics deals with the interconversions of various forms of energy and work. It does so by describing a *thermodynamic system* (which is a part of the universe separated from the rest) in terms of concepts and laws derived from the study of *macroscopic* phenomena such as pressure, volume, temperature, concentrations, etc. Any physical or chemical system can be described as a thermodynamic system or, simply, as a system.

A system is *isolated* when it can exchange neither energy nor matter with its surroundings; it is a *closed* system when it can exchange energy but cannot exchange matter with its surroundings; and it is an *open* system when it can exchange energy as well as matter with its surroundings. *Classical* or *equilibrium* thermodynamics describes systems which are in *equilibrium* or are undergoing *reversible* processes. A system is in equilibrium when it shows no internal tendency to change its properties with time. A system is in equilibrium when its *internal* parameters are completely determined by its *external* parameters. For example, consider a gas in a cylinder closed by a frictionless piston. This system is in equilibrium when the *external pressure* exerted by the piston on the gas is equal to the *internal pressure* of the gas, anywhere in the cylinder. If the gas is not in equilibrium we can still define the external pressure exerted by the piston, but this clearly is not a property of the system (the gas) itself; the internal pressure may vary from point to point in the gas before equilibrium is reached and its state cannot be determined by one value of the pressure equal to the external pressure of the piston. A system undergoes a reversible process when the path followed by such a process is one connecting intermediate states of equilibrium. For example, to reversibly expand the gas in a cylinder we have to release the pressure on the piston very slowly, in the limit, infinitely slowly, so that at any instant the pressure of the piston equals the pressure anywhere in the gas. If the piston is drawn back suddenly, the expansion of the gas occurs irreversibly. In that case the intermediate states of the gas are no longer equilibrium states.

Classical thermodynamics is a harmonious and self-consistent theory describing relations and correlations between the various parameters of a system in equilibrium or undergoing reversible processes. For irreversible processes, the laws of classical thermodynamics provide for a set of inequalities which only describe the *direction* of a change.

Reversible processes never occur in reality, since for changes to be reversible they must occur at an infinitely slow rate. Equilibrium thermodynamics, therefore, does not seem adequate for real processes. Indeed, time does not come into the formalism of equilibrium thermodynamics at all. In many cases, however, states which in reality were reached by processes moving along *irreversible* pathways can be realized, at least in "thought experiments," by *reversible* pathways as well. Thermodynamic quantities could thus be determined and this has in fact, proven to be very fruitful for the investigation of the properties and behavior of systems, including living ones.

The application of equilibrium thermodynamics to the description of the *course* of processes, especially of those in living systems, however, is limited. As we have mentioned before, equilibrium thermodynamics can only describe the direction of the changes and is unable to provide generalized quantitative relations. *Thermodynamics of irreversible processes* or *nonequilibrium thermodynamics* may make possible the application of the formulations of thermodynamics to quantitative investigations of the processes of living systems. This is particularly important for the description of transport processes such as those occurring through biological membranes (see, for instance, Katchalsky and Curran, 1967).

AII.2 First and Second Laws of Thermodynamics

If an *adiabatic system* (that is a system which cannot exchange energy or heat with its surroundings) is subject to external forces, *work* is performed. This work can be positive when it is done *by* the system *on* the surroundings, or negative when it is done *on* the system *by* the surroundings. If, for instance, the external force is a constant pressure P which causes a change in volume dV, the work is $dW = P\,dV$. Many experiments, particularly those of Joule, have confirmed that this amount of work, which changes the *state* of the system, is independent of the physical way in which the new state is reached. Thus, the compression work

$$W_{1\to 2} = \int_1^2 P\,dV$$

is independent of the values assumed by P during the change and is only and fully determined by the initial and final states of the system. There must, therefore, exist an energy function (a *state function*) whose decrease represents the amount of work,

$$-dU = dW$$

and whose change is dependent only on the initial and final states of the system

$$W_{1\to 2} = -\int_1^2 dU = U_1 - U_2 \tag{AII.1}$$

Equation (AII.1) implies that for a cyclic process (when the system goes from state 1 to state 2 and then back to state 1), the total change in U is zero:

$$\oint dU = \int_1^2 dU + \int_2^1 dU = U_2 - U_1 + U_1 - U_2 = 0 \tag{AII.2}$$

in which the *circular integral* $\oint$ indicates integration over a cyclic process. Relations (AII.1) and (AII.2) apply to all *thermodynamic state functions* whose total change depends only on the values in the initial and final states of the system, regardless of the way in which the change is accomplished. The function U is the *internal energy* of the system. When the adiabatic restriction is lifted, the system can take up energy (heat) from its surroundings. This energy dQ is used to increase the internal energy and to do work:

$$dQ = dU + dW$$

or in a more familiar arrangement

$$dU = dQ - dW \tag{AII.3}$$

Relation (AII.3) is the mathematical statement of the *first law of thermodynamics*. It states that energy can be neither created nor destroyed and thus expresses the impossibility of a *perpetuum mobile* (perpetual motion) of the first kind.

The first law provides an energy balance equation but it does not say anything about the *direction* in which natural processes spontaneously move. If two blocks of metal, one hot and the other cold, are placed in thermal contact with each other and isolated from their surroundings, we know by experience that the hot block becomes colder while the cold block becomes hotter; or in other words, that heat flows from the hot block to the cold block. This process continues until both blocks have the same temperature, somewhere between the two initial extreme temperatures. After they have reached this *equilibrium* temperature, no more heat flow takes place. We also know by experience that once in this state one of the blocks will *never* become hotter while the other becomes colder in a spontaneous way, although this would not at all violate the first law. Natural processes seem to proceed in a *unidirectional way*, and that direction seems to be the one in which driving power is *dissipated*. A mathematical pendulum may swing periodically forever but an actual pendulum will stop after a while due to friction in its bearings and frictional resistance of the air, and it will never thereafter start swinging by itself. The *second law of thermodynamics* makes it possible to predict the direction in which natural processes occur. It tells us that for a natural process to occur, energy must be dissipated or *degraded*.

The second law can be stated verbally in many equivalent ways. A useful one for our purpose is the one given by W. Thomson (Lord Kelvin) in 1853: "It is impossible by means of inanimate material agency to derive mechanical effect from any portion of matter by cooling it below the temperature of the coldest of the surrounding objects." This quotation states the impossibility of a *perpetuum mobile* of the second kind: it is impossible that the cooling of a single body, the ocean for example, can provide useful energy, for example, energy required to drive a ship, in a cyclic manner, even though the removal of thermal energy from the water and its cyclic transformation into mechanical work would not violate the first law.

Thus, no *positive* work can be obtained from a cyclic process when it occurs at a constant temperature:

$$\oint (dW)_T \leq 0 \tag{AII.4}$$

in which the subscript T means that the temperature is kept constant throughout the cycle. If a system produces a certain amount of work $W_{1\to 2}$ when it changes from a state 1 to a state 2 at a constant temperature, the amount of work $W_{2\to 1}$ required to bring the system back from state 2 to state 1 is *larger* or equal to $W_{1\to 2}$:

$$W_{1\to 2} \leq W_{2\to 1} \tag{AII.5}$$

The equalities of both (AII.4) and (AII.5) apply when the process occurs *reversibly*. In that case one can prove that

$$\oint \frac{dQ}{T} = 0 \tag{AII.6}$$

which means that when the system changes reversibly from state 1 to state 2 the value of the integral $\int_1^2 dQ/T$ depends only on the initial state 1 and the final state 2 and not at all on the way in which the change has taken place. T stands for the *absolute* temperature. For *reversible processes*, therefore, we can define a function S such that its change dS is

$$dS = \frac{dQ}{T} \tag{AII.7}$$

According to (AII.6), for a reversible process,

$$dS = 0$$

and, when the system changes from a state 1 to a state 2,

$$\int_1^2 dS = S_2 - S_1 \tag{AII.8}$$

thus indicating that the change in the function S when the system moves from state 1 to state 2 depends only on the initial and the final values of the function. The state function S is called the *entropy*. Its definition as described above was introduced by R. Clausius in 1865.

Although the entropy is defined above only for reversible processes, the fact that the entropy difference between two states of a system depends only on its initial and final value [Eq. (AII.8)] also makes it a useful function for natural irreversible processes. The change in entropy in going from a state 1 to a state 2 is always the same, that is, irrespective of the path between 1 and 2 and of whether or not the process is reversible. In order to investigate the change in entropy resulting from a natural irreversible transition from a state 1 to a state 2 by an isolated system, we can think of the system returning to its initial state 1 along a *reversible* path. We then can apply the definition for the entropy of reversible processes and arrive at an expression for the entropy difference between state 1 and state 2. If this is done we will discover that the second law requires that, for an irreversible process in an isolated system, *the entropy always increases*:

$$\Delta S = S_2 - S_1 > 0 \tag{AII.9}$$

Since all naturally occurring processes are irreversible, any change that actually occurs spontaneously in an isolated system is accompanied by a net increase in entropy. Stated the other way around, but equally valid, if there is any conceivable process for which the entropy can increase, it will occur spontaneously. For *irreversible* processes inequality (AII.9) expresses the second law, indicating the direction in which spontaneous processes will proceed. This direction is always towards a maximum entropy. When this maximum is reached no change will occur spontaneously and the system is *in equilibrium.*

At equilibrium, relation (AII.7) is valid. This relation represents the second law for *reversible* processes. The introduction of this equation into Eq. (AII.3) gives the expression for the combined first and second laws

$$dU = T\,dS - dW \tag{AII.10}$$

From this we derive that for an adiabatic process

$$dU = -dW$$

since $dQ = 0$ and hence $dS = 0$. For a reversible isothermic cycle Eq. (AII.10) becomes

$$\oint dU = T \oint dS - \oint dW$$

and, since both dU and dS depend only on final and initial states and therefore have zero circular integrals,

$$dW = 0$$

which mathematically expresses Lord Kelvin's formulation of the second law for an isothermic reversible cycle.

AII.3 Entropy

The physical meaning of entropy is not intuitively clear from Clausius' definition. It can be seen as that part of the heat-energy term that must be transported across a temperature difference in order to produce work. The entropy is an *extensive* quantity which has to be transported across a potential difference. The temperature is an *intensive* (potential) quantity which provides such a potential difference. All the energy terms, except the internal energy, consist of an extensive and an intensive (potential) part. Work is produced when there exists a difference of potential across which the conjugated extensive quantity can be transported. The work function, for instance, can contain many terms in addition to expansion work:

$$dW = P\,dV - f\,dl - \psi\,de - \sum_i \mu_i\,dn_i + \cdots \tag{AII.11}$$

in which changes in volume dV, length dl, charge de, and the number of moles of a species i in a chemical reaction dn_i are extensive quantities which can produce work when transported over "potential" differences provided by pressure P, mechanical force f, electric potential ψ, and chemical potential μ_i, respectively. Introducing (AII.11) in (AII.10) we obtain the *equation of Gibbs*

$$dU = T\,dS - P\,dV + f\,dl + \psi\,de + \sum_i \mu_i\,dn_i + \cdots \tag{AII.12}$$

which takes into account all possible changes in extensive properties (dS, dV, dl, de, dn_i) and relates the total change in internal energy to the sum of the product of intensive (potential) quantities (T, P, f, ψ, μ_i) and the changes in the extensive properties.

With the development of statistical mechanics, the concept of entropy became more tangible. Boltzmann has shown that entropy is directly related to the number of configurations in which a system can be realized. Consider, for example, a situation in which a large number (N) of apples has to be distributed over two baskets. Obviously, the most unlikely distribution is that in which there are no apples in basket 1 and N apples in basket 2. This distribution can be accomplished in only one way; the system can be realized by only one configuration. A distribution in which one apple is in basket 1 and $N - 1$ apples in basket 2 is a little more probable. This can be accomplished in N ways. The distribution in which 2 apples are in basket 1 and $N - 2$ in basket 2 can be achieved in even more ways by adding to any of the N apples which can be placed in basket 1 any of the remaining $N - 1$ apples, hence in $N\,(N - 1)$ ways; this system, thus, can be realized by $N\,(N - 1)$ configurations. By putting more and more apples in basket 1 the number of configurations in which the different distributions can be realized, and thus the probability or *randomness* of the distribution, will increase until each basket has $N/2$ apples.

If Ω is the number of configurations in which a system can be realized, the entropy is

$$S = k \ln \Omega \tag{AII.13}$$

in which k is the Boltzmann constant. The entropy is, thus, a measure of randomness, and to state that a spontaneous process always moves toward a maximum entropy is equivalent to saying that a process spontaneously moves toward a maximum randomness. The example of the apples in the two baskets is comparable to the two copper blocks of different temperature in thermal contact, or to a vessel in which a screen with a hole in it separates a zone containing a gas at high pressure from a zone containing gas at low pressure. In both cases a spontaneous process occurs from a state which is less probable, less random, to a state which is more probable, more random; heat flows from the hot block to the cold block until both blocks have the same intermediate temperature and gas molecules diffuse from the zone of high pressure to the zone of low pressure until the pressure throughout the vessel has the same intermediate value. In both cases, the process stops when the entropy has the maximum value of the equilibrium point.

AII.4 Thermodynamic Potentials

Linear combinations of thermodynamic state functions, such as the internal energy U and the entropy S, are also state functions; they are called *thermodynamic potentials*. Sometimes such combinations are more amenable to experimental determination. An example is the *enthalpy* H, defined as

$$H = U + PV \tag{AII.14}$$

from which it follows that

$$dH = dU + P\,dV + V\,dP$$

This function is particularly useful in situations where the pressure remains constant. In that case

$$dH = dU + P\,dV.$$

If the only work done in a system is expansion work $P\,dV$ (which is the case, for example, for a chemical reaction carried out in a test tube), application of the first law (AII.3) yields

$$dH = dQ. \tag{AII.15}$$

The enthalpy change dH, therefore, is the heat taken up (when positive) or given off (when negative) when a reaction occurs under *isobaric* (constant pressure) conditions and when no work other than expansion is done. It is often called the *reaction heat* or heat content. Many compounds have a characteristic reaction heat. In the case of combustions this heat is the *heat of combustion* when one gram molecule of the compound is completely burned in molecular oxygen. The heat of combustion is equal to the molar enthalpy change of the reaction. When heat is *given off* to the surroundings the enthalpy change is *negative* and the reaction is *exothermic*. When heat is taken up from the surroundings the enthalpy change is *positive* and the reaction is *endothermic*.

Another useful thermodynamic potential is the *Helmholtz free energy* F, defined by

$$F = U - TS \tag{AII.16}$$

from which it follows that

$$dF = dU - T\,dS - S\,dT$$

This potential function is useful for isothermal processes (processes occurring at a constant temperature) for which

$$dF = dU - T\,dS \tag{AII.17}$$

The maximum work can be obtained from a process when it occurs reversibly. In that case we obtain, by introducing expression (AII.10) into (AII.17),

$$-dF = dW_{\text{max}} \tag{AII.18}$$

The decrease in free energy, therefore, represents the maximum amount of work which can be performed under isothermal conditions. From Eqs. (AII.18) and (AII.16) we can deduce that the free energy is that part of the energy which, at a constant temperature, is useful for work. If a system is left alone, spontaneous processes which go in the direction of *increasing entropy* ($dS > 0$) also go in the direction of *decreasing free energy*. At equilibrium the entropy is at a maximum ($dS = 0$) and the free energy is at a minimum ($dF = 0$).

The Gibbs free energy is a potential function which is particularly useful for biochemical and biological processes, which usually occur not only at a constant temperature but also under constant pressure. The Gibbs free energy G is defined as

$$G = U - TS + PV \tag{AII.19}$$

At constant temperature and pressure

$$dG = dU - T\,dS + P\;dV = dH - T\,dS \tag{AII.20}$$

Using (AII.10) we obtain, for reversible processes,

$$-dG = dW_{\text{max}} - P\,dV = dW_{\text{net}} \tag{AII.21}$$

in which dW_{net} is all of the work other than expansion or compression work. Willard Gibbs called dW_{net} the *useful work*. The decrease in Gibbs free energy, therefore, represents the maximum amount of useful work (other than expansion or compression work) which can be obtained from a process occurring at constant temperature and pressure. Of course, at equilibrium ΔG is also equal to zero.

AII.5 The Chemical and the Electrochemical Potential

When we consider biological systems we must take into account that they are *open systems*; matter is brought into or taken away from the system by both transport processes and chemical reactions. It is important, therefore, that we generalize our treatment in such a way that such open systems are included. This can be done by including the change in composition brought about by the chemical potential μ in the work function [Eq. (AII.11)].

The variation of the Gibbs free energy at constant temperature and pressure, dG, is given by Eq. (AII.20). When we substitute the variation of the internal energy given by the Gibbs equation (AII.12), thereby ignoring all contributions to the work function except expansion work and "chemical" work, we get

$$dG = -S\,dT + V\,dP + \sum_i \mu_i\,dn_i \tag{AII.22}$$

The change in free energy by the addition (or removal) of component i to (or from) the system, keeping temperature, pressure, and the amount of all components other than component i constant, is given by

$$\left.\frac{\partial G}{\partial n_i}\right|_{T,P,n_{j\neq i}} = \mu_i \tag{AII.23}$$

The chemical potential is, thus, defined as the increase of the Gibbs free energy on the reversible addition of component i, holding temperature, pressure, and all other components constant.

For a pure one-component system the chemical potential is simply the Gibbs free energy per mole of the component. Therefore, μ_i is sometimes called the *molar free energy* of i when the concentration of i is expressed in moles. For a multicomponent system it can be mathematically shown that the total free energy at a given temperature and pressure is given by

$$G = \sum_i n_i \mu_i \tag{AII.24}$$

For an ideal gas the chemical potential can be calculated per mole by using the ideal gas equation

$$PV = RT \tag{AII.25}$$

in which R is the gas constant. Since on expansion the gas performs work given by

$$dW = P\,dV \tag{AII.26}$$

the change in chemical potential must be

$$d\mu_g = -P\,dV \tag{AII.27}$$

When we differentiate (AII.25), we obtain

$$d(PV) = P\,dV + V\,dP = d(RT) = 0 \tag{AII.28}$$

at constant temperature. Hence, $P\,dV = -V\,dP$ and

$$d\mu_g = V\,dP \tag{AII.29}$$

at constant temperature. Solving for V from Eq. (AII.25) and substituting it in (AII.29) gives

$$d\mu_g = RT(dP/P) \tag{AII.30}$$

at constant temperature. Integration of Eq. (AII.30) yields

$$\mu_g - \mu_g^\circ = RT \ln \frac{P}{P_0} \tag{AII.31}$$

The chemical potential μ_g is equal to μ_g° when the pressure of the gas is equal to P_0. We can use this as a standard against which the chemical potential can be measured by simply putting P_0 equal to 1 atm at any particular temperature. Equation (AII.31) then becomes

$$\mu_g = \mu_g^\circ + RT \ln P_g \tag{AII.32}$$

The calculation of the chemical potential of a solute in a solution proceeds analogously. We now use the van't Hoff equation

$$\pi_i = c_i RT \tag{AII.33}$$

which is valid for ideal (infinitely diluted) solutions. π_i is the osmotic pressure and c_i is the concentration of solute i. The work per mole performed by the osmotic pressure is $\pi_i\,d(1/c_i)$, and thus

$$d\mu_i = -\pi_i\,d(1/c_i) = -c_i RT\,d(1/c_i) = RT\,d(\ln c_i) \tag{AII.34}$$

Integration yields

$$\mu_i - \mu_i^\circ = RT \ln(c_i/c_0) \tag{AII.35}$$

Again we can use a value $c_0 = 1$ mole to define the standard potential μ_i°, giving

$$\mu_i = \mu_i^\circ + RT \ln c_i. \tag{AII.36}$$

We can apply the same procedure when calculating the chemical potential of a solvent. In a solution, the solvent also has a concentration; there is less solvent per unit volume in a solution than in the pure solvent. We prefer, however, to use the *mole fraction* instead of moles for the solvent concentration. The mole fraction is defined as

$$x_j = n_j/\sum_i n_i, \qquad i = 1, \ldots, j, \ldots \tag{AII.37}$$

The use of mole fractions for the solvent has the convenient property of being equal to 1 for the pure solvent. The chemical potential of the solvent in a solution is given by

$$\mu_s = \mu_s^\circ + RT \ln x_s \tag{AII.38}$$

whereby $\mu_s = \mu_s^\circ$ for the pure solvent.

Equations (AII.32), (AII.36), and (AII.38) are derived for ideal gases and ideal solutions. In such systems there are no molecular interactions: strictly speaking, the derived formulas are valid only for such systems. It is still formally correct, however, to use the equations, provided that we replace the gas pressure P_g in (AII.32) by a quantity called *fugacity* f_g, defined by

$$f_g = \alpha_g P_g \tag{AII.39}$$

and the concentration c_i (or mole fraction x_s) in (AII.36) and (AII.38) by the *activity* a_i, defined by

$$a_i = \alpha_i c_i \tag{AII.40}$$

The coefficients α_g and α_i are the fugacity coefficient and the activity coefficient, respectively. They are purely empirical coefficients. If the pressure or the concentration is sufficiently low (as in diluted systems) the use of pressure and concentration is a reasonable approximation.

If a solute bears an electric charge (for instance, an ionic species in a solution of an electrolyte) there is also an electrical potential that has to be taken into account. The ability to do useful work in that case is due to the chemical potential (related to concentration) *and* to the electrical potential (related to electric charge). For such a solute we define the *electrochemical potential*

$$\tilde{\mu}_k = \mu_k + z_k \mathscr{F} \psi = \mu_k^\circ + RT \ln c_k + z_k \mathscr{F} \psi \qquad \text{(AII.41)}$$

in which z_k is the valence of ionic species k, ψ is the electrical potential, and $\mathscr{F}$ is the *Faraday constant*, which is the electronic charge per gram equivalent (since the electronic charge ε is 1.6×10^{-18} coulomb and Avogadro's number A is 6.02×10^{23}, $\mathscr{F} = \text{A}\varepsilon = 96{,}489$ coulomb per gram equivalent; in calories per volt, a unit more useful for our purpose, $\mathscr{F} = 23{,}061$, since 1 cal = 4.184 joules).

AII.6 The Standard Free Energy of a Chemical Reaction

We can, in a fairly general way, represent a chemical reaction by the equation

$$a\text{A} + b\text{B} \rightleftharpoons p\text{P} + q\text{Q} \qquad \text{(AII.42)}$$

in which a, b, p, and q are the stoichiometric coefficients. For such a reaction the change of free energy is given by the difference between the sum of the chemical potentials of the products (the components on the right-hand side of the equation) and the sum of the chemical potentials of the reactants (the components on the left-hand side of the equation), each chemical potential being multiplied by the appropriate stoichiometric coefficient. Thus,

$$\begin{aligned}\Delta G &= p\mu_\text{P} + q\mu_\text{Q} - a\mu_\text{A} - b\mu_\text{B} \\ &= p\mu_\text{P}^\circ + pRT \ln c_\text{P} + q\mu_\text{Q}^\circ + qRT \ln c_\text{Q} \\ &\quad - a\mu_\text{A}^\circ - aRT \ln c_\text{A} - b\mu_\text{B}^\circ - bRT \ln c_\text{B}\end{aligned}$$

or

$$\Delta G = \Delta G^\circ + RT \ln c_\text{P}^{\,p} c_\text{Q}^{\,q} / c_\text{A}^{\,a} c_\text{B}^{\,b} \qquad \text{(AII.43)}$$

when we lump the standard potentials μ_i° together in one symbol ΔG°.

At equilibrium $\Delta G = 0$. Hence, since ΔG° is a constant, the logarithm in Eq. (AII.43) must be a constant at equilibrium. Thus,

$$\frac{c_\text{P}^{\,p} c_\text{Q}^{\,q}}{c_\text{A}^{\,a} c_\text{B}^{\,b}} = K \qquad \text{(AII.44)}$$

in which K is called the *equilibrium constant*. According to (AII.43)

$$\Delta G^\circ = -RT \ln K \tag{AII.45}$$

ΔG° is called the *standard free energy change* of the reaction and is a measure of the *direction* in which the reaction spontaneously proceeds toward equilibrium. It can be seen from (AII.44) and (AII.45) that ΔG° is negative when the concentrations of the products at equilibrium are higher than those of the reactants. A negative value of ΔG° means that the reaction proceeds spontaneously in the direction of the products. The reaction is said to be *exergonic*. A positive value of ΔG°, on the other hand, means that it will cost energy to push the reaction in the direction of the products. The reaction then is *endergonic*. The energy necessary to drive the reaction in the direction of the products can be obtained, for example, by *coupling* the reaction to an exergonic reaction which has a value of ΔG° at least as negative as the ΔG° of the original reaction is positive.

There are two important things to note about the standard free energy change. First, it is *not* an *actual change in free energy*; it is just a *measure* for the *direction* in which a reaction would go when it spontaneously proceeds to equilibrium and *how far* it would go. Second, its value *is not* a measure for the *rate* of the reaction. It is quite possible that a strongly exergonic reaction which has a large negative ΔG° proceeds with an infinite slow rate, for instance, because it has a high activation energy barrier. In many cases a catalyst, such as an enzyme in a biological system, can lower this barrier so that the reaction can proceed at a high rate toward equilibrium.

AII.7 Oxidation–Reduction Potential

Oxidation–reduction reactions are reactions which involve the transport of electrons. A substance is said to be reduced when it receives electrons. A substance becomes oxidized when it loses electrons. In aqueous media (such as we usually have in living cells) a reduced substance can pick up protons from the medium so that reduction becomes equivalent to the addition of hydrogen. An oxidized substance can bind to oxygen because of the strong electronegativity of oxygen. Essentially, the reaction is the transport of electrons; often, it does not go together with protonation or oxygenation, as in the cytochrome reactions of the respiratory chain.

Since oxidation–reduction reactions amount to the exchange of negative electric charge, it is convenient to express the free energy change of such reactions in units of electric potential. This can be done as follows.

We can represent a reaction in which substance **B** reduces a substance A by

$$A_{ox} + B_{red} \rightleftharpoons B_{ox} + A_{red} \tag{AII.46}$$

this reaction can be split into two half-reactions

$$A_{ox} \rightleftharpoons A_{red} \tag{AII.46a}$$

and

$$B_{red} \rightleftharpoons B_{ox} \tag{AII.46b}$$

The free energy change of the half-reaction (AII.46b) is the free energy change that is associated with the transition of B from the reduced to the oxidized state. According to (AII.43), this is equal to

$$\Delta G = \Delta G^\circ + RT \ln \frac{B_{ox}}{B_{red}} = \Delta G^\circ + RT \ln \frac{\alpha}{1 - \alpha} \tag{AII.47}$$

where α is the fraction of the amount of B which is oxidized. We can convert this to an electrical potential by dividing it by $-z\mathscr{F}$, in which z is the number of transported electrons in the reaction. (The negative sign comes in because it is a negative electric charge that is transported.) Thus,

$$E_B = E_B{}^\circ - \frac{RT}{z\mathscr{F}} \ln \frac{\alpha}{1 - \alpha} \tag{AII.48}$$

The potential E is the oxidation–reduction potential, or redox potential, and can be measured against a standard. A convenient standard for this purpose is the *hydrogen electrode.* A hydrogen electrode is a platinum electrode immersed in an acid solution against which hydrogen gas is bubbled. The half-reaction at the electrode, occurring when the electrode is part of a closed circuit, is

$$H_2 \rightleftharpoons 2H^+ + 2e \tag{AII.49}$$

and the potential is

$$E = E^\circ - \frac{RT}{2\mathscr{F}} \ln \frac{c_{H^+}^2}{c_{H_2}} \tag{AII.50}$$

When we use hydrogen gas at a pressure of 1 atm, $c_{H_2} = 1$ and (AII.50) becomes

$$E = E^\circ - \frac{RT}{\mathscr{F}} \ln c_{H^+} \tag{AII.51}$$

Since $\ln c_{H^+} = 2.3\ {}^{10}\log c_{H^+} = -2.3$ pH, $T = 298$ K (=25°C, the standard temperature used for electrochemical cells), $R = 1.98$ cal/mole degree, and $\mathscr{F} = 23{,}061$ cal/V,

$$E = E^\circ + 0.06 \text{ pH} \tag{AII.52}$$

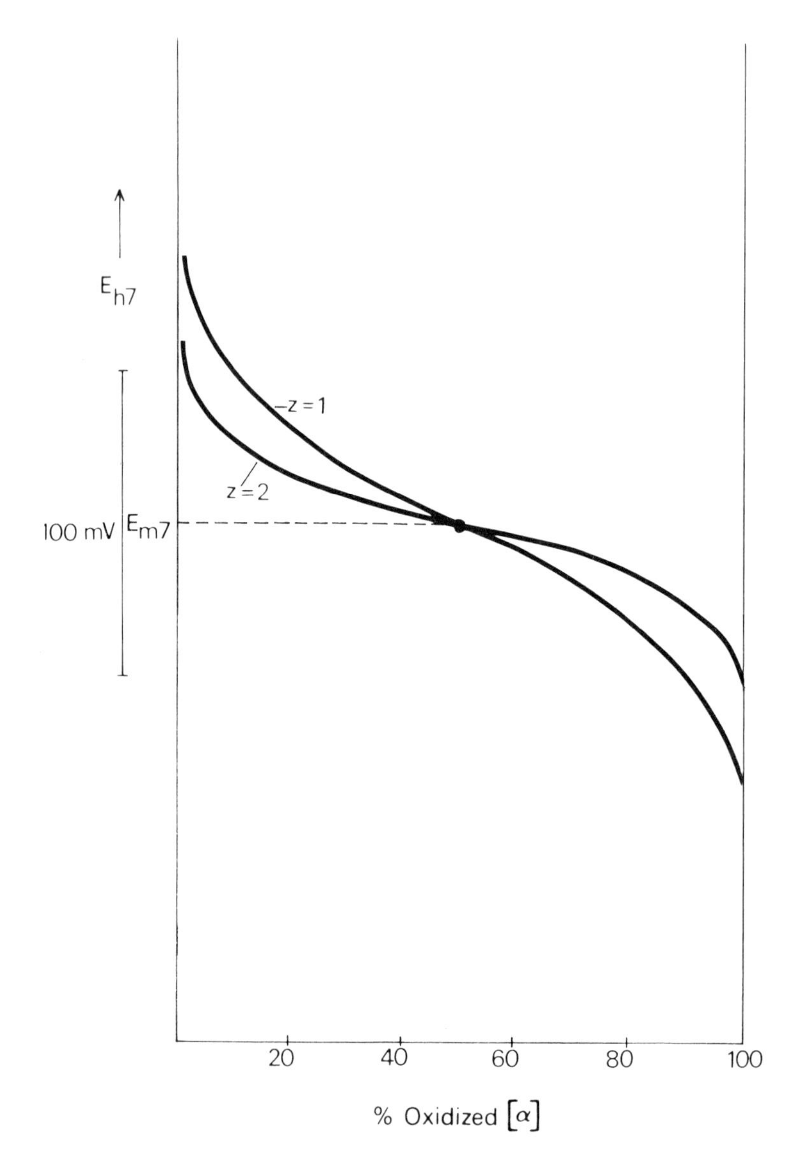

Fig. AII.1 Redox titration curves for a redox reaction involving one electron ($z = 1$) and a redox reaction involving two electrons ($z = 2$). E_{h7} is the redox potential standardized against the hydrogen electrode at pH $= 7$. E_{m7} is the midpoint potential.

The potential E measured by the hydrogen electrode emersed in a 1 N acid solution (pH = 0) is chosen *by convention* to be zero. Hence, $E = E^\circ = 0$ at pH = 0. Against this standard all redox potentials can be measured in an absolute way.

In living cells the ambient pH is not far from neutral. Thus, it is more convenient to take the standard at pH = 7, rather than at pH = 0. Therefore, where biologically important redox couples are concerned, the potential of the hydrogen electrode at pH = 7 is chosen to be zero. Thus,

$$E = 0 = E^\circ + 0.42 \tag{AII.53}$$

and the standard value (the redox potential of hydrogen)

$$E^\circ = -0.42 \quad \text{volt}$$

Using the hydrogen electrode at pH = 7 as a standard, we can rewrite Eq. (AII.48) as follows:

$$E_{h7} = E_{m7} - \frac{RT}{z\mathscr{F}} \ln \frac{\alpha}{1-\alpha} \tag{AII.54}$$

in which E_{h7} is the redox potential, measured with the standardized electrode, and E_{m7} is the *midpoint potential*, measured when half of the substance is in the oxidized state and half of it is in the reduced state. The midpoint potential is characteristic for a redox couple.

In Fig. AII.1, E_{h7} is plotted as a function of two values of z. Such curves are called *redox titration curves*. They can be obtained by titrating a redox reaction (either in the direction of reduction or in the direction of oxidation) with some indicator (a spectral change, for example) and simultaneously measuring the potential. The inflection point is at the midpoint potential.

References

Bohr, N. (1933). Light and Life, *Nature* **131,** 421–423.

Bremerman, H. (1967). Quantitative Aspects of Goal Seeking, Self-Organizing Systems, *Progr. Theoret. Biol.* **1,** 59–77.

Casadio, R., Baccarini-Melandri, A., Zannoni, D., and Melandri, B. A. (1974). *FEBS Lett.* **49,** 203–207.

Chen, S. H., and Yip, S. (1974). "Spectroscopy in Biology." Academic Press, New York.

Darmon, S. E., and Sutherland, G. B. B. M. (1947). *J. Amer. Chem. Soc.* **69,** 2074.

Delbrück, M. (1949). A Physicist Looks at Biology, *Trans. Conn. Acad. Arts Sci.* **38,** 175–190.

Duysens, L. N. M. (1952). "Transfer of Excitation Energy in Photosynthesis," Thesis, Univ. of Utrecht, The Netherlands.

Elsasser, W. M. (1958). "The Physical Foundation of Biology." Pergamon Press, Oxford.

Elsasser, W. M. (1966). "Atom and Organism." Princeton Univ. Press, Princeton, New Jersey.

Förster, Th. (1951). "Fluoreszenz Organische Verbindungen." Vanden Hoeck & Ruprecht, Göttingen.

Fröhlich, H. (1969). Quantum Mechanical Concepts in Biology, *in* "Theoretical Physics and Biology" (M. Marois, ed.), pp. 13–22. Wiley (Interscience), New York; North-Holland Publ., Amsterdam.

Gamow, G. (1955). Information Transfer in the Living Cell, *Sci. Am.* **193** (October), 70–84.

Harris, E. J. (1960). "Transport and Accumulation in Biological Systems." Academic Press, New York.

Heisenberg, W. (1962). "Physics and Philosophy." Harper & Row, New York.

Jackson, J. B., and Crofts, A. R. (1969). *FEBS Lett.* **4,** 185.

Jacob, F., and Monod, J. (1961). Genetic Regulator Mechanisms in the Synthesis of Protein, *J. Mol. Biol.* **8,** 318–356.

Kasha, M. (1963). Energy Transfer Mechanisms and the Molecular Exciton Model for Molecular Aggregates, *Rad. Res.* **20,** 55–71.

Katchalsky, A., and Curran, P. (1967). "Non-Equilibrium Thermodynamics in Biophysics." Harvard Univ. Press, Cambridge, Massachusetts.

Kornacker, K. (1972). Living Aggregates of Nonliving Parts, *Progr. Theoret. Biol.* **2,** 1–22.

Lehninger, A. L. (1970). "Biochemistry," p. 302. Worth, New York.

McMillan, W. S., and Mayer, J. E. (1945). The Statistical Dynamics of Multicompartment systems, *J. Chem. Phys.* **13,** 276–305.

Monod, J. (1970). "Le Hasard et la Nécessité." Ed. du Seuill, Paris.

Moore, W. (1972). "Physical Chemistry," 4th ed. Prentice-Hall, Englewood Cliffs, New Jersey.

Nason, A. C. (1965). "Textbook of Modern Biology." Wiley, New York.

Nirenberg, M. W., and Matthaei, H. (1961). The Dependence of Cell-free Protein Synthesis in *E. coli* upon Naturally Occurring or Synthetic Polyribonucleotides, *Proc. Natl. Acad. Sci. U.S.* **47,** 1588–1602.

Oosawa, F., and Higashi, S. (1967). Statistical Thermodynamics of Polymerization and Polymorphism of Protein, *Theoret. Biol.* **1,** 79–164.

Parsegian, V. L., Shilling, P. R., Monaghan, F. V., and Luchins, A. S. (1970). "Introduction to Natural Science, Part Two: The Life Sciences." Academic Press, New York.

Pauling, L., and Corey, R. B. (1951). Configuration of Polypeptide Chains. *Nature* **168,** 550–551.

Prigogine, I. (1955). "Introduction to the Thermodynamics of Irreversible Processes." Thomas, Springfield, Illinois.

Prigogine, I. (1969). Structure, Dissipation, and Life, *in* "Theoretical Physics and Biology" (M. Marois, ed.), pp. 23–52. North Holland Publ., Amsterdam: Wiley (Interscience). New York.

Pullman, B. (1965). Aspects of the Electronic Structure of Nucleic Acids, *in* "Molecular Biophysics" (B. Pullman and M. Weisbluth, eds.), pp. 117–189. Academic Press, New York.

Pullman, B., and Pullman, A. (1963). "Quantum Biochemistry," Wiley (Interscience), New York.

Schrödinger, E. (1944). "What Is Life?" Cambridge Univ. Press, London and New York.

Setlow, R. B., and Pollard, E. C. (1962). "Molecular Biophysics." Addison-Wesley, Reading, Massachusetts.

Snell, F. M., Shulman, S., Spencer, R. P., and Moos, C. (1965). "Biophysical Principles of Structure and Function." Addison-Wesley, Reading, Massachusetts.

Stanier, R. Y. (1970). Some Aspects of the Biology of Cells and Their Possible Evolutionary Significance, *in* "Organization and Control in Prokaryotic and Eukaryotic Cells" (H. P. Charles and B. C. J. G. Knight, eds.). Cambridge Univ. Press, London and New York.

von Bėkėsy, G. (1953). *J. Accoust. Soc. Am.* **25,** 786.

Vredenberg, W. J., and Duysens, L. N. M. (1963). *Nature* **197,** 355.

Watson, J. D., and Crick, F. H. C. (1953). A Structure for Deoxyribose Nucleic Acid. *Nature* **171,** 737–738.

Wiener, N. (1961). "Cybernetics," 2nd ed. MIT Press, Cambridge, Massachusetts.

Abbreviations

Angstrom	A	Kilocalorie	kcal
Atmospheric pressure (unit)	atm	Kilohertz	kHz
Centigrade	C	Micron	μm
Centimeter	cm	Millimeter	mm
Debye	D	Millivolt	mV
Decibel	dB	Nanometer	nm
Electron volt	eV	Radian	rad
Electrostatic unit	esu	Second	sec
Hertz	Hz	Volt	V
Kelvin	K		

Index

D

F

G

Q

R

S

W

X

Z

A
B 7
C 8
D 9
E 0
F 1
G 2
H 3
I 4
J 5